AF361637

Advances in Urban Groundwater and Sustainable Water Resources Management and Planning

Advances in Urban Groundwater and Sustainable Water Resources Management and Planning

Editors

Helder I. Chaminé
Maria José Afonso
Maurizio Barbieri

MDPI • Basel • Beijing • Wuhan • Barcelona • Belgrade • Manchester • Tokyo • Cluj • Tianjin

Editors

Helder I. Chaminé
Polytechnic of Porto, ISEP
Portugal

Maria José Afonso
Polytechnic of Porto, ISEP
Portugal

Maurizio Barbieri
Sapienza University of Rome
Italy

Editorial Office
MDPI
St. Alban-Anlage 66
4052 Basel, Switzerland

This is a reprint of articles from the Special Issue published online in the open access journal *Water* (ISSN 2073-4441) (available at: https://www.mdpi.com/journal/water/special_issues/ Urban_Groundwater_Management).

For citation purposes, cite each article independently as indicated on the article page online and as indicated below:

LastName, A.A.; LastName, B.B.; LastName, C.C. Article Title. *Journal Name* **Year**, *Volume Number*, Page Range.

ISBN 978-3-0365-6001-4 (Hbk)
ISBN 978-3-0365-6002-1 (PDF)

Medieval spring of Ourém (1434), Central Portugal. Cover image courtesy of Helder I. Chaminé.

Contents

About the Editors

Helder I. Chaminé

Helder I. Chaminé is a skilled Geologist (B.Sc., Ph.D., D.Sc.), with over 32 years of experience in multidisciplinary geoscience research and practice. He is a Coordinator Professor at the School of Engineering (ISEP), Polytechnic of Porto, Portugal. Additionally, he is the Head of the Laboratory of Cartography and Applied Geology at ISEP. Presently, he belongs to the directive board of the Portuguese Chapter of the International Association of Hydrogeologists (IAH) and the Technical Committee of Environmental Geotechnics of the Portuguese Geotechnical Society (SPG). In addition, he has been on the editorial board of the *Geosciences* MDPI, Hydrogeology Journal, Geotechnical Research ICE, Springer Nature Applied Sciences, Mediterranean Geoscience Reviews, Arabian Journal of Geosciences, Euro-Mediterranean Journal for Environmental Integration, and Discover Water. His main research interests are hard rock hydrogeological and GIS-based mapping for water resources; urban groundwater for sustainable water resources management and planning; vulnerability mapping and geohazards; hydrogeomechanics and underground environments; military geosciences and groundwater; higher education dissemination and geo-professional core values.

Maria José Afonso

Maria José Afonso is a skilled Hydrogeologist (B.Sc., M.Sc., Ph.D.) with over 32 years of experience in urban groundwater, hard-rock hydrogeology and applied geology research. She is an Adjunct Professor at the School of Engineering (ISEP), Polytechnic of Porto, Portugal. Her main research interests are urban groundwater for sustainable water resources management and planning; urban vulnerability mapping and geohazards; hard-rock hydrogeology and water resources; hydrogeochemistry in both natural and anthropogenic environments; as well as applied geology and hydrogeotechnics.

Maurizio Barbieri

Maurizio Barbieri is a skilled Geologist (B.Sc., Ph.D.) with over 27 years of experience and is a Full Professor at the Department of Chemical, Materials and Environmental Engineering of Sapienza University of Rome, Italy. He is the Head of the Geochemistry Laboratory at the Sapienza University of Rome. Presently, he is a Fellow of the Association of Applied Geochemists and European Chair (Italy) of the Society for Environmental Geochemistry and Health. He has been on the editorial board of the Arabian Journal of Geosciences, Chemie der Erde, Environmental Geochemistry and Health, Euro-Mediterranean Journal for Environmental Integration, Water, Geofluids and Discover Water. In addition, he is editor-in-chief (Hydrogeology Section) for *Geosciences* MDPI. His main research interests are environmental geochemistry, environmental hydrogeology; hydrogeochemistry; urban groundwater and geochemistry processes; water–rock Interaction; water resources management; geoenvironment and geo-hazards.

Preface to "Advances in Urban Groundwater and Sustainable Water Resources Management and Planning"

Water is a vital resource for the sustainable development and survival of humanity and ecosystems alike. The occurrence of this natural resource has always enabled the distribution of populations and biodiversity. In fact, early communities sought the proximity of springs, water courses, lakes, and rivers to establish settlements.

In nature, urban groundwater results in several processes, including climatic, geological, geomorphological, geochemical, ecotoxicological, and hydraulic processes, as well inducing sanitation. Urban development deeply impacts hydrological systems, particularly in the invisible component of the water cycle, the groundwater.

Urban and peri-urban areas continue to develop and expand uninterrupted, and urbanisation's extensive impact on groundwater is and associated with climate change and water quality degradation. The horizontal development of municipalities through urban expansion is a reality. In addition, the expansion in the underground environment must be considered, since this is the location of the infrastructures which relate to urbanisation and which have a significant impact on underground network areas (e.g., water supply, sewage, stormwater, metro, tunnels, and storage). Therefore, urban planning must consider the site's natural conditions (e.g., geology, geomorphology, groundwater, ecosystems) and the entire network of constructed infrastructure and underground structures.

This Special Issue emphasises the presentation and discussion of key studies on model urban and peri-urban areas. In addition, several papers describe the current state of the art on the challenges and emerging fields related to the mapping, characterisation, assessment, mitigation, and protection of sustainable groundwater systems and water resources in urban and peri-urban areas. In the current year, 2022, World Water Day has been committed to groundwater and the process making the invisible visible. Therefore, this Special Issue presents a set of papers that stimulate reflections, methodologies, and learned studies on the significance of fresh water in urban areas.

Helder I. Chaminé, Maria José Afonso, and Maurizio Barbieri

Editors

 *water*

Advances in Urban Groundwater and Sustainable Water Resources Management and Planning: Insights for Improved Designs with Nature, Hazards, and Society

Helder I. Chaminé [1,2,*], Maria José Afonso [1,2] and Maurizio Barbieri [3]

[1] Laboratory of Cartography and Applied Geology (LABCARGA), Department of Geotechnical Engineering, School of Engineering (ISEP), Polytechnic of Porto, 4200-072 Porto, Portugal
[2] GeoBioTec Centre, Georesources, Geotechnics, Geomaterials Research Group, University of Aveiro, 3810-193 Aveiro, Portugal
[3] Department of Chemical, Materials and Environmental Engineering (DICMA), La Sapienza University of Rome, 00184 Rome, Italy
* Correspondence: hic@isep.ipp.pt

1. Scope

"It appears therefore that, in early times, Man's interference with the natural flow of water consisted mainly in taking water from rivers and springs, and that this water would find its way back, in a polluted condition, into the rivers, having suffered some reduction in quantity by evaporation. The size of streams would, therefore, not be markedly interfered with, although the water would be greatly polluted. We have to remember, in this connection, that the population was considerably less, and the quantity of water used per head very much less in early times than is now the case (…)." R.L. Sherlock (1922, p. 272)

Citation: Chaminé, H.I.; Afonso, M.J.; Barbieri, M. Advances in Urban Groundwater and Sustainable Water Resources Management and Planning: Insights for Improved Designs with Nature, Hazards, and Society. *Water* **2022**, *14*, 3347. https://doi.org/10.3390/w14203347

Received: 11 October 2022
Accepted: 12 October 2022
Published: 21 October 2022

Publisher's Note: MDPI stays neutral with regard to jurisdictional claims in published maps and institutional affiliations.

In nature, urban groundwater results in several processes, including climatic, geological, geomorphological, geochemical, ecotoxicological, and hydraulic processes and sanitation, sustaining several ecological services. Urban development profoundly impacts hydrological systems, particularly in the invisible component of the water cycle: groundwater (e.g., [1–3]). That impact was noticed a long time ago and focused on societal roles in the development of urbanisation and the consequent contamination and pollution of hydrosystems (e.g., [4–7]). Additional issues in sustainable water resource management and hydrological cycle comprehension are added by urbanisation [8–10]. In addition, climate analysis is taking on an increasingly central role in the life of humanity. Climate greatly impacts many environmental issues and requires reliable, as well as complete, data [11,12].

An intricate network of pipes constitutes the anatomy of the urban underground, including, namely, conduits, channels, galleries, storm sewers, and other structures that alter the hydraulic conductivity of geomaterials (e.g., [9,13–18]). Consequently, these urban buried features act as favourable pathways for the fluid flow of urban-sourced contaminants into groundwater resources. In addition, the ground's surface is generally covered by several structures, such as buildings, asphalt, concrete, and bricks, that are perceived as effectively impervious. In addition, the increasing environmental pressures, such as overexploitation, contamination and/or pollution issues, and climate variability, affect urban groundwater systems (e.g., [1,19–21]).

The release of contaminants from urban infrastructures alters the chemistry of the surrounding environment and affects water quality. Urban streams (some also channelled and shallowly buried) receive dissolved and particulate chemical loadings from runoff, sewer connections, direct discharge from other waterways, and interactions with groundwater. The chemistry of urban runoff tends to be dominated by materials associated with or accumulated on impervious surfaces, such as heavy metals and deicing salt from roadways. Sewage treatment plants are typically designed to remove some but not all human-produced compounds and suspended material from water (e.g., [22,23]).

Source controls, i.e., pollution prevention, represent fundamental steps towards minimising the presence of pollutants in urban stormwater and the concomitant potentially adverse effects in receiving water bodies. Moreover, source control policies are the most cost-effective management tool for dealing with low-level diffuse pollution. Therefore, there is a strong need to advance this pollution control tool [24]. However, frequently, the opportunities for preventing pollution sources are limited or hard to achieve. Then, it may be more feasible to control the release activities rather than primary sources, such as atmospheric deposition, drainage surfaces, anthropogenic activities, and urban drainage systems [25].

Currently, more than half (54%) of the population lives in urban areas (ca. 4 billion people) at locations generally close to coastal areas, and it is projected that this proportion will grow to 68% by 2050 [26]. Urban groundwater is a relatively recent field of hydrological sciences (e.g., [3,15,19,27–30]). However, attention is focused on the relationship between urban development and water resource management that started in the 1950s–1960s, when the accelerated growth after World War II, especially in Europe and North America, started to create a wide range of hydrological problems. Most of these issues were related to urban runoff and flooding, so within a short period, urban hydrology was decisively established (e.g., [29,31,32]). Urban development impacted surface water resources, but the effects have also started on groundwater, hence the emergence of the urban groundwater concept (e.g., [9,19,30,31]). La Vigna [3] proposed a groundwater city classification that reflects geographical aspects, climate contexts, and hydrogeological settings, with a clear connection to groundwater dynamics with several categories: (i) coastal, lagoon, and delta groundwater cities (CGC-LDGC); (ii) volcanic groundwater cities (VGC); (iii) hard-rock and karst groundwater cities (HRGC-KGC); (iv) alluvial groundwater cities (AGC); (v) cold climate groundwater cities (CCGC); (vi) arid climate groundwater cities (ACGC).

A paradigm shift based on holistic management is required to design sustainable water systems. Thus, an urban water framework must be based on sustainable technical–scientific studies and embrace socioeconomic, cultural, heritage, and ethical dimensions [33]. Therefore, the solutions to be promoted should be planned and organised in an environmental balance and harmony with the natural environment (e.g., [6,21,34,35]) within a spirit of comprehensive studies and societal practices, eco-responsibility, and geoethics [36].

Lastly, researchers should continue identifying, mapping, and quantifying the contributing potential contamination and pollution sources and developing control strategies and treatment technologies in collaboration with the industry, local authorities and representatives of citizen groups. Furthermore, it is urgent to tackle innovative programs and actions visible in primary, secondary, and higher education about the crucial role of groundwater in society, geosystems, and ecosystems, particularly in urban areas (e.g., [2,37]). Moreover, enhanced public awareness of groundwater will inspire citizens and stakeholders to take advised, local action on water issues [38].

2. Articles

This Special Issue (SI) highlights the presentation and discussion of model urban studies, methodological approaches, and reflections that describe the current state-of-the-art methods on challenges and emerging fields related to the mapping, characterisation, assessment, mitigation, and protection of sustainable groundwater systems in peri-urban and urban areas.

The contributions unfold two major dimensions: (1) groundwater protection studies in urban areas and implications for sustainable groundwater resources management; (2) groundwater contamination investigations in model urban and peri-urban areas. The paper set includes model urban areas in Europe (Italy and Portugal, including the Azores), Asia (China, India, and Thailand), South America (Brazil), and Africa (Egypt, Kenya, and Zambia). The SI comprises 11 papers involving over 64 authors, all dealing with several groundwater methodologies and tools. In addition, the papers include six feature papers and one editor's choice paper.

Articles here shape many interesting approaches, such as the following:

(i) The feature paper from Foster et al. [39] analyses sustainable management drivers and policy demands related to urban self-supply from groundwater. In the last decades, the use of private water wells in developing cities increased enormously but in a chaotic manner. The authors outline this sensitive question based on ten globally selected urban cities from three continents. This insightful contribution highlights the following impressive thought: "it is thus very important that urban self-supply from groundwater must not be ignored by the public authorities and should be systematically included in city-wide surveys and monitoring of water-supply provisions. Policies need to be introduced that encourage municipal water utilities and local government offices to provide services to private water well users in return for formal water well registration and payment of a modest resource fee." [39].

(ii) The articles by Cai et al. [40], Andrei et al. [41], and Valente et al. [42] highlight groundwater hazard concerns in distinctive hydrogeological media, geological settings, and geoenvironmental issues. Cai et al. [40] point out a detailed study in Limin village, Nantong City, Jiangsu Province (SE China), regarding the analysis of the water level (pumping/recovery tests) forecast of groundwater sources using numerical modelling and the contamination impacts on the nearby environment. Andrei et al. [41] present an isotopic hydrology study for tracing municipal solid waste landfill contamination of groundwater in two urban areas, in Cagliari province, Sardinia (SW Italy), and Umbria region, Perugia province (Central Italy). The findings of two model regions confirm that the δ^2H isotope enrichment is a useful tracer for detecting contamination processes between leachate from municipal solid waste landfills and groundwater [41]. Finally, Valente et al. [42] describe an exploratory geo-hazard investigation in assessing the impact of volcanic eruptions on a groundwater-fed water supply system in the Ponta Delgada urban area, São Miguel Island (Azores, Portugal). This study offers key guidelines for other municipalities in the Azores or comparable volcanic islands, where the water supply issues during and after a volcanic event are similarly critical.

(iii) The papers of Afonso et al. [43], Mansilha et al. [44,45], and Zeferino et al. [46] report case studies in Portugal related to GIS mapping for environmental hydrogeology, hydrogeochemistry, and hydrodynamics assessment in peri-urban and urban areas. Afonso et al. [43] assessed the major urban hydrogeological processes and their dynamics, as well as anthropogenic interactions in groundwater systems in fissured media of the Porto city urban area (NW Portugal). Mansilha et al. [44] outline a study that identifies major effects of a large forest wildfire on groundwater quality from springs linked to a small supply system in a peri-urban forest area in Braga city's (NW Portugal) vicinities. In addition, the investigation concludes that an interlinkage between groundwater depletion and devastating wildfires might seem questionable, but the parametric drinking water values demonstrate the groundwater system's vulnerability to wildfires. Mansilha et al. [45] describe an environmental hydrogeology study on drained effluents from the abandoned colliery mine of São Pedro da Cova, located in the Porto peri-urban area (NW Portugal), examining their suitability for irrigation purposes. The results suggest a cost-effective methodology, minimising the pollution of natural streams and soils and increasing the potential use of effluents. Zeferino et al. [46] present a study to delineate the effectiveness of well-head protection areas after long-term applications on public supplies with continuous pumping located in a densely populated urban area of Montijo municipality (SW Portugal).

(iv) A set of papers [47–49] underlining several case studies on numerical analysis and modelling of groundwater resources management. Liu et al. [47] describe a model for analysing the development pattern of water resource's carrying capacity by examining the water conservancy in Jilin Province (NE China). The lessons learned could be applied to other regions. Krishan et al. [48] present a comprehensive study of hydrogeochemistry processes in the groundwater system salinity of the Mewat region, Haryana province (NW India). The outcomes of this study will be useful in managing and remedying groundwater systems. Finally, Abdelfattah et al. [49] outline a study on the coastal aquifer in the western area of Port Said (NE Egypt). The findings emphasise optimum withdrawing scenarios with sufficient groundwater but a smaller salinity.

3. Outlook

The urban water cycle provides a conceptual and unifying basis for a correct assessment of groundwater systems and leads to the foundation of studies on the sustainability of water resources. The role of climate, geology, geomorphology, land use and land cover, hydrogeochemistry, hydraulics, and human activities is important for an integrated assessment of water resources in urban areas. In addition, remote sensing provides valuable and up-to-date spatial information on terrain and natural resources. Recently a new focus has emerged, addressing questions on Geographic Information Systems (GIS) studies integrated into urban water supply systems, notably in historic cities. Sustainable urban groundwater systems are considered increasingly significant in global development issues such as management, protection, distribution, safety, and services (e.g., [2,3,19,27,29,39,50]). In addition, urban population growth and improved living patterns lead to increased water use and demand. While climate and environmental change raise several issues related to the accessibility of urban water resources, such an approach concerning all ecosystem characteristics in peri-urban and urban environments necessarily requires a transdisciplinary methodology that encompasses socioeconomic and cultural perspectives and technical-scientific solutions based on compatible designs with nature, and that is attentive to societal dynamics (e.g., [20,21,33]).

New challenges are emerging, and they are related to mapping, assessing, abstraction, and modelling the urban water cycle. The development of GIS-based inventories in urban areas is a challenge that is vital for planning and managing water resources, assessing water supply security, and defining asset strategies. Furthermore, the urban water cycle concept in urban environments should be highlighted, emphasising holistic and integrated sustainable management related to climatic, physiographic, hydraulic, environmental, and socio-cultural conditions. Indeed, the role of sustainability should be carried out by conducting a consistent analysis of the interlinkages between urban groundwater and the sustainable development goals, as highlighted, for example, by the UNESCO 2030 Agenda. Water issues will continue to be on the agenda of all societies as they cut across the balance of communities, ecosystems, energy transition and climate emergency, heritage, and societal issues and will mark the agenda of any society permanently. However, a correct understanding of the urban water cycle should be based on the sustainable protection and management of the resource, territorial planning, eco-responsibility, hydrogeoethics, and good practices.

In the current year, 2022, World Water Day was dedicated to groundwater, making the invisible visible. This SI offers a set of papers that promote reflections, methodologies, and learned studies on the importance of fresh water in urban areas.

Author Contributions: H.I.C. designed and drafted the editorial. M.J.A. and M.B. contributed to the editorial. All authors have read and agreed to the published version of the manuscript.

Funding: This research received no external funding.

Acknowledgments: Our thanks to the reviewers for their in-depth review support during the peer-reviewing process, contributing to the overall quality of the manuscripts. A word of appreciation to all authors for their valuable contributions to urban groundwater scope.

Conflicts of Interest: The authors declare no conflict of interest.

References

1. Foster, S.D.; Hirata, R.; Howard, K.W.F. Groundwater use in developing cities: Policy issues arising from current trends. *Hydrogeol. J.* **2011**, *19*, 271–274. [CrossRef]
2. Hibbs, B.J. Groundwater in urban areas. *J. Contemp. Water Res. Educ.* **2016**, *159*, 1–4. [CrossRef]
3. La Vigna, F. Urban groundwater issues and resource management, and their roles in the resilience of cities. *Hydrogeol. J.* **2022**, *30*, 1657–1683. [CrossRef]
4. Sherlock, R.L. *Man as a Geological Agent: An Account of His Action on Inanimate Nature*; H. F. & G. Witherby: London, UK, 1922.
5. Wittfogel, K.A. The hydraulic civilizations. In *Man's Role in Changing the Face of the Earth*; Thomas, W.L., Ed.; The University of Chicago Press: Chicago, IL, USA, 1956; pp. 152–164.
6. Leopold, L.B. Hydrology for urban land planning: A guidebook on the hydrologic effects of urban land use. *U.S. Geol. Surv. Circ.* **1968**, *USG*, S554.
7. Legget, R.F. *Cities and Geology*; McGraw-Hill: New York, NY, USA, 1973.
8. Marsalek, J.; Jiménez-Cisneros, B.; Karamouz, M.; Malmquist, P.; Goldenfum, J.; Chocat, B. *Urban Water Cycle Processes and Interactions*; UNESCO-HP, Urban water series; Taylor & Francis: Leiden, The Netherlands, 2008.
9. Hibbs, B.J.; Sharp, J.M. Hydrogeological impacts of urbanization. *Environ. Eng. Geosci.* **2012**, *18*, 51–64. [CrossRef]
10. van Leeuwen, K.; Frijns, J.; van Wezel, A.; van De Ven, F.H.M. Cities blueprints: 24 indicators to assess the sustainability of the urban water cycle. *Water Resour. Manag.* **2017**, *26*, 2177–2197. [CrossRef]
11. Gentilucci, M.; Barbieri, M.; Burt, P.; D'Aprile, F. Preliminary data validation and reconstruction of temperature and precipitation in central Italy. *Geosciences* **2018**, *8*, 202. [CrossRef]
12. Gentilucci, M.; Barbieri, M.; Lee, H.S.; Zardi, D. Analysis of rainfall trends and extreme precipitation in the middle Adriatic side, Marche region (Central Italy). *Water* **2019**, *11*, 1948. [CrossRef]
13. Lerner, D.N. Leaking pipes recharge ground water. *Ground Water* **1986**, *24*, 654–662. [CrossRef]
14. Vázquez-Suñé, E.; Carrera, J.; Tubau, I.; Sánchez-Vila, X.; Sole, A. An approach to identify urban groundwater recharge. *Hydrol. Earth Syst. Sci.* **2010**, *14*, 2085–2097. [CrossRef]
15. Vázquez-Suñé, E.; Sánchez-Vila, X.; Carrera, J. Introductory review of specific factors influencing urban groundwater, an emerging branch of hydrogeology, with reference to Barcelona, Spain. *Hydrogeol. J.* **2005**, *13*, 522–533. [CrossRef]
16. Wiles, T.J.; Sharp, J.M. The secondary permeability of impervious cover. *Environ. Eng. Geosci.* **2008**, *14*, 251–265. [CrossRef]
17. Attard, G.; Winiarski, T.; Rossier, Y.; Eisenlohr, L. Impact of underground structures on the flow of urban groundwater. *Hydrogeol. J.* **2016**, *24*, 5–19. [CrossRef]
18. Afonso, M.J.; Freitas, L.; Pereira, A.; Neves, L.; Guimarães, L.; Guilhermino, L.; Mayer, B.; Rocha, F.; Marques, J.M.; Chaminé, H.I. Environmental groundwater vulnerability assessment in urban water mines (Porto, NW Portugal). *Water* **2016**, *8*, 499. [CrossRef]
19. Custodio, E. Hidrogeología urbana: Una nueva rama de la ciencia hidrogeológica. *Boletín Geológico Y Min. Madr.* **2004**, *115*, 283–288.
20. Chaminé, H.I.; Afonso, M.J.; Freitas, L. From historical hydrogeological inventories, through GIS mapping to problem solving in urban groundwater systems. *Eur. Geol. J.* **2014**, *38*, 33–39.
21. Chaminé, H.I.; Teixeira, J.; Freitas, L.; Pires, A.; Silva, R.S.; Pinho, T.; Monteiro, R.; Costa, A.L.; Abreu, T.; Trigo, J.F.; et al. From engineering geosciences mapping towards sustainable urban planning. *Eur. Geol. J.* **2016**, *41*, 16–25.
22. Barbieri, M.; Nigro, A.; Sappa, G. Arsenic Contamination in groundwater system of Viterbo area (Central Italy). *Senses Sci.* **2014**, *1*, 101–106. [CrossRef]
23. Sappa, G.; Barbieri, M.; Andrei, F. Assessment of trace elements natural enrichment in topsoil by some Italian case studies. *SN Appl. Sci.* **2020**, *2*, 1409. [CrossRef]
24. Marsalek, J.; Viklander, M. Controlling contaminants in urban stormwater: Linking environmental science and policy. In *On the Water Front: Selections from the 2010 World Water Week in Stockholm*; Lundqvist, J., Ed.; Stockholm International Water Institute (SIWI): Stockholm, Sweden, 2011; Volume 101.
25. Müller, A.; Österlund, H.; Marsalek, J.; Viklander, M. The pollution conveyed by urban runoff: A review of sources. *Sci. Total Environ.* **2020**, *709*, 136125. [CrossRef]
26. UN-Habitat. *United Nations Human Settlements Programme. Envisaging the Future of Cities. World Cities Report 2022*; United Nations Human Settlements Programme: Nairobi, Kenya, 2022.
27. Foster, S.D. Impacts of urbanisation on groundwater. In *Hydrological Processes and Water Management In Urban Areas*; Massing, H., Packman, J., Zuidema, F.C., Eds.; International Association of Hydrological Sciences (IAHS): Wallingford, UK, 1990; pp. 187–207.
28. Chilton, J. *Groundwater in the Urban Environment: Selected City Profiles*; A.A. Balkema: Rotterdam, The Netherlands, 1999.
29. Howard, K.W.F. Sustainable cities and the groundwater governance challenge. *Environ. Earth Sci.* **2015**, *73*, 2543–2554. [CrossRef]
30. Schirmer, M.; Leschik, S.; Musolff, A. Current research in urban hydrogeology: A review. *Adv. Water Resour.* **2013**, *51*, 280–291. [CrossRef]
31. Howard, K.W.F. Urban groundwater issues: An introduction. In *Current Problems of Hydrology in Urban Areas, Urban Agglomerates and Industrial Centers*; Howard, K.W.F., Israfilov, R.G., Eds.; NATO Science Series; IV Earth and Environmental Sciences; Kluwer Academic Publishers: Dordrecht, The Netherlands, 2002; Volume 8, pp. 1–15. [CrossRef]
32. Howard, K.W.F. Urban Groundwater: Meeting the challenge. In *IAH Selected Papers on Hydrogeology*; Taylor & Francis Group, CRC Press: London, UK, 2007; Volume 8. [CrossRef]

33. Chaminé, H.I.; Carvalho, J.M.; Freitas, L. Sustainable groundwater management in rural communities in developed countries: Some thoughts and outlook. *Mediterr. Geosci. Rev.* **2021**, *3*, 389–398. [CrossRef]

34. McHarg, I.L. *Design with Nature, 25th-anniversary edition*; Wiley Series in Sustainable Design; Wiley: New York, NY, USA, 1992.

35. Bandarin, F.L.; van Oers, R. *Reconnecting the City: The Historic Urban Landscape Approach and the Future of Urban Heritage*; John Wiley & Sons: London, UK, 2015.

36. Peppoloni, S.; Di Capua, G. *Geoethics: Manifesto for an Ethics of Responsibility Towards the Earth*; Springer: Cham, Switzerland, 2022. [CrossRef]

37. Houben, G.J. Teaching about groundwater in primary schools: Experience from Paraguay. *Hydrogeol. J.* **2019**, *27*, 513–518. [CrossRef]

38. Cherry, J. Groundwater: The missing educational curriculum. *Ground Water* **2022**, 1–2. [CrossRef]

39. Foster, S.D.; Hirata, R.; Eichholz, M.; Alam, M.-F. Urban self-supply from groundwater: An analysis of management aspects and policy needs. *Water* **2022**, *14*, 575. [CrossRef]

40. Cai, J.; Wang, P.; Shen, H.; Su, Y.; Huang, Y. Water level prediction of emergency groundwater source and its impact on the surrounding environment in Nantong city, China. *Water* **2020**, *12*, 3529. [CrossRef]

41. Andrei, F.; Barbieri, M.; Sappa, G. Application of 2H and 18O isotopes for tracing municipal solid waste landfill contamination of groundwater: Two Italian case histories. *Water* **2021**, *13*, 1065. [CrossRef]

42. Valente, F.; Cruz, J.V.; Pimentel, A.; Coutinho, R.; Andrade, C.; Nemésio, J.; Cordeiro, S. Evaluating the impact of explosive volcanic eruptions on a groundwater-fed water supply system: An exploratory study in Ponta Delgada, São Miguel (Azores, Portugal). *Water* **2022**, *14*, 1022. [CrossRef]

43. Afonso, M.J.; Freitas, L.; Marques, J.M.; Carreira, P.M.; Pereira, A.J.S.C.; Rocha, F.; Chaminé, H.I. Urban groundwater processes and anthropogenic interactions (Porto Region, NW Portugal). *Water* **2020**, *12*, 2797. [CrossRef]

44. Mansilha, C.; Melo, A.; Martins, Z.E.; Ferreira, I.M.P.L.V.O.; Pereira, A.M.; Espinha Marques, J. wildfire effects on groundwater quality from springs connected to small public supply systems in a peri-urban forest area (Braga Region, NW Portugal). *Water* **2020**, *12*, 1146. [CrossRef]

45. Mansilha, C.; Melo, A.; Flores, D.; Ribeiro, J.; Rocha, J.R.; Martins, V.; Santos, P.; Espinha Marques, J. Irrigation with coal mining effluents: Sustainability and water quality considerations (São Pedro da Cova, North Portugal). *Water* **2021**, *13*, 2157. [CrossRef]

46. Zeferino, J.; Paiva, M.; Carvalho, M.D.R.; Carvalho, J.M.; Almeida, C. Long term effectiveness of wellhead protection areas. *Water* **2022**, *14*, 1063. [CrossRef]

47. Liu, T.; Yang, X.; Geng, L.; Sun, B. A three-stage hybrid model for space-time analysis of water resources carrying capacity: A case study of Jilin Province, China. *Water* **2020**, *12*, 426. [CrossRef]

48. Abdelfattah, M.; Abu-Bakr, H.A.-A.; Gaber, A.; Geriesh, M.H.; Elnaggar, A.Y.; Nahhas, N.E.; Hassan, T.M. Proposing the optimum withdrawing scenarios to provide the western coastal area of Port Said, Egypt, with sufficient groundwater with less salinity. *Water* **2021**, *13*, 3359. [CrossRef]

49. Krishan, G.; Sejwal, P.; Bhagwat, A.; Prasad, G.; Yadav, B.K.; Kumar, C.P.; Kansal, M.L.; Singh, S.; Sudarsan, N.; Bradley, A.; et al. Role of ion chemistry and hydro-geochemical processes in aquifer salinization—A case study from a semi-arid region of Haryana, India. *Water* **2021**, *13*, 617. [CrossRef]

50. Foster, S.D.; Hirata, R.; Custodio, E. Waterwells: How can we make legality more attractive? *Hydrogeol. J.* **2021**, *29*, 1365–1368. [CrossRef]

Article

Urban Self-Supply from Groundwater—An Analysis of Management Aspects and Policy Needs

Stephen Foster [1,2,*], **Ricardo Hirata** [1,3], **Michael Eichholz** [1,4] **and Mohammad-Faiz Alam** [1,5]

[1]   Groundwater Management Group, International Water Association, London E14-2BA, UK;
       rhirata@usp.br (R.H.); michael.eichholz@bgr.de (M.E.); mohammadfaizalam@yahoo.com (M.-F.A.)
[2]   Department of Earth Sciences, University College London, London WC1E-6BT, UK
[3]   CEPAS I USP Groundwater Research Center, Institute of Geosciences, University of São Paulo,
       São Paulo 05508-080, Brazil
[4]   Federal Institute for Geosciences & Natural Resources, D-30655 Hannover, Germany
[5]   International Water Association, Groundwater Management Group, International Water Management
       Institute, Delhi 110012, India
*   Correspondence: drstephenfoster@aol.com

Abstract: The use of private water wells for self-supply in developing cities has 'mushroomed' during recent decades, such that it is now an important component of total water-supply, but one all too frequently overlooked in official figures. Selected global experience of the phenomenon (from 10 cities in 3 continents) is succinctly summarized, and then analyzed from differing perspectives, before drawing recommendations on priorities for its improved management.

Keywords: urban water-supply; groundwater; private self-supply; management issues

Citation: Foster, S.; Hirata, R.; Eichholz, M.; Alam, M.-F. Urban Self-Supply from Groundwater—An Analysis of Management Aspects and Policy Needs. *Water* **2022**, *14*, 575. https://doi.org/10.3390/w14040575

Academic Editors: Helder I. Chaminé, Maria José Afonso, Maurizio Barbieri and Francesco De Paola

Received: 23 December 2021
Accepted: 10 February 2022
Published: 14 February 2022

Publisher's Note: MDPI stays neutral with regard to jurisdictional claims in published maps and institutional affiliations.

1. Scale of Phenomenon

The use of private water wells for urban self-supply (Figure 1) has 'mushroomed' over the past 20 years or so, especially in South Asia, Latin America, and Sub-Saharan Africa. Recent surveys suggest that private self-supply from groundwater in the urban areas of developing cities is an essential component of total water-supply [1–9], but one that is frequently overlooked in official figures. The phenomenon varies in level with the physical evolution and hydrogeological setting of any given city, but there is convincing evidence of an increasing dependence on private water wells in response to rapid urban population growth and escalating water demand, facilitated by the modest cost of water well drilling.

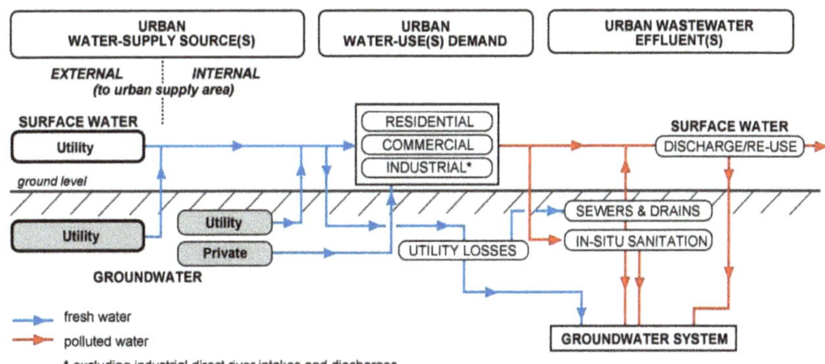

Figure 1. Schematic overview of urban water-supply sources and their interactions [10].

The term self-supply refers to water-supply investments that are financed by users themselves [11], and in developing economies, most self-suppliers use groundwater, and also sell water to neighbors. Self-supply from groundwater provides a rapid solution in areas where it is technically feasible and affordable. Nevertheless, private groundwater use tends to pass under the radar of national water-supply statistics [12], or the phenomenon is not recognized at all by government [3,13].

Private water well construction costs in most hydrogeological settings will be in the range of US$2000–20,000, but are considerably higher (US$30,000–50,000) where deep boreholes (of 200–300 m) are required. Private water well ownership will thus remain mainly the preserve of wealthy individuals or well-organized local communities. Though the practice reduces the pressure on the water-utility supplies, it can also have serious impacts on their cash flows and investment cycles [1,5], and result in inequalities around access to water-supply.

The initial private investment in water well construction is usually triggered by a highly inadequate urban utility water-service, and represents a 'coping strategy' to improve water-supply security by multi-residential users, individual properties, commercial premises, and industrial enterprises. In the longer term, the unit operational cost of private water wells is often lower than the cost of an equivalent municipal water-supply on an unsubsidized tariff, and thus, the practice continues as a 'cost-reduction strategy' even when the reliability of the public supply has improved.

2. Selected Global Experience

Profiles of 10 selected cities worldwide (Figure 2) will be included as illustrations of places with an important water-supply component derived by self-supply from groundwater. In some cases, significant progress has been made in exercising management over private water well users, but in the majority of cases, this has not yet been the case.

Figure 2. Location of selected cities whose water-supply has an important private component from water wells.

In Brazil, 172 million people have access to the public water mains, but only 30.4 million (17.7%) are provided by groundwater [14], due to the fact that the largest cities are supplied primarily by surface water. However, 52% of 5570 Brazilian municipalities are supplied either totally (36%) or partially (16%) by groundwater. During 2013–17, Brazil was hit by one of the most severe droughts of the last 80 years, which caused 48% of all cities to declare a water crisis [15], and those cities supplied exclusively by surface water were hit twice as often as those provided by groundwater.

Although official data suggest that surface water is the most used resource in larger cities, it is necessary to assess the role of complementary self-supply from private water wells, which are often irregular or illegal [16], to get the full picture. In Brazil, more than 90% of all groundwater is extracted by private water wells, highlighting the role that self-supply plays in national water security [17], with the following distribution among user types: domestic (30%), agricultural (24%), urban public (18%), and multiple-use (14%). These uses make the value of extracted groundwater US$12 billion, with the 2.5 million water wells valued at more than US$15 billion [18], equivalent to 6–7 years of government investments in sanitation.

For example, in São Paulo (Box 1), the largest urban conglomerate in South America (with over 21.5 million population), though groundwater supplies only 1% of the public water-utility, more than 12,000 private water wells provide 18% of its total demand, and more during crisis periods.

Box 1. São Paulo—Brazil.

> **São Paulo is home to 21.5 million people distributed across 39 municipalities, and occupies an area of 7946 km^2 with a GDP of US$237 billion/a. Public water-supply is provided to about 95% of the population, mainly by a complex surface-water system producing 5270 ML/d, of which only 1% is groundwater. However, there are more than 12,000 private water wells extracting 950 ML/d (18% of the public supply), and this increased to 25% during the last major water crisis of 2016–2019. Although private self-supply has increased water-supply security, large-scale uncontrolled water well drilling has caused problems, with both the lowering of water-table levels with conflict amongst users, and a significant risk of pollution with many private sources not having regular chemical analysis. The failure to manage groundwater resources is primarily attributed to a lack of appreciation of their importance for water-supply security, and a limited understanding of the conflicts between users. Thus, little pressure is exerted on the water management agencies, who have little incentive to try to regulate thousands of private water well owners.**

A similar situation occurs in Fortaleza (Box 2), where many multi-residential apartment blocks take a significant proportion of their water-supply from private water wells [19]. This phenomenon started as a 'coping strategy' during drought when the mains water-supply was inadequate, but continues as a 'cost-reduction strategy' to avoid having to pay the higher water tariffs.

The large city of Recife has a very poor public water-supply (Box 3), and some 6000–8000 private water wells serve 25% of the population of 1.6 million [20]. This situation is replicated in many cities, and even in medium-sized urbanizations such as Bauru (390,000 inhabitants), private water wells meet 18% of the total urban water demand. Private self-supply from water wells must be regarded as essential, given that the public water-supply is limited due to historical scarcity of infrastructure investments. Without the supply produced by private water wells, the water-supply of many cities would collapse in drought periods.

The invisibility of groundwater, when it comes to both human water-supply and environmental features, leads to very inefficient management of the resource nationally. Though laws and regulations are adequate, they are scarcely applied, which means that 88% of water wells are irregular (without an extraction license). This is the result of groundwater users, regulatory agencies, and water well contractors lacking a clear perception of legitimacy and fairness. There is both confusion due to changes in water law, and a predisposition towards law-breaking—perceiving that the benefits outweigh the risks.

Box 2. Fortaleza—Brazil.

Fortaleza, a city of over 3.0 million population, is situated on a coastal strip underlain by permeable aeolian and fluvial deposits (widely of 30+ m in saturated thickness, and the water-table at 2–15 m depth), which are locally susceptible to sea-water intrusion, and widely vulnerable to pollution. The climate is humid–tropical with an average rainfall of 800–1200 mm/a. Aquifer recharge results from diffuse rainfall recharge, infiltration of surface runoff along riverbeds, profuse leakage of water mains, and ground discharge of wastewater from septic-tanks and cesspits, with groundwater exhibiting some nitrate contamination (15–35 mg NO_3/L) and chloride concentrations of 100–150 mg/L. CAGECE (the water-service utility) provides some 60–70% of the population from surface-water sources (with a 'guaranteed yield' of 570 ML/d), but earlier drought periods (notably 1998) caused near collapse of the mains water-supply, and led some 40–50% of the population, together with commercial water-users, to construct water wells for direct self-supply. A groundwater survey in 2002–2003 inventorized 8950 fully-equipped water wells (compared to about 1700 in 1980), and concluded that:

- sunk capital in private water wells was at least US$19 million, and probably nearer to US$25 million;
- potential groundwater production is about 200 ML/d, representing 36% of total drought provision;
- more than 1500 water wells yielding more than 2 m^3/h (0.5 L/s) are effectively outside the law;
- groundwater use does not tax resources as a result of mains leakage and wastewater seepage.

Significant private domestic groundwater use arises as a result of consumers avoiding the use of mains water-supply at prices above the highly-subsidized 'social tariff' (equivalent to US$0.26/$m^3$). This has major financial implications for CAGECE in terms of loss of revenue from potential water sales, difficulties of increasing average tariffs, and resistance to recovering sewer-use charges.

Box 3. Recife—Brazil.

The rapidly-developing Recife Metropolitan Region, a humid region with rainfall of about 2000 mm/a, has a population of over 3.0 million, and a water demand of about 15,000 L/s (including high 'non-accounted for' losses). COMPESA (the state water-service utility) has progressively developed surface water sources, whose capacity reached over 8000 L/s by 2008, but which reduces in severe drought to less than 3000 L/s. Hydrogeologically, the region divides into two areas (north and south) by a major geological lineament, with sharply contrasting groundwater potential on either side. In the south, shallow groundwater is intensely exploited at an estimated rate of 2000 L/s as a low-cost supply for multi-residential properties and hotel facilities by some 6000–8000 private water wells, mainly drilled in response to extreme municipal water shortages during the droughts of 1993–1994 and 1999–2000. In the north, the Beberibe Aquifer of some 200 m thickness dips below younger strata as the coastline is approached, and since 1975, has been widely developed over a 20 km-wide strip by COMPESA to provide mains water-supply, reaching 1500 L/s by 2002. In addition, it is estimated that private industrial water wells are abstracting about 700 L/s, although the depth of the aquifer horizon is such as to have largely prevented residential self-supply from groundwater. The aquifer response has seen drawdowns to around −60 m MSL, although this was not accompanied by any rapid saline intrusion. The Beberibe Aquifer is a good example of a high-yielding groundwater system close to a major urban area, whose geographical extension and regional flow is not sufficient for it to become a 'sole source' of urban water-supply, but whose freshwater storage reserves are large and need to be conjunctively managed to provide increased water-supply security at minimum possible cost.

India is far-and-away the largest user and consumer of groundwater globally, with a total annual extraction estimated to be 245,000 Mm^3/a, but an increasing number (now 29%) of the aquifer units are excessively exploited [21]. Agricultural irrigation (over 75%) is the predominant use, but 85% of the population are believed to rely, directly or indirectly, on groundwater for domestic water-supply, with a total extraction in the range 55–60,000 Mm^3/a.

Virtually all Indian cities have insufficient municipal mains water-supply [22], and their residents have widely adopted self-supply from groundwater as at least a stop-gap solution [23]. The detailed situation in three cities (Box 4, Box 5 and Box 6)—Chennai (Tamil Nadu), Indore (Madhya Pradesh), and Aurangabad (Maharastra)—are presented to illustrate typical trends.

Box 4. Chennai—India.

> Chennai is the fourth-largest metropolitan area in India, and its 8.6 million population faced an acute water crisis during 2017–2019. Chennai's four main reservoirs almost dried-up as a result of persistent drought, and by June 2019, combined surface-water storage stood at only 0.1% of total storage, with the water utility only being able to supply 525 ML/d of the total demand of 830 ML/d. Much of the city became totally dependent on groundwater, which is abstracted extensively both within and outside the city limits. Within the city, there are about 420,000 private wells, but due to long-term overexploitation and limited recharge during poor monsoons, the water-table has fallen, causing wells to dry-up and groundwater quality to deteriorate through seawater intrusion. Thus, more than 5000 tankers of 9000 L capacity made 5–6 trips daily to supply groundwater from the surrounding rural areas for both the water utility and private operators at a total rate of 200–300 ML/d. However, a history of inadequate groundwater management has led to conflicts at the urban–rural interface.

Box 5. Indore—India.

> Indore is one of India's fastest-growing cities, with a population of about 1.9 million and some 0.5 million households. The Indore Municipal Corporation (IMC) faces major challenges in trying to meet escalating water demand, and currently only achieves 46% water-mains coverage with a supply of about 320 ML/d; it achieves an average supply reliability of about 1 h/day. Of this supply, some 40–60 ML/d is groundwater, but private water wells of residential, industrial, and commercial users extract a further 100 ML/d. Little is known about groundwater quality and any associated risks to public health, but a systematic survey must be an urgent priority for the city. It is estimated that 35–40% of all households rely on private water wells as their main source, with about 150,000 water wells producing some 90 ML/d for this purpose. The associated investment is the equivalent of US$190/household (US$87 million in total), and the unit cost of private groundwater supplies is estimated to be US$0.04/m^3 compared to the water-utility charge of US$0.23/m^3. The estimated operational cost of private water wells is US$1.2–2.0 million/year, which is equivalent to 36% of the annual water-utility revenue, demonstrating the magnitude of the private investment in securing an urban water-supply.

Box 6. Aurangabad—India.

> Aurangabad has grown rapidly in recent years to a population of about 1.2 million. In 1998, public water-supply became the responsibility of the Aurangabad Municipal Corporation (AMC), which takes 150 ML/d from a reservoir 45 km away and 180 m lower elevation. This supplements local groundwater sources that can only provide about 15 ML/d. However, shortages of electrical power needed for the high-lift pumping result in very poor service levels (widely, less than 1 h/day), and consequently, most residential properties and commercial premises have resorted to drilling private water wells. The local area has only limited groundwater resources, and the water-table falls from 5 m bgl to more than 10 m bgl in the dry season. Water well capital costs are very low (generally less than US$400), and an expenditure review revealed that private groundwater supplies cost US$0.15–0.25/m^3, compared with tinkered water at US$1.30–1.35/m^3. Consequently, residential and commercial consumers take their water-supply from different sources at different times of year: when available, the highly-subsidized AMC mains-supply at US$0.03/m^3; then, private water wells; and only when these fail in the dry season (April–June) do they use tankered supplies. This represents a coping strategy to reduce the period during which the most expensive water is needed, and as a result, about 60% of the total annual supply is from water wells, 35% from the AMC, and only 5% from tankers.

Bangkok (Box 7), the Thailand capital, is an excellent example of a city, and is driven by an increasing risk of tidal flooding seriously aggravated by land subsidence due to excessive water well pumping. It has struggled with urban groundwater over-exploitation for decades [24], but groundwater stability and an end to further land subsidence was finally achieved around 2005, after greatly enhancing the powers and capacity of the national Groundwater Department (GWD) in defined 'critical zones' as a proactive regulatory agency.

Box 7. Bangkok—Thailand.

Greater Bangkok occupies the lower part of the Chao Phrayh Basin, which is underlain by 500 m of interbedded alluvial and marine sediments containing eight semi-confined 'aquifer horizons' (recharged from the north), and overlain by a confining Holocene clay. By 1980, widespread exploitation of groundwater for urban water-supply, mainly by the Metropolitan Waterworks Authority (MWA), reached a level of about 500 ML/d, and caused a groundwater level decline to 40 m bsl, with evidence of significant related land subsidence. The initial approach to reducing groundwater abstraction was to require the MWA to close its water wells in favor of the development of distant surface-water sources, but increased water tariffs triggered a massive increase in private water well drilling. Abstraction reached over 2000 ML/d by the late 1990s, with a further 400 ML/d abstraction by three Provincial Waterworks Authorities (PWAs). The Groundwater Department (GWD) was then given increased power to reduce groundwater abstraction, with the definition of 'critical areas' where water well drilling must be banned, the sealing of water wells in areas with mains water-supply coverage, and licensing/charging for water well abstraction. Charges were raised under two separate components (a 'use fee' and 'conservation fee', each reaching US$0.21/m^3 by 2004). These measures had the positive outcome of controlling groundwater abstraction and reducing land subsidence. There are now just over 4000 licensed water wells operated by around 3000 owners, abstracting about 1600 ML/d (representing 15% of total water-supply). There has been conflict in some districts where the mains water-supply was extended but with high charges (US$0.60/m^3), and these were resolved by allowing private water well users to continue pumping up to 10 years (their next license renewal).

In the last decade or so, the construction of private water wells for urban self-supply has mushroomed in Sub-Saharan Africa [4,12,25–27], but with significant quality concerns in many places [28]. This has occurred equally in cities such as Lusaka (Box 8) and Douala (Box 9), where local aquifers are also in major use by the water-supply utility, as in cities such as Mombasa (Box 10), where the municipal water-supply is imported from distant sources. The driving factors are always the inadequate level and unreliability of the public supply, and the cost advantages to larger domestic water users of self-supply.

Box 8. Lusaka—Zambia.

Lusaka has grown rapidly from 0.5 million in 1978 to 2.8 million in 2018. It has long been dependent on local groundwater for its water-supply. In 2018, the water-utility operated 228 water wells to provide about 140 MI/d, with river treatment works providing a further 80 ML/d. The water-utility is still plagued by high water losses and poor revenue collection, but has taken a 'pro-poor initiative' by drilling stand-alone boreholes to supply water-kiosks at a subsidized tariff of US$0.25/m^3 (a 40–70% reduction). In addition, there are thousands of private water wells with a total abstraction of up to 300 ML/d. In low-income peri-urban areas, most households still rely on shallow dug-wells where the water-table is less than 3 m depth, but the dolomitic-limestone formation they tap (though high-yielding) is very vulnerable to pollution from urban wastewater and industrial effluents. Pit latrines are the predominant form of sanitation, and in these ground conditions, they are a serious hazard to groundwater quality, and the cause of frequent cholera outbreaks. Some large-scale projects to extend the main sewer network and wastewater treatment capacity are underway, but in the unplanned peri-urban slums, these are difficult and costly to implement.

Box 9. Douala—Cameroon.

> Douala, the Cameroon capital on the Gulf of Guinea, has grown very rapidly from about 1.0 million in 1995 to 3.8 million in 2015. The Doula Basin contains thick sedimentary formations, including a semi-confined coarse sandy Mio-Pliocene aquifer (up to 220 m thickness), and the consolidated deeper Continental Terminal aquifer. The climate is hyper-humid (with a precipitation of around 4000 mm/a), resulting in diffuse recharge approaching 1000 mm/a. The Cameroon Water Utility (CAMWater) provides a public supply to about 40% of the urban population, deriving 50% from water wells, and the balance from river treatment works. The rate of groundwater abstraction increased from about 55 ML/d in 1990 to 175 ML/d in 2010. Access to domestic water-supply remains a problem, and the demand is met by private self-supply from shallow water wells and purchase from water tankers. More than 70% of the urban population is served by latrine sanitation, and there is concern about groundwater quality degradation in the upper part of the Mio-Pliocene aquifer, which exhibits a patchy quality, with EC reaching 500–2000 μS/cm (compared to 200 μS/cm in the deeper horizons), but the generally reducing conditions result in the autoelimination of NO_3 and SO_4. In general terms, the hydrogeologic conditions are very favorable for developing new public water-supply wellfields upstream of the city in either the deeper horizons of the Mio-Pliocene and/or the Continental Terminal aquifers.

Box 10. Mombasa—Kenya.

> Mombasa is a major coastal city, whose population has grown rapidly from 0.4 million in 1989 to 1.5 million in 2018, and is set in a metropolitan area with a population of over 3.0 million. The total water demand is estimated to be in excess of 250 ML/d, but the currently available production capacity of the Mombasa Water Company (MWC) is only capable of meeting 20–25% of this demand. Groundwater provides most of the MWC supply, with the main sources being the Baricho-Sabaki wellfield, 100 km distant, with a design capacity of 95 ML/d (but only delivering 30 ML/d to Mombasa due to offtake by other towns); the Mzima springs via a 200 km pipeline with a design capacity of 35 ML/d, but currently delivering 20 ML/d; the much nearer Tiwi water wells, with a production of up to 10 ML/d; and Marere springs, providing another 5 ML/d. Although not adequately quantified and assessed, the very substantial urban water-supply deficit is met by large-scale water tankering, and by large numbers of uncontrolled shallow private water wells (mainly to 25 m depth in sands requiring well-screens), whose water quality is widely compromised by both wastewater percolation and saline-water encroachment, and is often virtually brackish in character.

3. Analysis from Different Perspectives

3.1. Private Water Well Users

In-situ private self-supply from groundwater is mainly practiced by urban dwellers who have sufficient financial resources (individually or communally) to act unilaterally to secure a more reliable water-supply [5]. Among higher income groups, ownership of a private water well is widely seen as enhancing personal water-supply security in the face of unreliable public water-supplies. Urban properties with water wells, or favorably located for access to groundwater, generally attract a higher market value.

There is a widespread perception that groundwater is of excellent quality, and private water wells are, therefore, reasonably safe. Though there may be some truth in this, there is also increasing evidence of significant pollution [28], and an adequately performing water-utility could offer a supply of more assured quality. On the other hand, the technological and financial inefficiencies of many water-utilities in the developing world leave questions as to whether this can currently be done, and at comparable cost [3,8].

Whether private domestic groundwater use for potable supply presents a serious risk to users themselves will depend on the type of anthropogenic pollution or natural contamination present, and interactions with systems of in-situ sanitation are a particular concern. Some pathogenic microbes, certain synthetic industrial chemicals, and soluble arsenic and fluoride are the greatest hazards. However, there are those who argue that, for some in society, reliable access to any low-cost, parasite-free water-supply is preferable to no water access at all (or use of scarce financial resources on expensive tankered supplies), and that advice on possible use, hazards, and provision of bottled drinking water is a sufficient public-health precaution [9].

3.2. Water-Service Utilities

The existence of groundwater self-supply is of significance to municipal utilities, since it both reduces pressure on (their often limited) water-supplies, and can meet demands whose location or temporal peaks present difficulty for mains water-supply. It is also especially appropriate for uses than are non-quality sensitive.

However, from a water-utility perspective, unregulated private access to groundwater in urban areas also usually means large numbers of richer residents opting to obtain most of their water-supply off-grid. Though this can free up utility water production to meet the needs of lower-income neighborhoods, it will substantially reduce utility revenue collection, and make it more difficult for water utilities to mobilize new investment for water infrastructure and maintain highly-subsidized social water tariffs [1,10]. Moreover, if mains sewerage is (or is planned to be) provided, private water wells will generate additional sewer-flows for which a way will be needed to collect charges (Box 2 and Box 3).

Where the municipal water-supply utility has 'excess developed resources', and is subject to commercial incentives (putting 'financial considerations' before 'social service'), it may well try to market substitution of mains water-supply for private self-supply (to multi-residential properties, and commercial and industrial users, rather than deploying the surplus to improve water-supply to low-income areas). Such action could distort a 'rational policy dialogue'.

3.3. Water Resource Management

Intensive private groundwater use does not necessarily cause serious resource exploitation problems (such as saline intrusion or land subsidence), because of abundant replenishment from water-mains leakage, in-situ sanitation seepage, and other urban sources [3]. The intensive use of private water wells can also incidentally recover a significant proportion of mains water-supply leakage to shallow unconfined aquifers. However, in those cities where the principal aquifers are significantly confined, sustainability problems may arise (Box 9).

Normally, sufficient groundwater resources are not available within the limits of larger cities to meet fast-growing urban water-demand sustainably (Figure 1). Where high-yielding aquifers are present in the hinterland of cities, the development of 'external wellfields' is an attractive option for water utilities compared to long-distance import of surface-water resources. However, groundwater also remains widely exploited by water utilities through individual water wells scattered around urbanized areas, many without specific protection measures.

4. Balanced Public Policy Formulation

4.1. Resource Stocktaking and Risk Assessment

There is an urgent need to take stock of urban private water well use, and better understand the dynamics of overall investment in water-supply provision and its socioeconomic implications. Public policy needs to be formulated to reconcile the widely differing perspectives of private users and public utilities, and establish how private water well use can be better harmonized with public utility water-supply and sanitation investments [1,7,13,26]. A key question is whether 'off-grid solutions' to water-supply provision have an increasing role to play, despite their potential health risks and lack of economy of scale.

It is important that groundwater resources are used efficiently in developing cities, since they can play a key role in climate change adaptation. In this context, it will be important to manage the large groundwater storage of most aquifers to improve water-supply security. To achieve this the monitoring, the assessment, management, and protection of groundwater must be undertaken, and political and public awareness of the resource must be greatly improved.

The public administration will also need to assess the benefits of private groundwater use in terms of relieving pressure on municipal resources (especially for non-sensitive uses, such as garden irrigation, laundry and cleaning, cooling systems, recreational facilities, etc.), and of guarding against the possibility of groundwater table rebound and associated urban drainage problems should groundwater abstraction radically reduce.

A risk assessment of current private urban water well use practices is a pre-requisite for policy development, and will have various components:

- evaluation of the state of aquifer resources, the risk of saline intrusion or continuous water-level decline, and loss of access;
- appraisal of groundwater quality status, including the types of any aquifer pollution and the hazard associated with any natural contamination (such as arsenic or fluoride);
- audit of sanitary construction standards of private water wells, their mode of use, and the implications in terms of health hazard;

If the assessment indicates a high-level of risk to either groundwater resources and/or water well users, certain actions should be implemented as appropriate to local conditions, including use metering and charging (directly or indirectly) to constrain water well abstraction [3], or issuing health warnings and use advice to private water well operators by declaring sources unsuitable for potable and sensitive uses [10].

Certain tools are useful to guide private investment in groundwater development:

- maps of water well yield potential and reliability, depth to main aquifer horizons, static groundwater levels, groundwater pollution vulnerability, potential contaminant loads, and natural quality hazards;
- protocols for water well design, construction, and operation, and the design and operation of in-situ sanitation (septic tanks and improved latrines);
- order-of-magnitude assessment of the status of groundwater resource abstraction, levels of sustainability, and seriousness of risks associated with persistent excessive abstraction;
- guidelines on groundwater use precautions in relation to quality risks, and procedures for rainwater harvesting and aquifer recharge enhancement at the individual urban plot level.

4.2. Improving Regulation and Monitoring

Some regulation of private self-supply from water wells in urban areas is needed, so as to be able to keep the phenomenon in check [3], but if it is not enforced systematically, it is likely to further distort private use. Moreover, regular quality monitoring would be highly advisable, since, without it, the use of private water wells will be a 'risky business', especially from shallow aquifers as a result of the hazard of their significant pollution from in-situ sanitation, industrial discharges, and chemical spillages.

The large majority of private urban water wells are unregulated or completely illegal [17]. If this situation can be regularized, taking advantage of advances in geographical positioning, data capture, and storage systems, it will have a number of benefits:

- urban groundwater users can receive sound information and advice relevant to their use (water well construction standards, pollution risks/alerts, use precautions), and can be protected against the impacts of excessive total abstraction and/or inadequate water well spacing;
- sanitary completion standards of water wells can be improved, and their potential interaction with in-situ sanitation units (latrine, cesspools, and septic tanks) reduced;
- the public administration will be in possession of much better data on private use, which will, in turn, feed into more realistic groundwater resource assessment and municipal water-supply provision;
- public authorities could undertake periodic water analysis as a service to legal urban groundwater users (who, ideally, in return would pay a modest annual 'water resource fee').

The most forthright attempts to regularize the private use of urban groundwater have been in places such as Recife (Box 3) and Fortaleza (Box 2) in Brazil, where the municipal utility has argued for levying of a volumetric water charge in respect to mains sewer use by private water well users. This has resulted in municipal utilities drawing up comprehensive inventories of private water wells on multi-residential, commercial, and industrial properties.

In Fortaleza, charging for sewer use on properties with private water wells is by type/size of property, but in Recife, metering is being introduced for this purpose.

A persistent policy question is under what circumstances do the risks or inconveniences of private residential self-supply in urban areas justify an attempt to ban such use of groundwater. Historically, 'urban groundwater use bans' have been necessarily introduced to address certain specific problems. However, attempts at such bans can be justly criticized because they:

- are unrealistic and, in effect, unimplementable at present;
- could impose intolerable strain on municipal water-supply in some cities;
- do not represent good use of scarce water resources, including the recovery of high levels of physical mains leakage losses;
- run the risk of promoting abandonment of groundwater pumping with water-table rebound, which, in low-lying cities, could imply major disruption and cost.

5. Concluding Summary

Hundreds of millions of urban dwellers worldwide rely on self-supply from private water wells every day. Thus, for local government and municipal water-utilities, finding modalities through which they can find a way of working with, and improving the public health dimensions of, self-supply makes sound sense.

The 'pros' of urban domestic self-supply from groundwater are that it considerably improves water-supply access for some user groups, albeit this only includes the poor, where community action occurs and/or the saturated aquifer occurs at very shallow depth. It also both mobilizes significant private investment in water-supply provision, and incidentally recovers a significant proportion of mains water-supply leakages to the ground.

At a large scale, the 'contras' are that the phenomenon compromises revenue collection, and undermines the investment of municipal water utilities, making it more difficult for them to improve service coverage and quality. Shallow self-supply water wells are also prone to pollution in urban areas from in-situ sanitation, industrial discharges, and accidental spillages.

It is thus very important that urban self-supply from groundwater must not be ignored by the public authorities, and should be systematically included in city-wide surveys and monitoring of water-supply provisions. Policies need to be introduced that encourage municipal water utilities and local government offices to provide services to private water well users in return for formal water well registration and payment of a modest resource fee. Such services need to include periodic chemical analyses to identify pollution hazards or natural contamination, and provide advice on the suitability of the supply for specific uses.

Author Contributions: Conceptualization S.F. 100%; M.E.thodology S.F. 100%; validation S.F. 100%; formal analysis S.F. 40%, R.H. 20%, M.E. 20%, M.-F.A. 20%; investigation S.F. 40%, R.H. 20%, M.E. 20%, M.-F.A. 20%; data curation S.F. 70%, R.H. 10%, M.E. 10%, M.-F.A. 10%; writing—original draft preparation S.F. 100%; writing—review and editing S.F. 70%, R.H. 10%, M.E. 10%, M.-F.A. 10%; supervision S.F. 100%; project administration S.F. 100%. All authors have read and agreed to the published version of the manuscript.

Funding: This research received no external funding.

Institutional Review Board Statement: Not appropriate for review of this type.

Informed Consent Statement: Not appropriate for review of this type.

Data Availability Statement: All data used are given in paper.

Acknowledgments: The first author generously acknowledges the excellent dialogue on this topic maintained during 2018–2021 with members of the IWA Groundwater Management Group Steering Committee (which he Chairs) that enriched the content of this paper.

Conflicts of Interest: The authors declare no conflict of interest.

References

1. Foster, S.; Hirata, R.; Misra, S.; Garduno, H. *Urban Groundwater Use Policy–Balancing the Benefits and Risks in Developing Nations*; GW-MATe Strategic Overview Series; World Bank: Washington, DC, USA, 2010.
2. Grönwall, J.T.; Mulenga, M.; McGranahan, G. *Groundwater, Self-Supply and Poor Urban Dwellers—A Review with Case Studies of Bangalore and Lusaka*; IIED Human Settlements-Water & Sanitation Paper; IIED: London, UK, 2010.
3. Foster, S.; Hirata, R. Groundwater use for urban development: Enhancing benefits and reducing risks. *Water Front.* **2011**, *2011*, 21–29.
4. Grönwall, J. Self-supply and accountability: To govern or not to govern groundwater for the (peri-) urban poor in Accra, Ghana. *Environ. Earth Sci.* **2016**, *75*, 1163. [CrossRef]
5. Foster, S. Global Policy Overview of Groundwater in Urban Development—A Tale of 10 Cities! *Water* **2020**, *12*, 456. [CrossRef]
6. Alam, M.F.; Foster, S. Policy priorities for the boom in urban private wells. *IWA—Source* **2019**, *16*, 54–57.
7. Sutton, S.; Butterworth, J. *Self-Supply: Filling the Gaps in Public Water-Supply Provision*; Practical Action Publishing: London, UK, 2021.
8. Foster, T.; Priadi, C.; Kotra, K.K.; Odagiri, M.; Rand, E.C.; Willetts, J. Self-supplied drinking water in low and middle income countries of the Asia-Pacific. *NPJ Clean Water* **2021**, *4*, 37. [CrossRef]
9. Carrard, N.; Foster, T.; Willetts, J. Groundwater as a source of drinking water in southeast Asia and the Pacific. *Water* **2019**, *11*, 1605. [CrossRef]
10. Foster, S.; Vairavamoorthy, K. *Urban Groundwater–Policies and Institutions for Integrated Management*; GWP Perspectives Paper; Global Water Partnership: Stockholm, Sweden, 2013.
11. Oluwasanya, G.; Smith, J.; Carter, R. Self-supply systems: Urban dug wells in Abeokuta, Nigeria. *Water Sci. Technol. Water Supply* **2011**, *11*, 172–178. [CrossRef]
12. Danert, K.; Healy, A. Monitoring groundwater use as a domestic water source by urban households: Analysis of data from Lagos State, Nigeria and Sub-Saharan Africa with implications for policy and practice. *Water* **2021**, *13*, 568. [CrossRef]
13. Grönwall, J.; Danert, K. Regarding groundwater and drinking-water access through human rights lens: Self-supply as a norm. *Water* **2020**, *12*, 419. [CrossRef]
14. ANA. *Atlas Brasil de Abastecimento Urbano de Água: Panorama Nacional*; Agência Nacional de Águas Engecorps/Cobrape: Brasilia, Brazil, 2010.
15. IBGE. *Perfil dos Municípios Brasileiros*; Instituto Brasileiro de Geografia e Estatística Publication: Brasilia, Brazil, 2017.
16. Foster, S.; Hirata, R.; Custodio, E. Waterwells—How can we make legality more attractive? *Hydrogeol. J.* **2021**, *29*, 10–17. [CrossRef]
17. Foster, S.; Hirata, R. Sustainable groundwater use for developing country urban populations: Lessons from Brazil. *Water* **2012**, *21*, 44–48.
18. Hirata, R.; Suhogusoff, A.; Marcellini, S.; Villar, P.; Marcellini, L. *As Águas Subterrâneas e sua Importância Ambiental e Socioeconômica para o Brasil*; Instituto de Geociências Universidade de São Paulo: São Paulo, Brazil, 2019.
19. Foster, S.; Garduño, H. *Groundwater Use in Metropolitan Fortaleza-Brazil: Evaluation of Strategic Importance and Potential Hazards*; GW-MATe Case Profile Collection; World Bank: Washington, DC, USA, 2006.
20. Hirata, R.; Montenegro, S. Eaux souterraines et sécurité hydrique dans la RMR. In *Affronter le Manque d'Eau dans une Métropole, le cas de Recife, Brésil*; Septentrion Presses Universitaires: Paris, France, 2018; Volume 1, pp. 67–74.
21. CGWB. *National Compilation on Dynamic Groundwater Resources of India*; Central Ground Water Board Publication: Faridabad, India, 2021.
22. Buurman, J.; Santhanakrishnan, D. Opportunities nad barriers in scaling-up of 24/7 urban water-supply: The case of Karnataka-India. *Water Policy* **2017**, *19*, 1189–1205. [CrossRef]
23. World Bank. *Deep Wells and Prudence, towards Pragmatic Action for Addressing Groundwater Overexploitation in India*; World Bank Publication: Washington, DC, USA, 2010.
24. Buapeng, S.; Foster, S. *Controlling Groundwater Abstraction and Related Environmental Degradation in Metropolitan Bangkok–Thailand*; GW-MATe Case Profile Collection; World Bank: Washington, DC, USA, 2008.
25. Foster, S.; Bousquet, A.; Furey, S. Urban groundwater use in tropical Africa—A key factor in enhancing water security? *Water Policy* **2018**, *20*, 982–994. [CrossRef]
26. Foster, S.; Eichholz, M.; Nlend, B.; Gathu, J. Securing the critical role of groundwater for the resilient water-supply of urban Africa. *Water Policy* **2020**, *22*, 121–132. [CrossRef]
27. Matsa, M.M.; Chokuda, F.; Mupepi, O.; Dzawanda, B. An assessment of groundwater quality in Zimbabwe's urban areas: The case of Mkoba-19 suburb, Gweru. *Environ. Monit. Assess.* **2021**, *193*, 7. [CrossRef] [PubMed]
28. Lapworth, D.J.; Nkhuwa, D.C.W.; Okotto-Okotto, J.; Pedley, S.; Stuart, M.E.; Tijani, M.N.; Wright, J.J.H.J. Urban groundwater quality in Sub-Saharan Africa; current status and implications for water security and public health. *Hydrogeol. J.* **2017**, *25*, 1093–1116. [CrossRef] [PubMed]

Article

Water Level Prediction of Emergency Groundwater Source and Its Impact on the Surrounding Environment in Nantong City, China

Jinbang Cai [1], Ping Wang [2], Huan Shen [2,*], Yue Su [2] and Yong Huang [2]

[1] Nanjing Institute of Environmental Sciences, Ministry of Ecology and Environment, Nanjing 210042, China; cjb@nies.org

[2] School of Earth Science and Engineering, Hohai University, Nanjing 210098, China; pingwang@hhu.edu.cn (P.W.); suyue@hhu.edu.cn (Y.S.); hyong@hhu.edu.cn (Y.H.)

* Correspondence: shenh@hhu.edu.cn; Tel.: +86-1529-575-7958

Received: 2 November 2020; Accepted: 14 December 2020; Published: 16 December 2020

Abstract: Based on the geological and hydrogeological conditions, and in situ hydrogeological tests of the emergency groundwater source in Nantong City, China, a 3D numerical model of the heterogeneous anisotropy in the study area was established and calibrated using data from pumping and recovery tests. The calibrated model was used to simulate and predict the water level of the depression cone during the emergency pumping and water level recovery. The results showed that after seven days of pumping, the water level in the center of the depression cone ranged from −51 m to −55 m, and compared with the initial water level, the water level dropped by 29 m to 32 m. The calculated water level has a small deviation compared with that of the analytical solution, which indicates the reliability and rationality of the numerical solution. Furthermore, during water level recovery, the water level of pumping wells and its surroundings rose rapidly, which was a difference of about 0.28 m from the initial water level after 30 days, indicating that the groundwater level had recovered to the state before pumping. Due to the emergency pumping time is not long, the water levels of Tonglu Canal, surrounding residential wells, and other aquifers will not be affected. After stopping pumping, the water level recovers quickly, so the change of water level in a short time will not lead to large land subsidence and has little impact on the surrounding environment.

Keywords: emergency groundwater source; numerical model; drawdown; in situ hydrogeological tests

---

1. Introduction

With the rapid economic development and the acceleration of urbanization, large-scale water pollution in cities has occurred from time to time in recent years, such as the Songhua River, Taihu Lake and Beijiang River and other surface water pollution incidents. The safety of urban water supply is seriously threatened [1–4]. Because the groundwater is less vulnerable to be polluted, it has become an important part of the urban water supply source [5,6]. The emergency water source, as a temporary water supply source that guarantees the basic living water of residents, plays an important role in dealing with large-scale water shortages or water pollution in cities that cause water supply difficulties [7–9].

As many sudden water safety issues are frequently discovered, humans' awareness of prevention of secondary drinking water crises has also been continuously improved. For many countries and regions, the emergency water source has become an important part of the safety guarantee system and the construction of strategic emergency water source has also become an urgent problem to be solved [10–14].

International experts and researchers have done a lot of theoretical and applied research work on the development and utilization of groundwater emergency water source so far [15–23]. Praveen Kumar Amar [24] proposed that water resources in emergency water sources need to be dynamically adjusted and allocated reasonably. Lan et al. [25] established a groundwater flow and migration model based on regional hydrogeological conditions and monitoring data from 52 boreholes, and then carried out water level prediction and environmental impact assessment on the withdrawal plan of the emergency groundwater source in Jiujiang City along the Yangtze River. The results showed that short-term emergency water supply had little impact on the environment. According to the characteristics of emergency water supply, Dai et al. [26] divided water sources into emergency water sources and backup water sources. Taking a northern city in China as an example, a systematic study was conducted on the factors affecting groundwater sources and the environmental effects caused by emergency withdrawal. Zhu et al. [27] established a migration model for the Dawu water source through MODFLOW, which was used to simulate and predict the migration of pollutants in the water source under the action of water interception wells and extraction wells. Song et al. [28] gave a detailed overview of the optimal allocation of urban water supply from the perspectives of changes, modeling methods and challenges and it provided a reference for the construction of urban groundwater sources. In order to identify the influencing factors and environmental impacts of reservoir-type water sources, the reservoir-type water source vulnerability (WSV) assessment method was established by Zhang et al. and was successfully applied to Tianjin Yuqiao Reservoir [29]. In order to evaluate the feasibility of using groundwater as an emergency water source to alleviate the effects of drought, the groundwater model in arid areas was established by Mussá et al. [30] In order to use limited groundwater resources more efficiently and rationally, Bozek et al. [31] proposed a semi-quantitative risk assessment method for emergency groundwater source to improve the distribution and dispatch of groundwater resources in the Czech Republic. Capelli et al. [32] discussed an approach for identifying "strategic groundwater resources" for human consumption and the approach was illustrated in the northern Latium volcanic complexes of Italy. Perfler et al. [33] studied and analyzed the factors affecting the safety and quality of drinking water in Austria, and put forward detailed requirements. In order to better respond to urban emergencies, Zhang et al. [34] analyzed the feasibility of karst groundwater as an emergency groundwater source from the perspective of environmental geology. They believed that the emergency withdrawal will not cause problems such as ground collapse and water pollution. Guo et al. [35] analyzed and studied the groundwater dynamics and flow field evolution characteristics of Huairou emergency water source during the recent 30 years of withdrawal. Wu et al. [36] built a regional groundwater numerical model. The groundwater level recovery capacity of the emergency groundwater source area in Ningbo was predicted under the condition of emergency withdrawal and the environmental effects of groundwater fall funnel, ground subsidence and salt water intrusion caused by emergency withdrawal were evaluated. Ye et al. [37] discussed the concept, characteristics and selection of emergency water source and they also demonstrated the feasibility of constructing emergency water sources in Xinchengzi District of Shenyang. Based on the current water demand and groundwater characteristics of Jilin City in China, Liu et al. [38] carried out a systematic modeling of groundwater emergency source determination, emergency capability analysis and emergency withdrawal plan design. The prediction results of the model showed that the drawdown of the five emergency withdrawal schemes (30 days, 60 days, 90 days, 180 days, and 360 days) was 2.0–8.0 m, which had little impact on the overall groundwater level of Jilin City and the environmental risk was little.

In summary, most of the existing research on emergency groundwater source mainly focuses on the qualitative analysis of emergency water source locations, dynamic characteristics, plan optimization, and protection countermeasures [39–43]. Only a few researchers have explored the environmental impact of groundwater exploitation. The research on water level prediction and environmental impact during the operation period of the groundwater emergency source area is not enough [44–46]. Taking the emergency groundwater source in Nantong as an example, the purpose of this paper is to predict the dynamic changes of the groundwater level and evaluate the impact on the surrounding environment. Firstly, the general situation, topographic features, stratum situation and hydrogeological conditions of the emergency groundwater source in Nantong are described. Then, the numerical method and model calibration method are introduced. Finally, the groundwater level during pumping and recovery is predicted and the impact of the emergency groundwater source on the surrounding environment is analyzed in detail from multiple perspectives based on the situation of the project area, which provides some references for the construction and operation of emergency groundwater source.

2. Study Site

2.1. Study Area

The emergency groundwater source of Nantong is located in Limin Village, Tongzhou District, Jiangsu Province, the length is 801.77 m, the width is 190.1 m, and the total area is 138,115 m^2. The project involves thirteen pumping wells, of which two wells are test wells J_1 and J_2. Groundwater pumping volume for a single well is 1920 m^3/day, and the total pumping volume is 25,000 m^3/day. There are seven wells on the side close to the Tonglu Canal, with a distance of 127–130 m between the wells. The other six wells are on the side far from the riverbank, with a distance of 134–148 m between the wells. The annual average temperature in this area is 15.1 °C, and that of January and July are 2.5 and 27.3 °C, respectively. The annual average precipitation and evaporation are 1084 mm and 857 mm, respectively. The spatial and temporal distribution of precipitation is uneven and mainly concentrated in the summer and autumn. Among them, the plum rain and typhoon rain are mainly from June to July, August to September, respectively. The location of study area and pumping wells is shown in Figure 1.

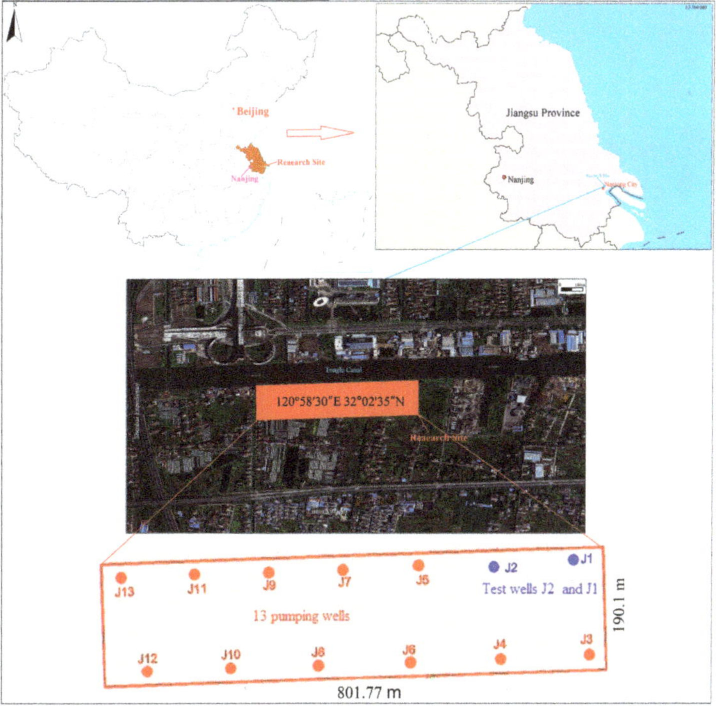

Figure 1. Location of study area and pumping wells.

2.2. Topographic Features

The terrain of Nantong is flat, and the ground elevation is about between 3.2 m and 4.5 m over the sea level. The northwest area is slightly higher than in the southeast. The landforms mainly include three plain areas, namely, (1) delta plain: it is the area extending outward from the ancient barrier spit on the north bank of the Yangtze River. It is also the area where the estuary sandbank contacted the land in the early stage. This landform belongs to the range of fluvial marine deposits and starts from north Fangong dike in the south to Changsha town in the west. There are dish-shaped pits in some areas, such as Zhanghuangdang, Changshadang, and so on; that is, the elevation is less than 3 m. Since the soil in this area was formed earlier, it developed into fluvo-aquic soil after artificial, dry plowing, and maturation; (2) water net plain: this area is distributed in the middle, centered on the sediment zone of ancient river branch in Nantong, extending south to Tonglu Canal sediment zone. The eastern area is higher, with an elevation of about 5 m. The water network here is dense and mostly composed of river and lake silt. The water net polder plain is located in the northern part of Haian and is a low-lying plain area in the Lixia River lagoon sediment zone. The elevation is below 4 m, the farming conditions are better, and the soil fertility is high; and (3) marine deposition plain: the eastern estuary of the mainstream of the Yangtze River, the ancient Hengjiang River, namely the marine deposition plain. The main terrain trend of this area is that the levee on both sides is titled toward the center. Nowadays, the marine sediments outside Fangong dike are mostly artificially modified soils that have formed fluvo-aquic saline soil. The salt content in the 1 m soil is also low at 0.6%, the groundwater salinity ranges from 3.0 to 5.0 g/L, and part of the soil is transitioning to fluvo-aquic soil. The topographic map of the study area is shown in Figure 2.

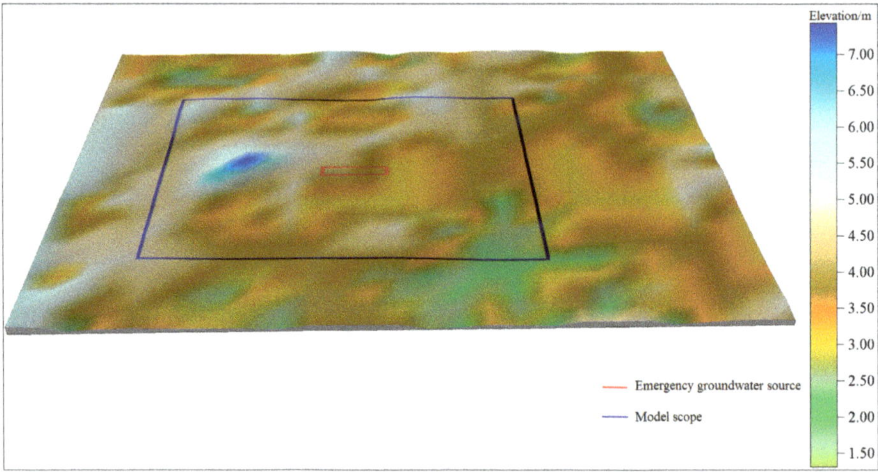

Figure 2. Topographic map of the study area.

2.3. Geological Setting

The area is mainly overlain by Quaternary alluvial deposits which consist of loose sediments with rhythmic changes alternately appearing in sand and clay layers [47]. The thickness of the Quaternary aquifer ranges from 240 m to 300 m. The sedimentary sequence is complex and phase changes are frequent. According to the difference of sedimentary sequence, the Quaternary system can be divided into Lower, Middle, and Upper Pleistocene and Holocene deposits. (1) Lower Pleistocene Series: The overall sequence is a set of fluvial sequences, and the middle section is affected by transgression in some areas and shows the characteristics of estuary sedimentation. The complete basic sequence consists of gravel-bearing coarse sand and medium-fine sand at the bottom, and transitions upward to silty sand and clay loam, showing a complete fluvial facies dual-texture. The thickness of the strata is between 100 m and 150 m. (2) Middle Pleistocene Series: The lithology is divided into upper and lower sections. The lower section is mainly silty sand, fine sand, medium-coarse sand, and gravel-bearing medium-coarse sand, and the upper section is mainly silty clay, silty sand, and fine sand. The thickness of the strata is between 30 m and 60 m. (3) Upper Pleistocene Series: The lithological characteristics are mainly gray, gray-green silty-fine sand. The sedimentary environment is dominated by fluvial facies, and later it appears as floodplain facies. The thickness of the strata is about 100 m. (4) Holocene Series: It is composed of delta facies deposits consist of gray silt and gray-black mucky clay, and a combination of fine clastics mainly composed of silty sand, sandy loam and mucky clay. The thickness of the strata is between 15 m and 50 m.

2.4. Hydrogeological Conditions

(1) Aquifer types

According to the nature of aquifers and the prospecting data, the aquifers are divided into four categories: the phreatic aquifer, the first confined aquifer, the second confined aquifer, and the third confined aquifer [48]. The hydrogeological map of the study area is shown in Figure 3.

a. The phreatic aquifer

The aquifer formation is composed of Holocene Series (Q_4) Yangtze River delta facies which consist of silty clay, silty sand, and silty-fine sand. The buried depth is above 50 m. In the vertical section, the mineral particles in the upper and lower sections are coarser, while in the intermediate section, the mineral particles become finer. The area is divided into two water-bearing sections, namely, the upper and lower water-bearing sections. The upper section is suitable for residential wells, with a small depth, mainly used for washing and irrigation of residents. The lower section is more suitable for shallow wells, with a depth of about 20 m, mainly used for water source extraction in phreatic aquifers during dry periods. The burial depth of the water level fluctuates up and down with the change of seasons, and the water level varies within the range of 1 to 3 m. The lower aquifer has a nature of semiconfined aquifer, and certain sections are connected to the first confined water. The quantity of groundwater pumped of a single well in the upper water-bearing section is not less than 10 m^3/day and the lower above 100 m^3/day. The water quality is freshwater-brackish water, with a salinity of 1 to 1.5 g/L. Due to the low quantity of groundwater pumped, poor water quality, and lack of stability, there is currently no large-scale exploitation in the area. The dynamic variation of groundwater level in the phreatic and the first confined aquifer are shown in Figure 4.

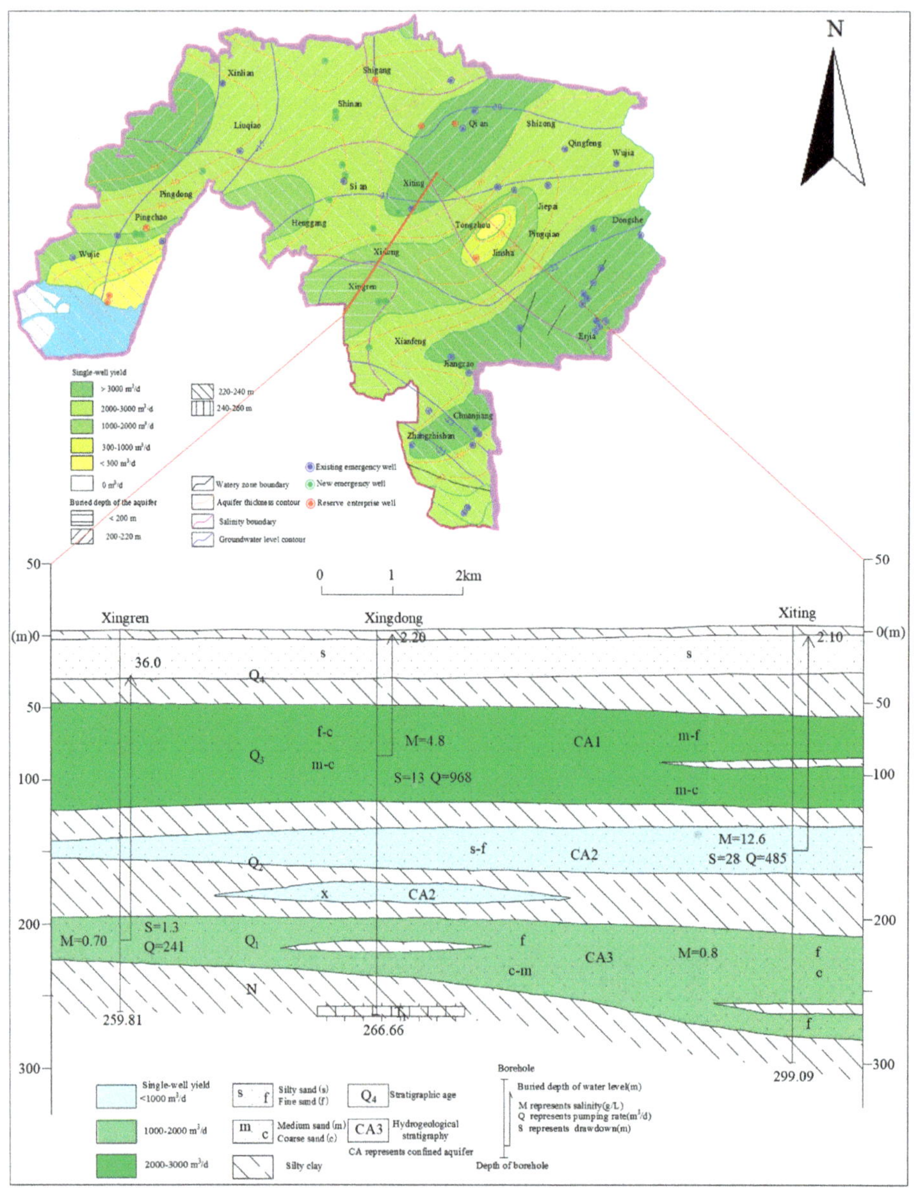

Figure 3. Hydrogeological map of the study area.

b.   The first confined aquifer

The first confined aquifer is mainly composed of the Upper Pleistocene alluvial and alluvial-marine loose sand layer, and the secondary transgression has had a great impact on it. The burial depth of the upper interface of the aquifer ranges from 50 m to 60 m. The lithology is mainly mucky clay or silty clay, and there is silty sand inter-bedded silt in some local areas. The thickness of the aquifer ranges from 10 m to 20 m. The impervious layer composed of gray-yellow, brown-yellow silty clay, clay, and mucky soil forms the lower interface of the aquifer formation. The aquifer in this area is composed of sand pebbles, gravel layers, gravel-bearing coarse sand, medium-coarse sand, medium-fine sand, fine sand, silty sand. The thickness of the aquifer is more than 60 m. The elevation of water level in the first confined aquifer is between 0 and 2 m, but the groundwater level may have dropped to the lowest value of −0.84 m due to the increase in artificial exploitation around in July 2013 (Figure 4). This aquifer is highly permeable and has abundant recharge sources. The discharge rate of a single well is usually between 2000 and 3000 m^3/day. The water quality is relatively poor and is brackish water with a salinity of more than 3.0 to 5.0 g/L.

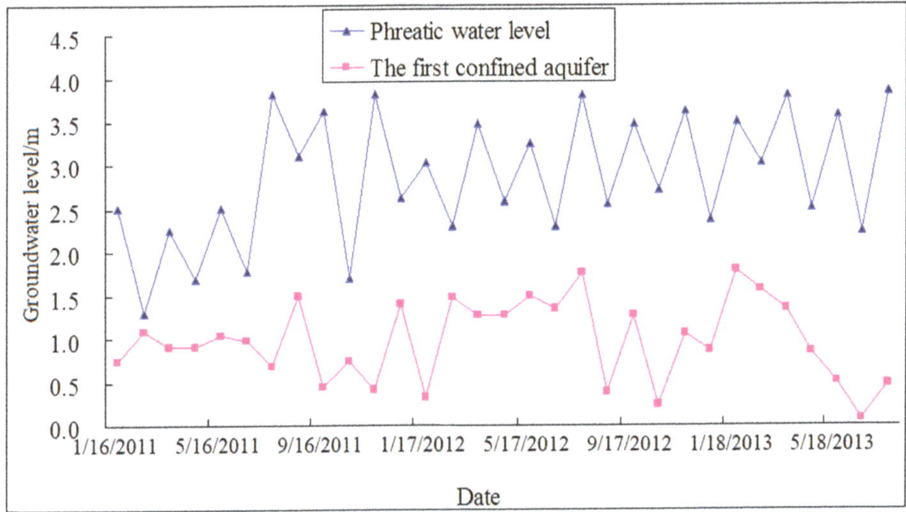

Figure 4. Dynamic variation of groundwater level.

c.   The second confined aquifer

The second confined aquifer is composed of blanket sand formed near the estuary of the Yangtze River during the Middle Pleistocene. The buried depth of the upper interface of the aquifer ranges from 130 m to 150 m. The lithology is mainly gravel-bearing fine and medium-coarse sand, which turns into silty sand in the upper part. The thickness of the aquifer ranges from 25 to 30 m. The quantity of groundwater pumped of a single well is 1000 m^3/day, and the buried depth of static level is in the range of 3 m to 5 m. The salinity is more than 1.0 g/L. At this stage, there is no exploitation activity. The dynamic variation of groundwater level in the second confined aquifer is shown in Figure 5.

d.   The third confined aquifer

The depositional age of the third confined aquifer is the Lower Pleistocene Series (Q_1). The buried depth of the upper interface of the aquifer is 200 m. The lithology is gravel-bearing fine and medium-coarse sand. The thickness of the aquifer ranges from 20 to 25 m. This aquifer is the main exploitation horizon due to its better water richness, wide distribution, and guaranteed water quality. The lithology is mainly gravel-bearing medium-coarse sand, coarse sand, fine sand or gravel-bearing coarse sand, medium-fine sand, silty sand. The quantity of groundwater pumped from the single well ranges from 1000 and 2000 m^3/day. Because the third confined aquifer has good burial conditions, the clay loam of the upper interface is dense and brown, with a relatively stable distribution, and a large thickness, effectively blocking the saltwater from the upper first confined aquifer and second confined aquifer. Therefore, the water quality of the confined aquifer in this area is better than that of other aquifers and is suitable for residents, with salinity less than 1.0 g/L and hardness less than 50 mg/L. The water type is mainly freshwater type, and the water temperature ranges from generally

22 to 23 °C. Due to the deep burial, it is relatively difficult to recharge the groundwater and to fully recover to the static level in a short time. Therefore, the water level fluctuates greatly and water level elevation ranges from −22 to −30 m. The water level change is mainly caused by artificial exploitation. The dynamic variation of groundwater level in the third confined aquifer is shown in Figure 6.

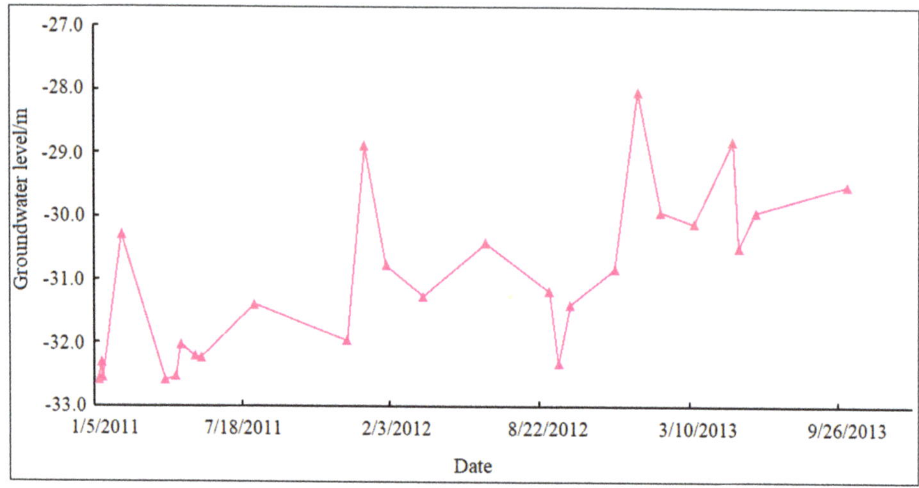

Figure 5. Dynamic variation of groundwater level in the second confined aquifer.

(2) Recharge, runoff, and discharge of groundwater

The phreatic aquifer is closely related to surface water due to its small buried depth and distributes a wide range, and there is a complementary relationship between them. The aquifer is recharged by atmospheric precipitation infiltration, lateral infiltration of surface water bodies, and return of farmland irrigation water. Since this area is located in the estuary area of the Yangtze River delta, the lithological particles of the phreatic aquifer are relatively coarse and the permeability is relatively high, so the runoff conditions are relatively good. The aquifer is discharged by vertical and horizontal direction from phreatic water evaporation, lateral infiltration of surface water bodies and leakage recharge to deep aquifers, and artificial exploitation. Among them, artificial exploitation and phreatic evaporation are the main discharge pathways.

The phreatic aquifer is recharged by atmospheric precipitation infiltration, and then recharged to the first confined aquifer by the leakage, so there is a certain hydraulic connection between the first confined aquifer and phreatic aquifer in this area. Between the first confined aquifer and second confined aquifer, the thickness of the impermeable layer in some sections is very small, only 5–15 m. Therefore, the two aquifers have a relatively close hydraulic connection and a complementary relationship. This area is located downstream of the regional seepage field. The regional seepage recharge comes from the upstream of the northwest and is mainly recharged by atmospheric precipitation infiltration. The confined water-bearing sand layer has high water permeability. Under the effect of hydraulic head, the groundwater produces regional runoff from west to east, which makes the confined water recharge laterally. The runoff and discharge of the first confined aquifer and second confined aquifer are mainly restricted by two factors: regional circulation and artificial exploitation. The former is mainly controlled by the groundwater flow field in the Yangtze River delta and the seepage velocity depends on the hydraulic gradient of the groundwater. Generally speaking, it flows from west to east and from north to south slowly, and the discharge direction is from upstream to downstream. Secondly, artificial exploitation and leakage recharge are also one of the ways of runoff and discharge.

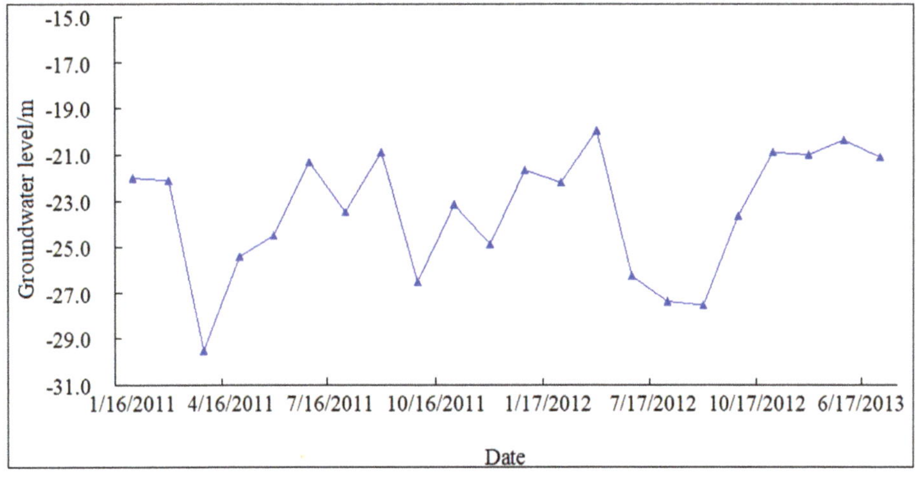

Figure 6. Dynamic variation of groundwater level in the third confined aquifer.

Recharge sources of the third confined aquifer are from the upstream recharge area, lateral recharge of the aquifer in the adjacent area and the elastic and plastic storage of the aquifer. The third confined water runoff in this area mainly flows from north to south, namely from the weak exploitation area to the strong exploitation area. As of the end of 2012, the number of third confined water extraction wells in this area was 105, most of which were located in the southern region, accounting for about 70% of the total. Land subsidence had already occurred in some areas due to overexploitation of groundwater. From north to south, the groundwater level gradually dropped from −15 m to −30 m. The main way of discharge in the third confined water was artificial exploitation.

3. Methods

3.1. Numerical Method

In this study area, numerical methods were used to simulate the groundwater flow movement by using the FEFLOW software (Finite Element Subsurface Flow System). It is developed by Wasy GmbH, Institute for Water Resources Planning and Systems Research Ltd. Germany [49]. It can simulate the regional seepage field and the planning and management plan of groundwater resources, and simulate the impact of underground exploitation on regional groundwater and its optimal countermeasures. It combines the precipitation-runoff model to dynamically simulate the "precipitation-surface water-groundwater" water resources system, analyzes the interdependence between the various components of the water resources system, and studies the rational use of water resources and the impact of ecological environment protection. In this study, the software is used to simulate the change of groundwater level in the third confined aquifer under pumping conditions. The numerical results can provide a relevant basis for the operation of emergency groundwater sources.

Based on the hydrogeological conditions of the study area, a mathematical model of the three-dimensional groundwater flow movement in the study area can be expressed as [50,51]:

$$\begin{cases} \frac{\partial}{\partial x}\left(K_{xx}\frac{\partial H}{\partial x}\right) + \frac{\partial}{\partial y}\left(K_{yy}\frac{\partial H}{\partial y}\right) + \frac{\partial}{\partial z}\left(K_{zz}\frac{\partial H}{\partial z}\right) + W = S_s\frac{\partial H}{\partial t} & (x,y,z) \in \Omega \\ H(x,y,z,t)\big|_{t=0} = H(x,y,z,t_0) & (x,y,z) \in \Omega \\ H(x,y,z,t)\big|_{\Gamma_1} = H(x,y,z,t) & (x,y,z) \in \Gamma_1 \\ K_{xx}\frac{\partial H}{\partial x}\cos(n,x) + K_{yy}\frac{\partial H}{\partial y}\cos(n,y) + K_{zz}\frac{\partial H}{\partial z}\cos(n,z)\big|_{\Gamma_2} = q(x,y,z,t) & (x,y,z) \in \Gamma_2 \\ H(x,y,z,t) = (x,y,z) \in \Gamma_3 \end{cases} \tag{1}$$

where K_{xx}, K_{yy} and K_{zz} are the hydraulic conductivities of the principal directions for the anisotropic aquifers. $H(x, y, z)$ is the hydraulic head value at the instant t. W is the source and sinks term. t is the time. S_s is the specific storage. Ω is the computational domain. $H_0(x, y, z, t_0)$ is the initial water level. The recharge capacity per unit area for the second boundary condition is given by $q(x, y, z, t)$ and $\cos(n, x)$, $\cos(n, y)$, and $\cos(n, z)$ are the directional cosines for the normal to the body surface. Γ_1, Γ_2 and Γ_3 are the first, second and free surface boundary conditions, respectively.

The finite element discretization of Equation (1) can be given as an algebraic equation for the seepage flow domain:

$$[G]\{H\} + [P]\left\{\frac{dH}{dt}\right\} = \{F\} \tag{2}$$

$$\frac{dH}{dt} = \frac{H(t+\Delta t)-H(t)}{\Delta t} \tag{3}$$

$$[G]\{H_{t+\Delta t}\} + [P]\left\{\frac{H(t+\Delta t)-H(t)}{\Delta t}\right\} = \{F\} \tag{4}$$

where $[G]$ is the seepage matrix, $\{H\}$ is the hydraulic head matrix, $[P]$ is the storage matrix, and $\{F\}$ is the known right-hand member.

3.2. Model Calibration Method

The difference between the calculated and observed values in the model is estimated using root mean squared error (RMSE) as the mean average of the squared differences, and the minimum sum of squares of the groundwater level (H) residual is used as the objective function: $\min E = \| H_c - H_0 \|^2$. Thus,

$$E\left(K_j^i\right) = \sum_{k=1}^{M} \sqrt{(H_k^c - H_k^o)^2} \tag{5}$$

where K_j^i is the pending parameters for different material partitions. H_k^c and H_k^o are the calculated and observed groundwater level values, respectively.

3.3. Conceptual Model

(1) Model Scope

The aquifer system of this simulation mainly includes the phreatic aquifer in the Quaternary loose rock pervious aquifer, the first confined aquifer, second confined aquifer, and third confined aquifer, and three aquitards between each aquifer. Considering the geological conditions of the study area and the data from the pumping tests, the aquifer system is generalized as heterogeneous anisotropy, and the horizontal permeability (1.0–5.0 m/day) is greater than the vertical permeability (0.1–0.5 m/day). According to the borehole-log lithology records and the in situ pumping tests, the influence radius is between 320.15 and 565.25 m. Considering the hydrogeological conditions and the errors in the pumping test, the simulation scope is 1 km around the emergency groundwater source. The model calculation area is shown in Figure 7.

(2) Initial and Boundary Conditions

The stream was usually set as the boundary condition, that is, the model scope should not exceed the Tonglu River. The pumping was conducted in the third confined aquifer, but there was no hydraulic connection with the stream, so the stream should not be considered as the evaluation scope. In addition, the recharge between the stream and the phreatic aquifer was through the streambed. The boundary around the study area was the first boundary type. Considering that the influence scope of pumping was small, the water head around the model was treated as a constant. The top boundary of the model was a recharge boundary for precipitation, and also a discharge boundary for evaporation so it was treated as a free surface boundary. The bottom boundary was treated as the impervious boundary, which was the lower interface in the third confined aquifer. The initial water level of the simulation prediction was determined by the water level of the monitoring well in each aquifer, and the initial water level of the entire simulation area was obtained by interpolation. Let the center of the water source be the origin of coordinates in the calculation area, the positive direction of the y-axis along the due north, the positive direction of the x-axis along the due east, and the positive direction of the z-axis along the vertical upward. Considering the geological setting and the relatively large thickness of the aquifer, the model was divided into seven layers in the vertical direction, namely, three aquitards and four aquifers. The model domain was discretized using triangular elements with 97,352 nodes in each slice and 169,358 elements in each layer.

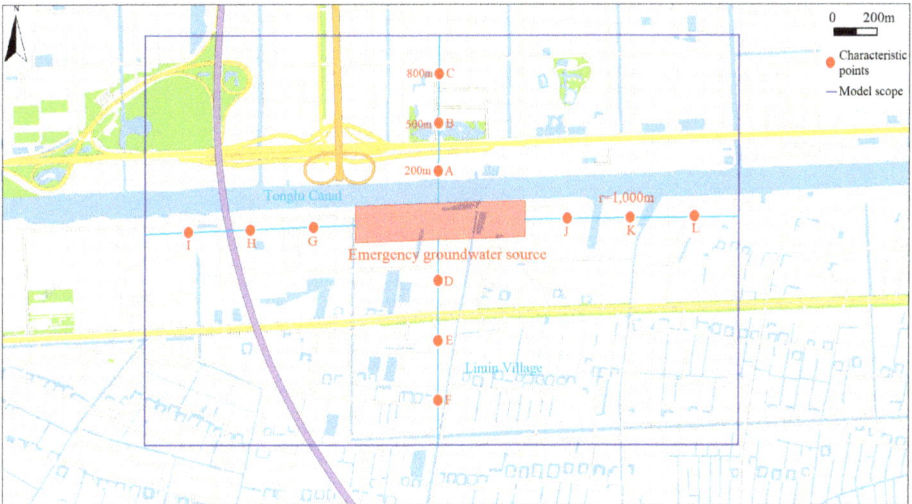

Figure 7. Model calculation area.

3.4. Field Tests and Parameters Determination

(1)   Permeability test in streambed sediments

The permeability test was mainly used to determine the hydraulic connection between the surface water (Tonglu Canal water) and the phreatic aquifer and to quantitatively calculate the hydraulic conductivity of the streambed sediments in the study area. The Figure 8 showed a vertical standpipe in the stream channel, where the lower part of the pipe had been pressed into the streambed and was filled with the unconsolidated sediment. Water was poured into the pipe to fill the rest of the pipe. Then, head losses in the pipe were measured at different times. The water level at the bottom of the sediment column was approximately equal to the water level in the stream. Because of the difference in water levels at the two ends of the sediment column in the pipe, water flowed through the sediment column, and the water table in the pipe fell. The standpipe test used polyvinyl chloride (PVC) pipe with a length of 2 m. Due to the test lasted for a long time, to ensure that the water level in the pipe would not change due to other conditions and cause inaccurate data, the top of the pipe was sealed with a thin film to prevent evaporation or precipitation.

According to the research result of Chen [52], K_v can be determined as

$$K_v = \frac{L_v}{t_1 - t_2} \ln \frac{h_1}{h_2} \tag{6}$$

where K_v is the vertical hydraulic conductivity of the streambed sediments. L_v is the thickness of the measured streambed sediments in the pipe. h_1 and h_2 are the water head in the pipe at time t_1 and t_2, respectively. It was assumed that the river stage was constant during the test.

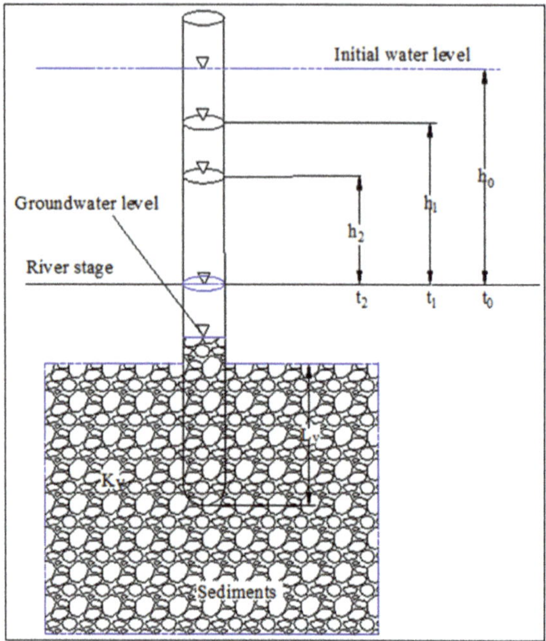

Figure 8. Schematic diagram of the permeability test in streambed sediments.

On the north side of the study area, a vertical standpipe was inserted to conduct two permeability tests in Tonglu Canal, one for K_{v1} and the other for K_{v2}. The position coordinates of vertical standpipe for the first test were 32°2′39.32″ N, 120°58′43.31″ E. The upper part of the pipe was 21 cm away from the river stage and L_{v1} was 41 cm. The observation time was 45 h from 2 November to 4 November. The calculated value of hydraulic conductivity was between 3.83×10^{-7} and 1.11×10^{-6} m/s, and the average value was 6.95×10^{-7} m/s, namely 0.06 m/day. The position coordinates of vertical standpipe for the second test were 32°2′39.22″ N, 120°58′40.22″ E. The upper part of the pipe was 91 cm away from the river stage and L_{v2} was 50 cm. The observation time was 45 h. The calculated value of hydraulic conductivity was between 6.23×10^{-8} and 5.01×10^{-7} m/s, and the average value was 4.68×10^{-7} m/s, namely, 0.04 m/day. According to the results of two tests, the average hydraulic conductivity was 0.05 m/day.

(2) Pumping tests

Pumping tests were conducted to determine the hydrogeological parameters of the aquifer and the hydraulic relationship between the aquifers in test boreholes, numbered as J_1 and J_2. The borehole depth of J_1 was 294 m, the range of depth of pumping level was from 270 m to 290 m, and the borehole depth of J_2 was 300 m, and the pumping level ranges from 204 to 212 m and 276 to 288 m. The observation and recording time of pumping test data was 1 min, 3 min, 5 min, 10 min, 20 min, 30 min, 1 h, and then recorded every other hour. After three tests, if the data obtained was the same or the water level difference did not exceed 2 cm within 4 h, the pumping test was considered stable and could be stopped. After that, the water level recovery observation was needed. The time interval was 1 min, 3 min, 5 min, 10 min, 20 min, 30 min, until the complete recovery. The pumping and recovery test data are shown in Table 1. The calculated hydraulic conductivities were 1.46 and 2.65 m/d, respectively. Since the third confined aquifer was mainly composed of medium-fine sand and silt fine sand, the hydraulic conductivity of the third confined aquifer was between 1.0 and 5.0 m/d.

(3)   Determination of Influence Radius

Table 1. Statistics parameters of pumping and recovery water tests.

Well Number	Static Level (m)	Dynamic Level (m)	Drawdown (m)	Pumping Rate (m^3/day)	Water Temperature (°C)	Stable Time (h)	Recovery of Water Levels (m)	Recovery Time of Water Level (h)
J_1	−22.85	−28.92	5.07	1920	22	25	−22.95	3.5
J_2	−22.65	−26.95	4.0	1920	22	25	−22.85	3.5

The third confined aquifer was divided into upper and lower sections. The upper section was composed of fine sand and medium-fine sand, with a thickness between 13 and 21 m; the lower section was gravel-bearing medium-coarse sand with a thickness between 26 and 30 m. The permeability and water yield properties of the two sections were relatively high. Using J_1 as a pumping well and J_2 as an observation well to conduct in situ pumping tests, and then using the results of the pumping tests to calculate the influence radius when a single well is pumped, as follows:

$$\lg R = \frac{s_w \lg r_1 - s_1 \lg r_w}{s_w - s_1} \tag{7}$$

where R is the radius of influence. s_w is the drawdown of the pumping well. r_1 is the distance from the observation well to the pumping. r_w is the radius of the pumping well. s_1 is the drawdown of the observation well. The radius of the well in this pumping test is 0.3 m. When the pumping rate is 80 m^3/h, the steady drawdown of the pumping well is 5.07 m, the drawdown of the observation well is 0.47 m, and the distance between wells is 127.3 m, so the calculated influence radius is 148.65 m.

Considering that 13 wells working at the same time during the operation of the groundwater source, which the influence radius of them was larger than that of a single well for pumping tests, so the "large-well method" could be used to calculate the influence radius for multi-well tests. The "large-well method" was to simplify the area occupied by the gallery pattern system into a large well in the prediction of discharge rate, and then apply the method of groundwater dynamics to calculate the influence radius and predict the discharge rate of the gallery pattern system. The r_0 can be expressed as

$$r_0 = \eta \frac{a + b}{4} \tag{8}$$

where r_0 is the reference radius, a is the length of the multi-well area, b is the width of the multi-well area, and the values of η, a, and b are shown in. The length and width of the multi-well area are 801.77 and 190.17 m, respectively. According to the Table 2, η is 1.12, that is, r_0 is 277.74 m. The influence radius of large well is calculated using the formula for steady flow to a confined aquifer, and the influence radius of R_0 is between 320.15 and 565.25 m.

Table 2. Relationship of η and a/b.

a/b	0	0.2	0.4	0.6	0.8	1.0
η	1.00	1.12	1.14	1.16	1.18	1.18

3.5. Model Calibration

According to the characteristics of heterogeneity and anisotropy of the study area, combined with the stratum lithology distribution, the water yield property of the aquifer, and other conditions, the study area was divided into four aquifers on the profile (phreatic aquifers, first confined aquifer, second confined aquifer, and third confined aquifer) and three impermeable aquifers (clay loam layer). The hydraulic conductivity obtained from the pumping water tests, recovery water tests and the empirical hydraulic conductivity of each soil layer was taken as the initial values of the model parameters. The initial values of hydraulic conductivity for different aquifers were shown in Table 3. The initial value of the model was verified using Equation (5) in Section 3.2. By comparing the experimental drawdown of the pumping wells with the calculated drawdown, and the flow model parameters were obtained by numerical reverse analysis. The water level fitting curve was shown in Figure 9. It can be seen in Figure 9 that the measured value of the drawdown in 4 to 12 h of pumping test fits well with the calculated value. The initial stage of the pumping test was poor, but the overall variation tendency remained consistent. The inversion parameters can reflect the permeability of the aquifer. The inversion values of hydraulic conductivity for different aquifers were shown in Table 4.

Table 3. Initial values of hydraulic conductivity for different aquifers.

Aquifer Types	Hydraulic Conductivity K (m/day)
Phreatic aquifer	0.5–5.0
1st CA [1]	1.0–5.0
2nd CA	0.5–1.0
3rd CA	1.0–5.0
Aquitard	0.02–0.5

[1] CA means confined aquifer.

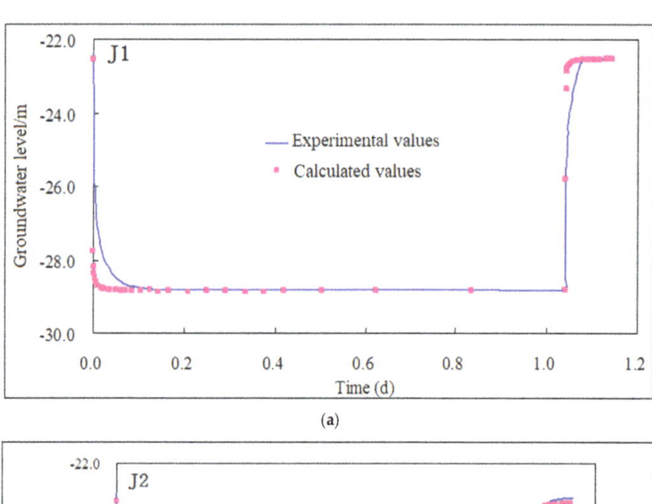

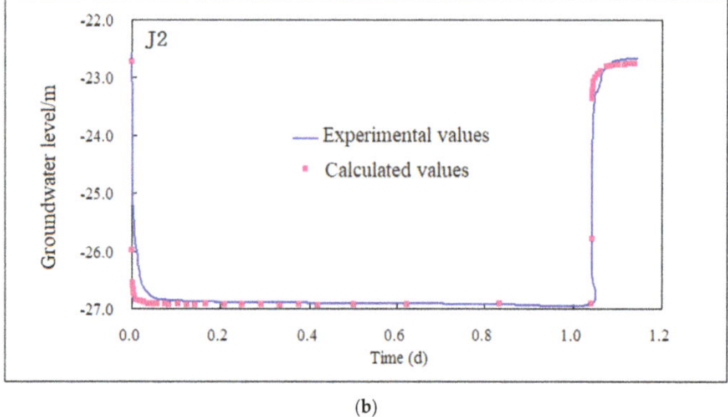

Figure 9. Fitting curves of experimental and calculated groundwater levels from pumping test ((**a**) Pumping test well J1, (**b**) Pumping test well J2).

Table 4. Inversion values of hydraulic conductivity for different aquifers.

Aquifer Types	Hydraulic Conductivity(m/day)		
	K_{xx}	K_{yy}	K_{zz}
Phreatic aquifer	1.5	1.5	0.15
Aquitard	0.03	0.03	0.003
1st CA [1]	3.5	3.5	0.35
Aquitard	0.02	0.02	0.002
2nd CA	5.0	5.0	0.5
Aquitard	0.03	0.03	0.003
3rd CA	4.8	4.8	0.48

[1] CA means confined aquifer.

4. Results Analysis and Discussion

4.1. Prediction of Groundwater Level during Pumping and Recovery Tests

During the operation of the emergency groundwater source, in addition to the existing 2 test wells, 11 new wells were added, numbered as J_3 to J_{13}. The pumping rate of each well was about 1920 m^3/day, and the total pumping rate was about 25,000 m^3/day. All emergency wells exploited groundwater in the third confined aquifer.

(1)　Prediction of groundwater level during pumping

After surface water was contaminated, the time from emergency treatment to water supply resumption generally did not exceed seven days. Therefore, the maximum time for emergency pumping was set to seven days, 13 wells were working at the same time, and the total pumping rate was about 25,000 m^3/day. The calibrated groundwater flow numerical model was used to predict groundwater level change after pumping as shown in Figure 10.

It can be seen from Figure 10 that the drawdown was the most at the multi-well, and a cone of depression was formed there. With the increase of pumping time, the cone of depression gradually spread to the surrounding area. After seven days of emergency pumping, the cone of depression expanded to the boundary. It can be seen from Figure 7 that after 1, 3, 5, and 7 days of pumping, the lowest water level in the center of the depression cone was about −42.11, −49.17, −52.64, and −55.82 m, and the maximum drawdown 19.11, 26.17, 29.64, and 32.82 m, respectively.

The water level changes around the emergency groundwater source can be analyzed by taking characteristic points (Figure 7). Three characteristic points (A–L) were taken from the north, south, west, and east of the water source. The distances from the boundary of the water source are 200, 500, and 800 m, respectively. The calculated water level is shown in Table 5. It can be seen the closer to the water source, the more the drawdown and vice versa. When pumping for five days, the dynamic water level change at 200 m from the water source boundary exceeded −35 m, and the dynamic water level change at 500 m and 800 m from the water source boundary was much lower than the warning stage, which is −35 m.

To understand the changing trend of the groundwater level in the center of the depression cone in the profile, three different profiles were taken for research. The groundwater level change with time was shown in Figure 11. It could be seen that the groundwater level in the center of the depression cone ranged from −51 m to −56 m after seven days of pumping. Compared with the initial water level, the drawdown was about 29–32 m. Among them, the J5-J9 pumping wells located in the center of multi-well had the largest drawdown, with a value of −55 m. The water level of the remaining wells was about −52 m. The water level between the two pumping wells was generally −42 m to −45 m. Based on the principles of groundwater dynamics and using the analytical formula for pumping test from multi-well, the drawdown of 13 wells during the pumping could be calculated, which was between 28.5 m and 31.26 m. The results were consistent with the numerical results, which indicated that the prediction results were reasonable.

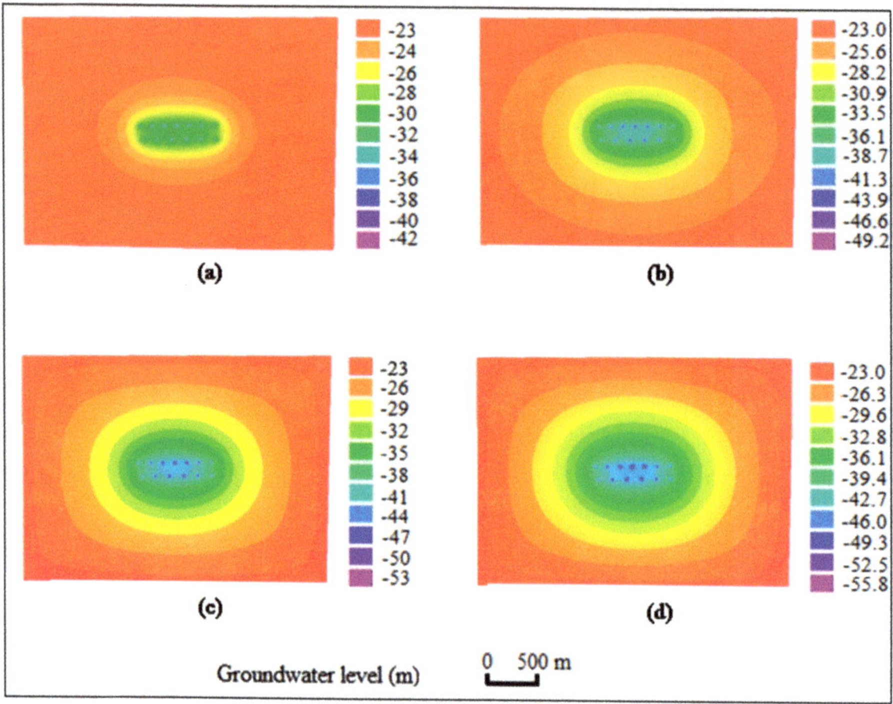

Figure 10. Contour maps of groundwater level in the third confined aquifer during pumping. (**a**) emergency groundwater supply day 1, (**b**) emergency groundwater supply day 3, (**c**) emergency groundwater supply day 5, and (**d**) emergency groundwater supply day 7.

Table 5. Statistics of groundwater level changes around the water source during pumping.

Point Position	The Distances from the Boundary of the Water Source (m)	Groundwater Level after Pumping for a Certain Time (m)			
		1 day	3 days	5 days	7 days
A	200	−26.41	−32.21	−36.90	−38.56
B	500	−24.16	−26.65	−29.87	−29.61
C	800	−23.65	−24.11	−24.53	−25.48
D	200	−26.21	−32.75	−36.78	−38.24
E	500	−24.21	−26.76	−29.70	−29.56
F	800	−23.84	−24.36	−24.68	−25.26
G	200	−25.35	−29.89	−32.16	−34.32
H	500	−24.14	−26.01	−27.76	−28.68
I	800	−23.92	−24.18	−24.49	−24.71
J	200	−25.61	−30.13	−32.96	−34.16
K	500	−24.12	−26.14	−27.69	−28.54
L	800	−23.67	−24.14	−24.26	−24.58

(2)   Prediction of groundwater level during recovery tests

When the emergency groundwater source operated for seven days, the pumping was stopped, and the water level of the pumping well and its surroundings rose rapidly (Figures 12 and 13). On the 1st, 5th, 10th, 15th, 20th, and 30th day during water level recovery, the water level in the center of the depression cone were −42.07, −33.13, −30.16, −26.58, −24.51, and −23.28 m, respectively. Compared with the seventh day of pumping, the water levels increased by 13.75, 22.69, 25.66, 29.24, 31.31, and 32.54 m, respectively. It can be seen that the rising process of water level was fast and then slow.

When the water level recovery was for five days, the lowest water level was −33.13 m. After 30 days of water level recovery, the difference from the initial water level was about 0.28 m. It indicated that the groundwater recovered to the state before pumping (Table 6).

Table 6. Statistics of groundwater level changes around the water source during water level recovery.

Point Position	The Distances from the Boundary of the Water Source (m)	Groundwater Level Recovery for a Certain Time (m)					
		1 day	5 days	10 days	15 days	20 days	30 days
A	200	−37.68	−32.67	−28.90	−26.48	−24.37	−23.26
B	500	−28.95	−26.93	−25.53	−24.98	−23.89	−23.16
C	800	−24.72	−24.29	−23.96	−23.74	−23.22	−23.04
D	200	−37.79	−32.56	−28.76	−26.36	−24.16	−23.24
E	500	−29.04	−26.83	−25.52	−24.85	−23.73	−23.12
F	800	−24.93	−24.26	−23.91	−23.57	−23.24	−23.05
G	200	−33.44	−30.32	−28.20	−26.14	−24.34	−23.26
H	500	−27.35	−26.55	−25.97	−24.59	−23.90	−23.18
I	800	−24.23	−24.02	−23.84	−23.71	−23.19	−23.07
J	200	−33.48	−30.16	−28.36	−26.15	−24.26	−23.24
K	500	−27.17	−26.53	−25.86	24.42	−23.74	−23.17
L	800	−24.29	−24.63	−24.36	−23.69	−23.16	−23.09

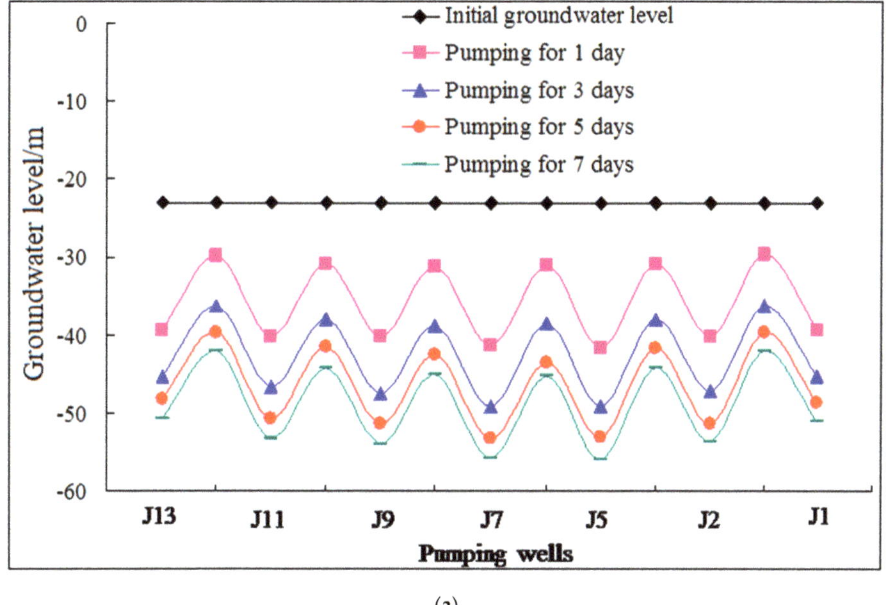

(a)

Figure 11. *Cont.*

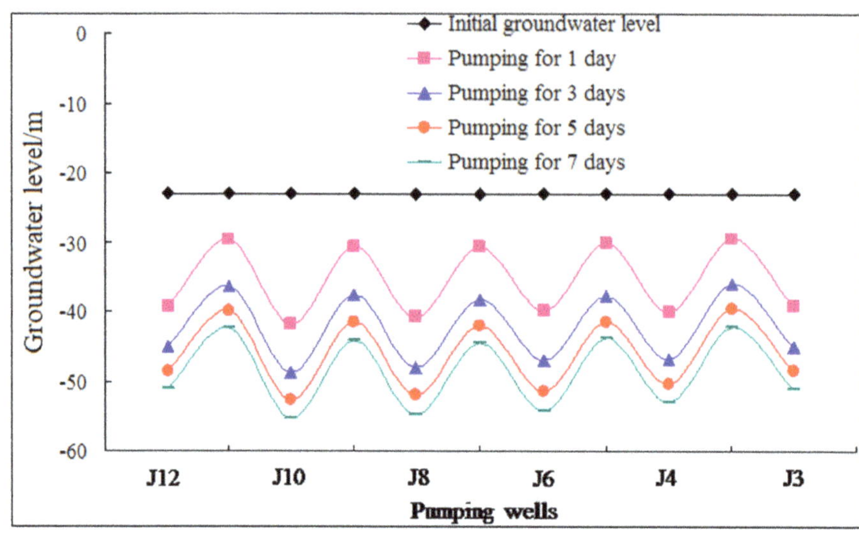

(**b**)

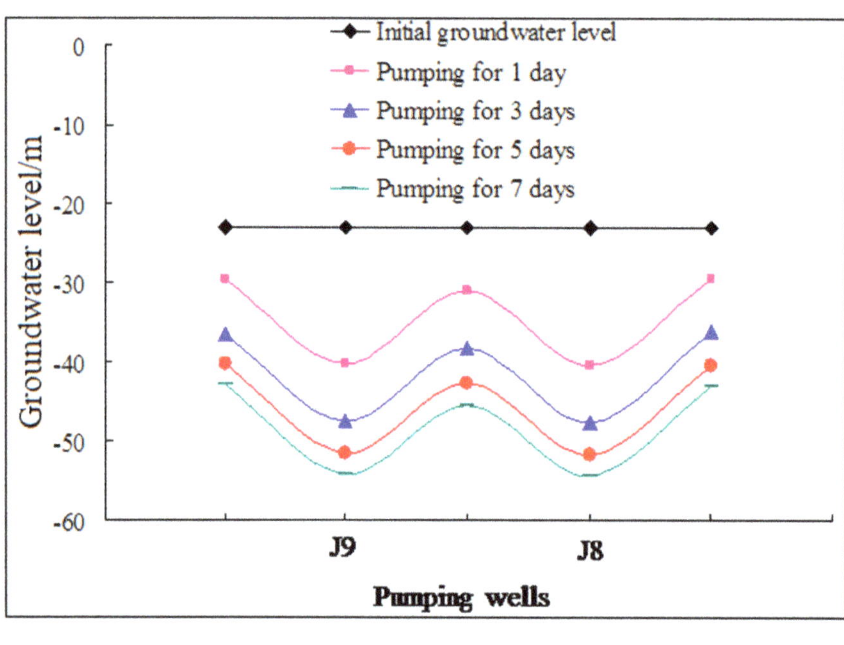

(**c**)

Figure 11. Curves of the cone of depression during pumping ((**a**) Section J13-J1, (**b**) Section J12-J3, (**c**) Section J9-J8).

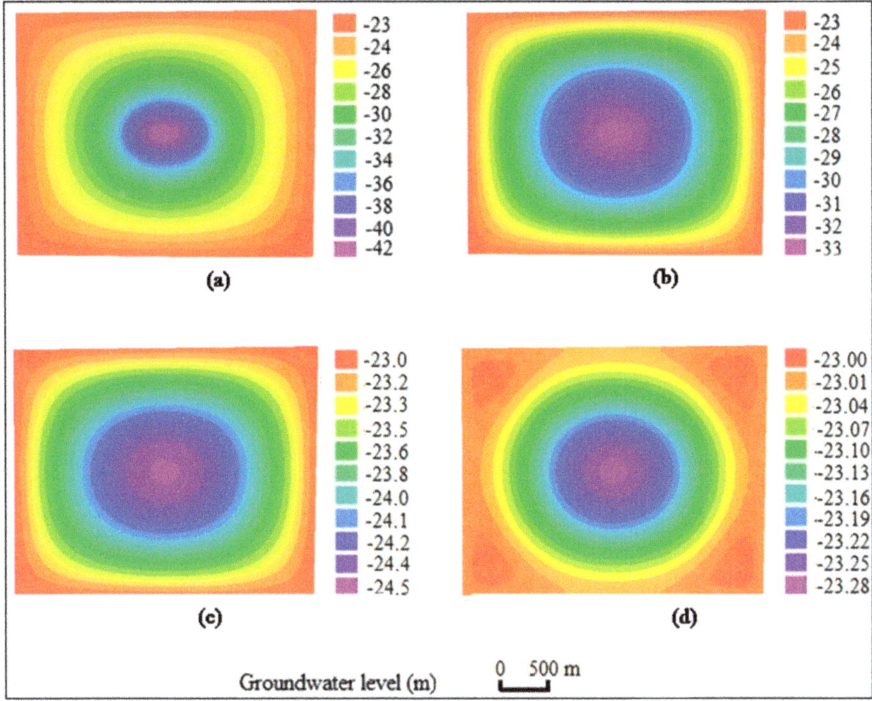

Figure 12. Contour maps of groundwater level in the third confined aquifer during water level recovery ((**a**) water level recovery day 1, (**b**) water level recovery day 5, (**c**) water level recovery day 20, and (**d**) water level recovery day 30).

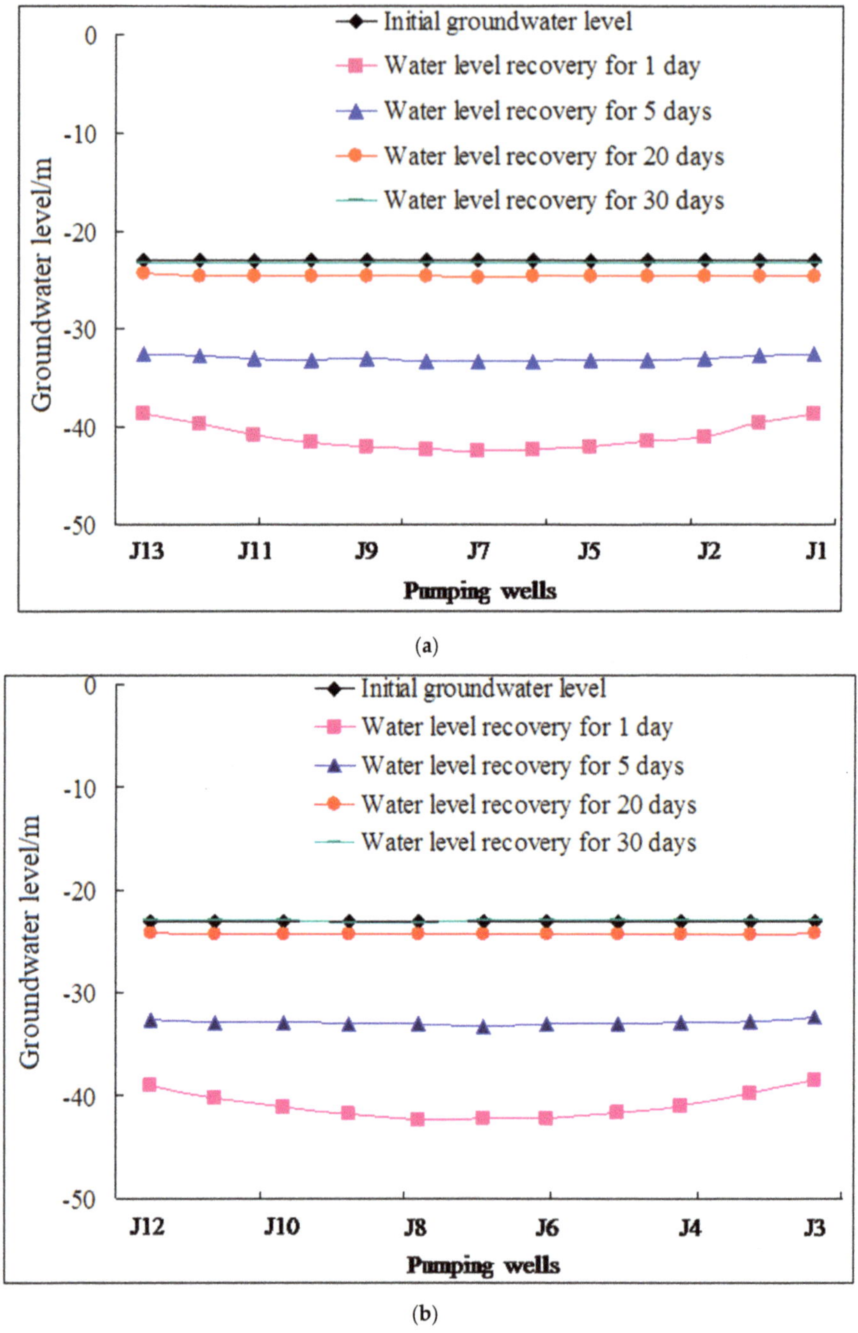

Figure 13. Cone depressions during pumping ((**a**) Section J13-J1, (**b**) Section J12-J3).

4.2. Impact of Emergency Water Sources on the Surrounding Environment during Operation

(1)   Analysis of impact on Tonglu Canal

Tonglu Canal flows near the emergency groundwater source, and the distance from the pumping well is about 20 m. Tonglu Canal is at the proposed location of the Limin village project. The normal water level, the warning water level, the highest water level on records, and the floodwater level in a one-hundred-year return period of the canal were 2.21, 2.61, 3.35, and 3.65 m, respectively. The streambed elevation was −2.10 m and the thickness of streambed sediments ranges from 40 to 100 cm. The permeability was low and the hydraulic conductivity was 0.05 m/day. There was a certain hydraulic connection between Tonglu Canal and phreatic water, and the recharge between them had been considered in the numerical simulation. The emergency groundwater source was located at the third confined aquifer, which is deeply buried. There were three aquitards between phreatic water and third confined water. Therefore, pumping in the third confined aquifer for a short time would hardly affect the phreatic aquifer. That was to say, pumping in the third confined aquifer had no impact on Tonglu Canal.

(2)   Analysis of impact on the surrounding residential wells

At present, the drinking water of the surrounding residents in the project area mainly is from surface water, but some villagers still keep residential wells, which are mainly used for washing clothes, watering vegetables, and livestock water. These residential wells are shallow and are generally distributed in the phreatic aquifer and the first confined aquifer. When pumping in the third confined aquifer, the hydraulic connection between the phreatic water, the first confined aquifer and third confined aquifer is weak, so it will not cause the drawdown in the residential wells. Moreover, there are thicker, low permeability clay or clay loam impervious layers between the aquifer formation in the deep groundwater. The hydraulic connection and the leakage effect between the aquifers are weak. Therefore, groundwater exploitation in the third confined aquifer will not have a significant impact on the phreatic aquifer and the first confined aquifer. It should be pointed out that this prediction came from the model results and analysis of hydrogeological conditions. In the future, more observation wells are needed to confirm our findings.

(3)   Analysis of impact on the second confined aquifer

The depth of the static level and the upper interface of the second confined aquifer range from 3.0 to 5.0 and 130 to 150 m, respectively. The lithology of the aquifer is fine sand with a thickness of 25–30 m. According to subsurface data [53], there is a stable distribution of tight silty clay impervious layer with a thickness of 40–50 m between the second confined aquifer and the third confined aquifer, so the hydraulic connection between the two aquifers is relatively weak. The results of numerical simulation also show that the water level in the second confined aquifer has not changed after seven days of pumping, and it is not affected by the pumping results of the third confined aquifer (Figure 14). It should be pointed out that this conclusion is based on the discussion of numerical simulation. The second confined aquifer did not have the observation well, so more groundwater level changes in the second confined aquifer needed to be verified through observation wells. In addition, due to the barrier effect of the silty clay, the saltwater in the second confined aquifer is also difficult to enter the third confined aquifer, so the water quality of the third confined aquifer will not be affected.

(4)   Analysis of impact on land subsidence

Regional data show that the stratum in this area is relatively soft and easily compressed to cause land subsidence. The total settlement of soil layers is equal to the sum of the settlement of each soil layer. The subsidence at the wall of the pumping well is caused by the drop in the groundwater level. There is no drawdown above the groundwater level, so the land subsidence is neglect. However, the sand or clay layer below the initial water level can cause land subsidence. Therefore, the land subsidence caused by the pumping well occurs in the part of the initial water level to the lower interface of the aquifer. According to the rock formation type and data, the empirical value is selected for calculation. When the maximum drawdown is 32 m, the total settlement is about 27.33 mm (where the total thickness of the water-bearing sand layer and clay layer are 90 and 184 m, respectively; the initial water level is about above 26 m). The third confined aquifer formation is composed of the Lower Pleistocene series (Q_1) strata and has a relatively compact structure and low compressibility. Therefore, during the operation of the emergency groundwater source, a large area of land subsidence will not occur due to larger drawdown.

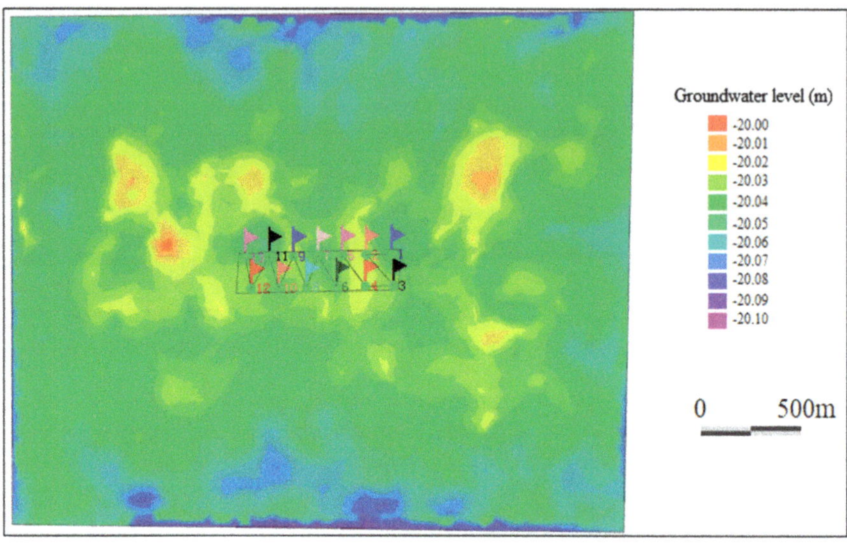

(**a**) Initial groundwater level

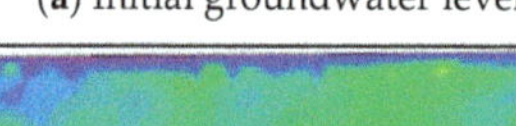

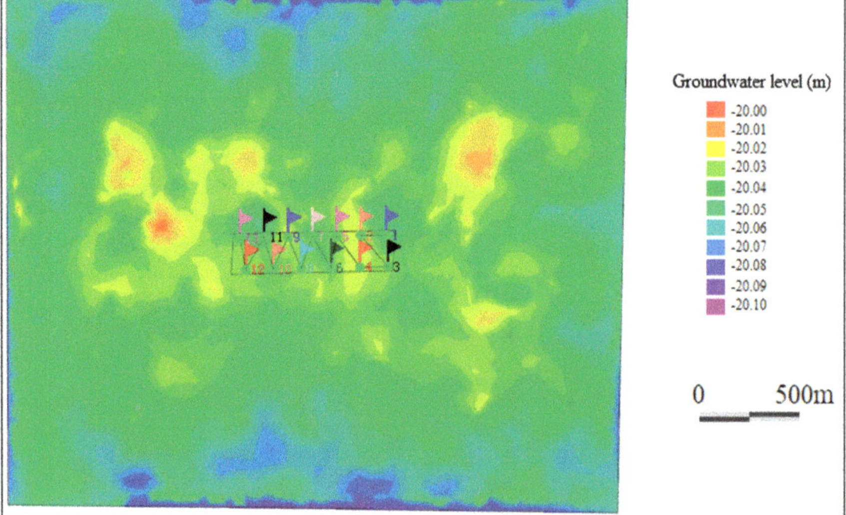

(**b**) Pumping for 7 days

Figure 14. Groundwater level of the second confined aquifer.

In addition, due to local governments restricting groundwater exploitation in recent years, the groundwater level has risen significantly. This measure has effectively controlled the problem of land subsidence. The results showed that during the operation of the emergency groundwater source for seven days, the groundwater level in the center of the depression cone dropped to about −55.82 m (Figure 15), which exceeded the warning water level (namely −35 m) by 20.82 m. If this is maintained for a long time, the water level will be difficult to recover and may cause land subsidence. However, the emergency groundwater source of this project has been operating for less than seven days, and the water level rises quickly during water recovery, so the water level change in a short time will not cause land subsidence.

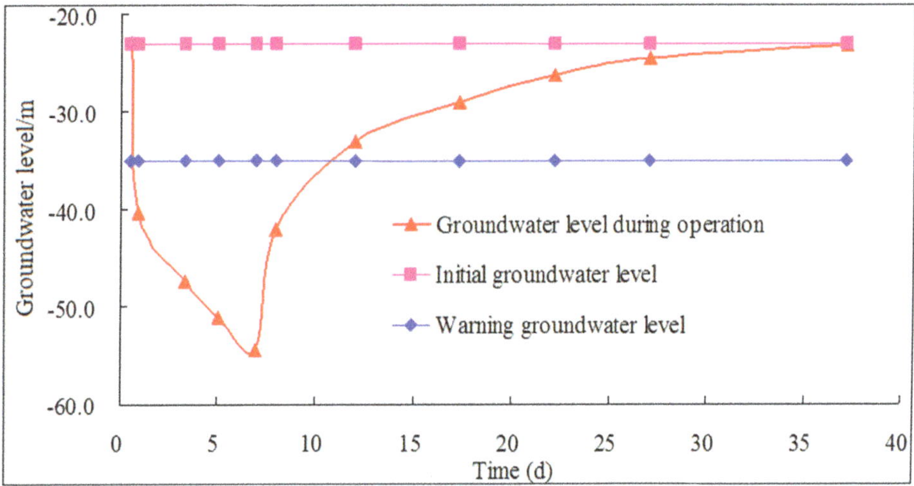

Figure 15. Change curves of groundwater level in the center of depression cone during the operation of emergency groundwater source.

5. Conclusions

The numerical methods are used to predict the change of groundwater level during the operation of emergency groundwater sources and the analysis of its impact on the surrounding environment. The main conclusions drawn from this study are as follows:

(1) According to the vertical standpipe test, the vertical hydraulic conductivity of the streambed sediments was 0.05 m/day, and the hydraulic conductivity of the aquifer was 1.0–5.0 m/day obtained through the pumping test. These parameters were taken as the initial values of the model parameters. Based on the water level of the pumping test and the calculated water level, the established numerical model was calibrated. The calibrated model was used to predict the changes in the groundwater level during pumping and water level recovery.

(2) Numerical results show that the groundwater level at the pumping well had the largest drawdown and a cone of depression was formed there. After seven days of pumping, the water level in the center of the depression cone ranged from −51 to −55 m and compared with the initial water level, the water level dropped by 29 to 32 m. Among them, the J5-J9 located in the center of the multi-well had the largest drawdown, and the water level at the other pumping wells had a small drawdown, which was consistent with the results of the analytical solution; In addition, during water level recovery, the water level of pumping wells and its surroundings rose rapidly, which was a difference of about 0.28 m from the initial water level after 30 days, indicating that the groundwater level had recovered to the state before pumping.

(3) The vertical hydraulic conductivity of the streambed sediments calculated by the vertical standpipe test was small, indicating that the Tonglu Canal had a certain hydraulic connection with the phreatic water. However, the emergency groundwater source was located at the third confined aquifer which was buried depth, and there were three aquitards between phreatic water and

third confined water; therefore, the pumping in the third confined aquifer had no impact on the Tonglu Canal. The residential wells around the emergency water source are shallow and generally distributed in the phreatic aquifer. When pumping in the third confined aquifer, the hydraulic connection between phreatic water and the third confined aquifer was weak, so it would not cause any apparent drawdown in the residential wells. The numerical results also showed that the water level in the second confined aquifer had not changed after 7 days of pumping, and was not affected by the pumping results of the third confined aquifer. During the operation of the emergency groundwater source for seven days, the groundwater level in the center of the depression cone dropped to about −55.82 m, which exceeded the warning water level (namely, −35 m) by 20.82 m. If this is maintained for a long time, the water level will be difficult to recover and may cause land subsidence. However, this project has been operating for less than seven days, and the water level recovers quickly, so the change of water level in a short time will not lead to large land subsidence and has little impact on the surrounding environment.

Overall, the change of groundwater level of the emergency water source in Nantong during the emergency pumping period and recovery period was predicted accurately by numerical method. The normal operation of the emergency water source had low impact on the surrounding environment. At the same time, the impact on the surrounding environment need to be confirmed by setting up more observation wells in the future. In addition, due to the different geological conditions, climatic conditions and economic development conditions of each emergency water source, our research results were limited and preliminary. Further research is needed to assess the impact of emergency pumping on the migration of pollutants.

Author Contributions: H.S. and Y.H. conceived and designed the research theme; Y.H. designed the analytical methods; P.W. and Y.S. contributed to the field investigation and data collection; J.C. and Y.H. analyzed the data and drew the related figures; J.C. wrote the paper and H.S. revised the paper. All authors have read and agreed to the published version of the manuscript.

Funding: This research was funded by The National Natural Science Foundation of China, grant number 41572209.

Acknowledgments: The authors would like to acknowledge the support provided by Pei Tian, Quanjie Zhong and Zhaoning Du of Hohai University.

Conflicts of Interest: The authors declare no conflict of interest.

References

1. Carrard, N.; Foster, T.; Willetts, J. Groundwater as a Source of drinking water in southeast Asia and the Pacific: A multi-country review of current reliance and resource concerns. *Water* **2019**, *11*, 1605. [CrossRef]
2. Zhou, Y.; Wu, F.P.; Chen, Y.P. Emergency reserved water demand estimation for public events of accidental water source pollution. *J. Nat. Resour.* **2013**, *28*, 1426–1437.
3. Scawthorn, C.; Ballantyne, D.B.; Blackburn, F. Emergency water supply needs lessons from recent disasters. In *Water Supply*; Tokyo, Japan, 15–18 November 1998; IWA Publishing: London, UK, 2000; Volume 18, pp. 69–77.
4. Vizintin, G.; Ravbar, N.; Janez, J.; Koren, E.; Janez, N.; Zini, L.; Treu, F.; Petric, M. Integration of models of various types of aquifers for water quality management in the transboundary area of the Soca/Isonzo river basin (Slovenia/Italy). *Sci. Total Environ.* **2018**, *619*, 1214–1225. [CrossRef] [PubMed]
5. Zhao, S.; Huang, B.; Guo, B. Water Environment Prediction and Evaluation on Reservoir Dredging and Source Conservation Project-Case Study of Duihekou Reservoir. In *Communications in Computer and Information Science, Proceedings of the 3rd Annual International Conference on Geo-Informatics in Resource Management and Sustainable Ecosystem (GRMSE), Wuhan, China, 16–18 October 2015*; Springer: Cham, Switzerland, 2016; Volume 569, pp. 520–524.
6. Sheriff, J.D.; Lawson, J.D.; Askew, T.E.A. Strategic resource development options in England and Wales. *Water Environ. J.* **1996**, *10*, 160–169. [CrossRef]
7. Su, H.B.; Zhang, T.M.; Hu, C.Y.; Long, L.Y. Calculation of ecological compensation for water sources for water diversion projects. *IOP Conf. Ser. Earth Environ. Sci.* **2016**, *39*, 12004. [CrossRef]

8. Yang, G.Q.; Li, Y.; Jiang, Y.H.; Liu, H.Y.; Jin, Y. A discussion of the emergency groundwater supply mode in Ningbo City-Dasong River Basin as an example. *Yangtze River* **2020**, *9*, 1–7.
9. Wu, Z.W.; Li, Z.C.; Wang, H.L. Simulation of seepage field under an emergency condition of groundwater supply at Maanshan City. *J. Water Resour. Archit. Eng.* **2020**, *18*, 244–249.
10. Zhu, H.H.; Dong, Y.N.; Xing, L.T.; Lan, X.X.; Yang, L.Z.; Liu, Z.Z.; Bian, N.F. Protection of the Liuzheng water source: A Karst WATER system in Dawu, Zibo, China. *Water* **2019**, *11*, 698. [CrossRef]
11. Tu, L.Q. Estimate of urban emergency water source in Xincai County. *Ground Water* **2020**, *42*, 91–93.
12. Zhang, Y.; Liu, Y.; Mao, L.; Gong, X.L.; Ye, S.J.; Liu, Y.; LI, J. Risk prediction of groundwater emergency water sources in Nantong Binhai new area. *J. Water Resour. Water Eng.* **2019**, *30*, 100–106.
13. Tu, S.B. Study on groundwater emergency water source in Huadu District, Guangzhou. *Ground Water* **2019**, *41*, 11–13.
14. Sun, Y.; Yue, Y.H.; Liu, Z.G.; Cai, Z.C.; Zheng, T. Evaluation on the groundwater resource in riverside emergency water source based on GMS. *Geol. Resour.* **2019**, *28*, 72–77.
15. De Melo, M.T.C.; Fernandes, J.; Midoes, C.; Amaral, H.; Almeida, C.C.; Da Silva, M.A.M.; Mendonca, J.J. Identification and management of strategic groundwater bodies for emergency situations in Portugal. In Proceedings of the 33rd International Geological Congress, Oslo, Norway, 6–14 August 2008.
16. Michalko, J.; Kordík, J.; Bodiš, D.; Malík, P.; Černák, R.; Bottlik, F.; Veis, P.; Grolmusová, Z. Identification and management of strategic groundwater bodies for emergency situations in Bratislava District, Slovak Republic. In Proceedings of the Biennial Conference of the Ground-Water-Division of the Geological-Society-of-South-Africa, Pretoria, South Africa, 10–12 October 2011; CRC Press: Leiden, The Netherlands, 2014; Volume 19, pp. 165–178.
17. Schwecke, M.; Simons, B.; Maheshwari, B.; Ramsay, G. Integrating alternative water sources in urbanized environments. In Proceedings of the 2nd Conference on Sustainable Irrigation Management, Technologies and Policies, Univ. Alicante, Alicante, Spain, 11–13 June 2008; WIT Press: Southampton, UK, 2008; pp. 351–359.
18. Verjus, P. *Albian-Neocomien Underground Water Resources. How to Safeguard and Manage an Emergency Strategic Resource for Drinking Water*; Societe Hydrotechnique de France: Paris, France, 2003; pp. 51–56.
19. Lowry, T.S.; Bright, J.C.; Close, M.E.; Robb, C.A.; White, P.A.; Cameron, S. Management gaps analysis: A case study of groundwater resource management in New Zealand. *Int. J. Water Resour. Dev.* **2003**, *19*, 579–592. [CrossRef]
20. Merz, C.; Lischeid, G. Multivariate analysis to assess the impact of irrigation on groundwater quality. *Environ. Earth Sci.* **2019**, *78*, 1–11. [CrossRef]
21. Carrillo-Rivera, J.J.; Cardona, A.; Huizar-Alvarez, R.; Graniel, E. Response of the interaction between groundwater and other components of the environment in Mexico. *Environ. Geol.* **2008**, *55*, 303–319. [CrossRef]
22. Wang, S.F.; Li, J.; Liu, Y.Z.; Liu, J.R.; Wang, X. Impact of South to North Water Diversion on groundwater recovery in Beijing. *China Water Resour.* **2019**, *7*, 26–30.
23. Meng, Y.; Zheng, X.; Qi, S.; Mingtang, L.; Zhuojun, L.; Long, J. Safe pumping in areas prone to karst collapses: A case study of the urban emergency water source of the Guanghua basin in the Pearl River Delta. *Carsologica Sin.* **2019**, *38*, 924–929.
24. Amar, P.K. Ensuring safe water in post-chemical, biological, radiological and nuclear emergencies. *J. Pharm. BioAllied Sci.* **2010**, *2*, 253–266. [CrossRef]
25. Lan, Y.; Jin, M.G.; Yan, C.; Zou, Y.Q. Schemes of groundwater exploitation for emergency water supply and their environmental impacts on Jiujiang City, China. *Environ. Earth Sci.* **2015**, *73*, 2365–2376. [CrossRef]
26. Dai, C.L.; Chi, B.M.; Liu, Z.P. City's emergency water source field in north China with the example of Changchun City. *Procedia Environ. Sci.* **2012**, *12*, 474–483. [CrossRef]
27. Zhu, H.H.; Zhou, J.W.; Jia, C.; Yang, S.; Wu, J.; Yang, L.Z.; Wei, Z.R.; Liu, H.W.; Liu, Z.Z. Control effects of hydraulic interception wells on groundwater pollutant transport in the Dawu water source area. *Water* **2019**, *11*, 1663. [CrossRef]
28. Song, P.B.; Wang, C.; Zhang, W.; Liu, W.F.; Sun, J.H.; Wang, X.Y.; Lei, X.H.; Wang, H. Urban multi-source water supply in China: Variation tendency, modeling methods and challenges. *Water* **2020**, *12*, 1199. [CrossRef]
29. Zhang, Y.; Zhang, K.; Niu, Z. Reservoir-Type water source vulnerability assessment: A case study of the Yuqiao Reservoir, China. *Hydrol. Sci. J.* **2016**, *61*, 1291–1300. [CrossRef]

30. Mussá, F.E.F.; Zhou, Y.; Maskey, S.; Masih, I.; Uhlenbrook, S. Groundwater as an emergency source for drought mitigation in the Crocodile River catchment, South Africa. *Hydrol. Earth Syst. Sci.* **2015**, *19*, 1093–1106. [CrossRef]

31. Bozek, F.; Bumbova, A.; Bakos, E.; Bozek, A.; Dvorak, J. Semi-Quantitative risk assessment of groundwater resources for emergency water supply. *J. Risk Res.* **2015**, *18*, 505–520. [CrossRef]

32. Capelli, G.; Salvati, M.; Petitta, M. Strategic groundwater resources in Northern Latium volcanic complexes (Italy): Identification criteria and purposeful management. In Proceedings of the International Symposium on Integrated Water Resources Management, Unvi. Calif, Davis, CA, USA, 21–23 April 2000; IAHS Press: Wallingford, UK, 2001; Volume 272, pp. 411–416.

33. Perfler, R.; Unterwainig, M.; Mayr, E.; Neunteufel, R. The security and quality of drinking water supply in Austria–Factors, present requirements and initiatives. *Österreichische Wasser Abfallwirtsch* **2007**, *59*, 125–130. [CrossRef]

34. Zhang, X.L.; Li, F.; Liu, H.Z. Analysis on the emergency-type groundwater source fields of Qujing City in Yunnan. *Adv. Mater. Res* **2013**, *610–613*, 2653–2657. [CrossRef]

35. Guo, G.X.; Shen, Y.Y.; Zhu, L.; Li, Y.; Xu, L. Evolution of groundwater flow field in Huairou emergency groundwater well field and its surrounding area under impacts of multiple factors. *South North Water Transf. Water Sci. Technol.* **2014**, *12*, 160–164.

36. Wu, B.H.; Zhou, Q.S.; Pan, X.Q.; Lin, D.H. Research of emergency water source area construction and environmental effect evaluation for Shepan Island of Ningbo City. *Water Resour. Prot.* **2017**, *33*, 41–48.

37. Ye, Y.; Xie, X.M.; Chai, F.X.; Zhao, Q.S. Research on groundwater emergency water source field of city. *Water Resour. Power* **2010**, *28*, 47–49.

38. Liu, J.F. Research on Groundwater Emergency Source Field for Drinkingwater in Urban Jilin City. Ph.D. Thesis, Jilin University, Changchun, China, June 2017.

39. Lu, C.J. Resources Evaluation on Groundwater of the Emergency Wellfield in the NORTH Wuqing District of Tianjin. Master's Thesis, China University of Geosciences (Beijing), Beijing, China, June 2017.

40. Wu, X.Y.; Gu, J.H. Advance in research on urban emergency management capability assessment at home and abroad. *J. Nat. Disasters* **2007**, *16*, 109–114.

41. Polomcic, D.; Gligoric, Z.; Bajic, D.; Cvijovic, C. A hybrid model for forecasting groundwater levels based on fuzzy C-mean clustering and singular spectrum analysis. *Water* **2017**, *9*, 541. [CrossRef]

42. Seyam, M.; Alagha, J.S.; Abunama, T.; Mogheir, Y.; Affam, A.C.; Heydari, M.; Ramlawi, K. Investigation of the influence of excess pumping on groundwater salinity in the Gaza Coastal Aquifer (Palestine) using three predicted future scenarios. *Water* **2020**, *12*, 2218. [CrossRef]

43. Zhu, F. Numerical Simulation of Ordovician Karst Water and Analysis of Environmental Effect of Groundwater Exploitation in Xishan, Beijing. Ph.D. Thesis, Capital Normal University, Beijing, China, May 2014.

44. Yu, G.C.; Zhou, Z.Y.; Yang, L.H.; Huang, W.P.; Wang, X.H. Anticipation of environmental effects and delimitation of emergency groundwater sources in Jiaxing. *Bull. Sci. Technol.* **2017**, *33*, 53–57.

45. Zhu, M.J.; Wang, S.Q.; Kong, X.L.; Zheng, W.B.; Feng, W.Z.; Zhang, X.F.; Yuan, R.Q.; Song, X.F.; Sprenger, M. Interaction of surface water and groundwater influenced by groundwater over-extraction, waste water discharge and water transfer in Xiong'an new area, China. *Water* **2019**, *11*, 539. [CrossRef]

46. Liu, S.X.; Shen, H.T.; Zhao, J.K.; Wu, M.J. Geo-Environmental effects since the beginning of groundwater exploitation restriction in the Zhejiang coastal plain. *J. Geol. Hazard Environ. Preserv.* **2013**, *24*, 37–44.

47. Li, C.X.; Chen, Q.Q.; Zhang, J.Q.; Yang, S.Y.; Fan, D.L. Stratigraphy and paleoenvironmental changes in the Yangtze Delta during the Late Quaternary. *J. Asian Earth Sci.* **2000**, *18*, 453–469. [CrossRef]

48. Ma, Q.; Luo, Z.; Howard, K.W.F.; Wang, Q. Evaluation of optimal aquifer yield in Nantong City, China, under land subsidence constraints. *Q. J. Eng. Geol. Hydrogeol.* **2018**, *51*, 124–137. [CrossRef]

49. Diersch, H.-J. *FEFLOW: Finite Element Modeling of Flow, Mass and Heat Transport in Porous and Fractured Media*; Springer: Berlin/Heidelberg, Germany, 2014; p. 996.

50. Huo, Z.L.; Feng, S.Y.; Kang, S.Z.; Cen, S.J.; Ma, Y. Simulation of effects of agricultural activities on groundwater level by combining FEFLOW and GIS. *N. Z. J. Agric. Res.* **2007**, *50*, 839–846. [CrossRef]

51. Xue, Y.Q.; Xie, C.H. *Numerical Simulation for Groundwater*; China Science Publishing & Media Ltd.: Beijing, China, 2007.

52. Chen, X.L. Measurement of streambed hydraulic conductivity and its anisotropy. *Environ. Geol.* **2000**, *39*, 1317–1324. [CrossRef]

53. Zhu, S.F.; Sheng, J.; He, T.J. The application of geophysical well logging in a hydrogeological survey in the Nantong area. *Shanghai Land Resour.* **2016**, *37*, 89–91.

Publisher's Note: MDPI stays neutral with regard to jurisdictional claims in published maps and institutional affiliations.

water

Article

Application of ^{2}H and ^{18}O Isotopes for Tracing Municipal Solid Waste Landfill Contamination of Groundwater: Two Italian Case Histories

Francesca Andrei [1], Maurizio Barbieri [2] and Giuseppe Sappa [1,*]

1. Department of Civil Building and Environmental Engineering (DICEA), Sapienza University of Rome, 00184 Rome, Italy; francesca.andrei@uniroma1.it
2. Department of Earth Sciences (DST), Sapienza University of Rome, 00185 Rome, Italy; maurizio.barbieri@uniroma1.it
* Correspondence: giuseppe.sappa@uniroma1.it; Tel.: +39-064-458-5010 or +39-345-280-8882

Abstract: Groundwater contamination due to municipal solid waste landfills leachate is a serious environmental threat. During recent years, the use of stable isotopes as environmental tracers to identify groundwater contamination phenomena has found application to environmental engineering. Deuterium (^{2}H) and oxygen (^{18}O) isotopes have successfully used to identify groundwater contamination phenomena if submitted to interactions with municipal solid waste landfills leachate, with a significant organic amount. The paper shows two case studies, in central and southern Italy, where potential contamination phenomenon of groundwater under municipal solid waste landfills occurred. In both cases, isotope compositions referred to ^{2}H and ^{18}O highlight a δ^2H enrichment for some groundwater samples taken in wells, located near leachate storage wells. The δ^2H enrichment is probably caused by methanogenesis phenomena, during which the bacteria use preferentially the hydrogen "lighter" isotope (^{1}H), and the remaining enriched the "heavier" isotope (^{2}H). The study of the isotope composition variation, combined with the spatial trend of some analytes (Fe, Mn, Ni) concentrations, allowed to identify interaction phenomena between the municipal solid waste landfills leachate and groundwater in both case histories. Therefore, these results confirm the effectiveness of ^{2}H isotopes application as environmental tracer of groundwater contamination phenomena due to mixing with municipal solid waste landfills leachate.

Keywords: environmental isotope; δ^2H; municipal solid waste; leachate contamination; natural tracers

Citation: Andrei, F.; Barbieri, M.; Sappa, G. Application of ^{2}H and ^{18}O Isotopes for Tracing Municipal Solid Waste Landfill Contamination of Groundwater: Two Italian Case Histories. *Water* **2021**, *13*, 1065. https://doi.org/10.3390/w13081065

Academic Editor: Giorgio Mannina

Received: 3 March 2021
Accepted: 9 April 2021
Published: 13 April 2021

Publisher's Note: MDPI stays neutral with regard to jurisdictional claims in published maps and institutional affiliations.

1. Introduction

Municipal solid waste landfills leachate can cause serious environmental problems for groundwater quality, due to the organic and inorganic pollutants of the leachate plumes [1]. Landfill leachate is defined as a liquid effluent containing contamination materials percolating through deposited waste and released within a landfill [2]. It is one of the most complex waste liquids to manage because it contains many organic and inorganic compounds such as nutrients, chlorinated organics, dissolved organic matter, inorganic compounds (e.g., ammonium, calcium, magnesium, sodium, potassium, iron, sulfates, chlorides) and heavy metals (e.g., cadmium, chromium, copper, lead, zinc, nickel) [3–7]. In fact, leachate coming from biological and physico-chemical decomposition of waste can cause damage to the environment when it percolates to the surrounding groundwater and stream water systems [3,7,8]. The leachate composition, in terms of chemical and microbiological matters, varies among different landfill types and depends on the characteristics of the solid wastes, age of the landfill, climate, environmental conditions, landfill operational mode and decomposition mechanism of the organic matter [3,7,9,10]. The spatial distribution and variation of the landfill leachate depend on the nature and partitioning of the waste properties (contouring degree and solid wastes compacting), moisture content, temperature, pH, oxygen level, microbial activity, groundwater inflow, surface water runoff, local rainfall regime, hydrogeological settings and characteristics of landfill (age, design, such as size, depth and lining system) [6].

Generally, waste decomposing processes by bacteria in landfill involve four stages: (i) hydrolysis, (ii) acid fermentation, (iii) acetogenesis and methanogenesis and (iv) settlement phase. This entire system is dynamic, and every phase create a suitable environment by the preceding stage, leading directly towards the gas and leachate production [6,11]. The first phase of waste decomposition depends on the oxygen amount of organic matter and this phase continue until the available oxygen is deplete. This phase is followed by reactions such as oxidation, hydrolysis and anaerobic acidification. In particular, the third phase, acetogenesis and methanogenesis, produces a decrease in acetic acid (CH_3COOH) and determines the production of methane (CH_4) and carbon dioxide (CO_2). CH_4 and CO_2 are the major landfill gases and their formation is influenced by bacterial decomposition, waste composition, organic matter availability, moisture content, pH, temperature and possible chemical reactions in landfill [7,11].

Therefore, it is fundamental the assessment of environmental risks associated with landfill leachate going to water resources, specially to groundwater. These phenomena need environmental monitoring programs that have to detect the potential leachate influence in groundwater near to the landfill area. The monitoring is important for the leachate characterization and to avoid or mitigate environmental damages [12]. Several studies [1,7,8,13–16] have proposed multidisciplinary approaches to supply information about landfill environmental impacts. In order to assess the landfill leachate environmental impact, in addition to conventional parameters such as total dissolved solids (TDS), total suspended solids (TSS), hardness, alkalinity, chloride, chemical oxygen demand (COD), biochemical oxygen demand (BOD), total organic carbon (TOC), and common inorganic ions, the use of additional tools such as stable isotopes can provide more information.

Generally, environmental tracers can be defined as physical properties and chemical components of water, whose spatial and temporal differentiation can be used to give information about pathways and dynamics of water through the environment [17]. The feature of environmental tracers (physical, chemical and isotopic) is that they are part of environmental e not must be artificially introduced [18]. The traditional isotope ratios of hydrogen (^2H) and oxygen (^{18}O) are recognizable tracers related to the water cycle and have widely used in hydrology [19]. In particular, the ^2H and ^{18}O isotopes are used in environmental hydrogeology not only because they are the main constituents of the water molecule but, also, they are stable isotopes with a nuclear composition not changed over time.

However, they can be subjected to isotopic fractionation phenomena. The isotopic fractionation is defined as the partitioning of heavy isotopes versus light isotopes in exchange reactions. This phenomenon will be more evident if the mass difference between the isotopes is sensitive. In environmental hydrogeology, evaporation and condensation processes highly influence the isotopic content: the "lighter" molecules tend to evaporate more quickly than the "heavier" ones. Instead, the opposite phenomenon is noted during the condensation processes [20–23].

Several studies [4,8,15,22] have observed that stable isotopes by landfill leachates, such as $\delta^{13}C$, δ^2H and $\delta^{18}O$, are influenced by processes within municipal solid waste (MSW) landfills, mainly on methanogenesis phase of the landfill. Landfill gases (CO_2 and CH_4) and landfill leached products (water and inorganic carbon) have a characteristic isotopic, in terms of δ^2H e $\delta^{18}O$, respect to the surrounding environment [24,25].

Methanogenesis is the process that determines the production of methane (CH_4) and carbon dioxide (CO_2) in a landfill by some microorganisms, known as methanogens. They, under anaerobic conditions, use organic and inorganic compounds to produce CH_4 and CO_2 [26–28]. The first two phases of the waste decomposing processes, hydrolysis and acid fermentation, cause a substantial reduction in the oxygen quantity in the landfill [11]. It is necessary to consider that isotope fractionation is a phenomenon strongly influenced by state change, such as condensation and evaporation processes [20–23]. Therefore, the changes in oxygen quantity due to hydrolysis and acid fermentation processes may not necessarily cause also variations in the 18-oxygen isotopic content. However, there aren't yet solidified scientific results that allow to prove the hydrolysis and acid fermentation influence on the isotopic content, in terms of δ^2H e $\delta^{18}O$. On the contrary, several studies [29–31] have shown that the methanogenesis phase is the cause of the enrichment of the dissolved inorganic carbon ($\delta^{13}C$) and hydrogen (δ^2H) stable isotopes in the leachate.

During the oxidation of the organic substance under anoxic environment, the methanogens can follow two metabolic pathways, are shown below [32]:

i. $CO_2 + 4H_2 \rightarrow CH_4 + 2H_2O$ (CO_2 reduction);

ii. $CH_3COOH \rightarrow CH_4 + CO_2$ (acetate fermentation).

In case of CO_2 reduction (i), microorganisms first use the "lighter" molecule of carbon dioxide (^{12}C) to produce CH_4, leaving the remaining CO_2 enriched in the "heavier" one (^{13}C). When acetate fermentation occurs (ii), CH_4 produced by the reaction is isotopically depleted in ^{13}C, and, on the contrary, CO_2 produced by the reaction (ii) is isotopically enriched in ^{13}C [23,32]. During the acetate fermentation process (ii), the hydrogen comes partly (¾) from CH_3COOH and partly (¼) from water present in the leachate. On the contrary, during CO_2 reduction process (i), the hydrogen is assumed to come completely from leaching water and therefore water is significantly more enriched in the "heavier" isotopes (^2H). Therefore, isotopic fractionation occurring is due to the preferential use of "lighter" hydrogen (^1H) isotope with the gas phase, CH_4, leaving the "heavier" isotope (^2H) in the liquid phase [8]. Due to the methane production, δ^2H ratios can be located outside of the traditional ranges associated with natural waters of direct meteoric origin meteoric water line, undergoing a shift to positive range. The δ^2H enrichment in groundwater surrounding landfill sites can be used as an indicator of leachate migration from landfill to groundwater and other activities near the study area [22].

Moreover, studies [2,15,24] have demonstrated that the landfill leachate is highly enriched in δ^2H respect to local average precipitation values (by δ^2H: +30‰ to 60‰) due to the production of microbial methane within the reservoir of a landfill. Another factor that can determine the variation of δ^2H content in the leachate produced by landfills is the alternation between dry and wet seasons. Some studies [8] have also shown that δ^2H in leachate is equal to +16‰ in dry season and +6‰ in wet season. This variation between dry and wet seasons is plausibly caused by the rainwater dilution, making less evident the isotopic signature of leachate [8]. On the contrary, it has also been observed that methanogenesis does not affect the ^{18}O isotopic composition [8,33]. Studies highlight [24,34] that ^{18}O isotopic composition variations in leachate are due to seasonal variations of precipitation, reflecting a rapid movement of water through the pile with minimal storage in the waste. In fact, Hackley et al. (1996) [24] have showed that $\delta^{18}O$ was depleted in seepage water, related to local precipitation value (by $\delta^{18}O$: -1‰ to -3‰) [15].

However, the degradation of organic matter is not constant, during the landfill operational time. In fact, the burial of wastes in a landfill has been developed, depending on some complex series of chemical and biological reactions. During the methanogenesis, the accumulated acids are converted to CH_4 and CO_2 by bacteria, and the methane production rate will increase. During this phase, the methane production rate will reach its maximum, and decrease thereafter as the pool of soluble substrate (carboxylic acids) decreases [35,36]. Therefore, this behavior could affect the ^2H isotopic contents, but, on the contrary, it doesn't involve ^{18}O, at all. Despite experimental results [37] showing that the δ^2H isotopic content increases continuously from the waste deposition [15], even if there aren't yet consolidated scientific results that can confirm this phenomenon.

This paper purpose is to show the effectiveness of the application of the hydrogen (^2H) as tracer of the leachate presence in groundwater near to municipal solid waste landfills, when it is present a significant organic amount. As a matter of fact, here it is shown as the leachate presence in groundwater can influence δ^2H ratios without determining an $\delta^{18}O$ enrichment. Two case studies are presented: the former in Sardinia, southern Italy, and the latter in Umbria, central Italy. The two case histories show how the use of the hydrogen

(δ^2H) and oxygen (δ^{18}O) isotopes, combined with other investigation tools, can highlight leachate contamination phenomena in groundwater.

2. Geological and Hydrogeological Setting

2.1. Case History I in Sardinia (Southern Italy)

The study area is located in Sardinia, in the province of Cagliari, on the border between two different municipalities (Figure 1). The landfill plant is part of a complex geological setting, characterized by the presence of the Nurallao deposits, consisting mainly of sandstones (Serra Longa sandstones) (Geological Map of Sardinia (1:25,000.00)). However, on the north-west side of the plant, it is also possible to check for the presence of alluvial deposits, dates to the Holocene epoch, and the anthropic deposits. The anthropic deposits are from both the landfill and the quarries near the study area.

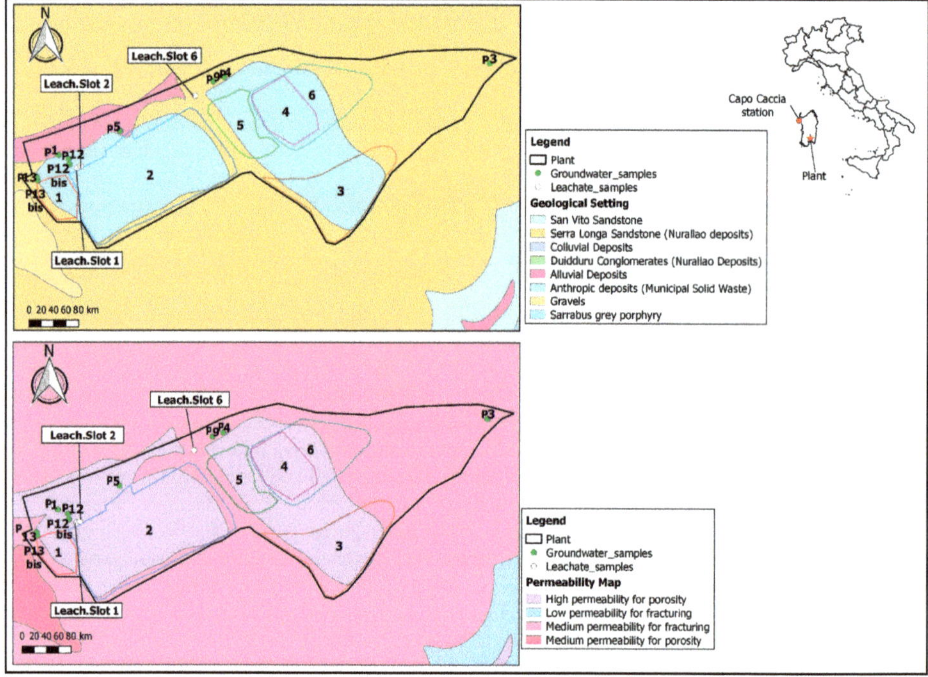

Figure 1. Geological (**above**) and Permeability (**below**) Map of Case History I.

The geological setting (above in Figure 1) is confirmed by the hydrogeological framework (below in Figure 1), with permeability ranging from medium for fracturing, in correspondence with the Nurallao deposits, up to high levels for porosity, in alluvial deposits, in west part of the study area (Permeability Map of Sardinia (1:25,000.00)).

The study area is characterized by two artesian aquifers, as shown in Figure 2:

- Aquifer 1 (shallow): represented by conglomerates and sand of the Nurallao deposits;
- Aquifer 2 (deep): characterized by conglomerates and gravels of the Nurallao deposits.

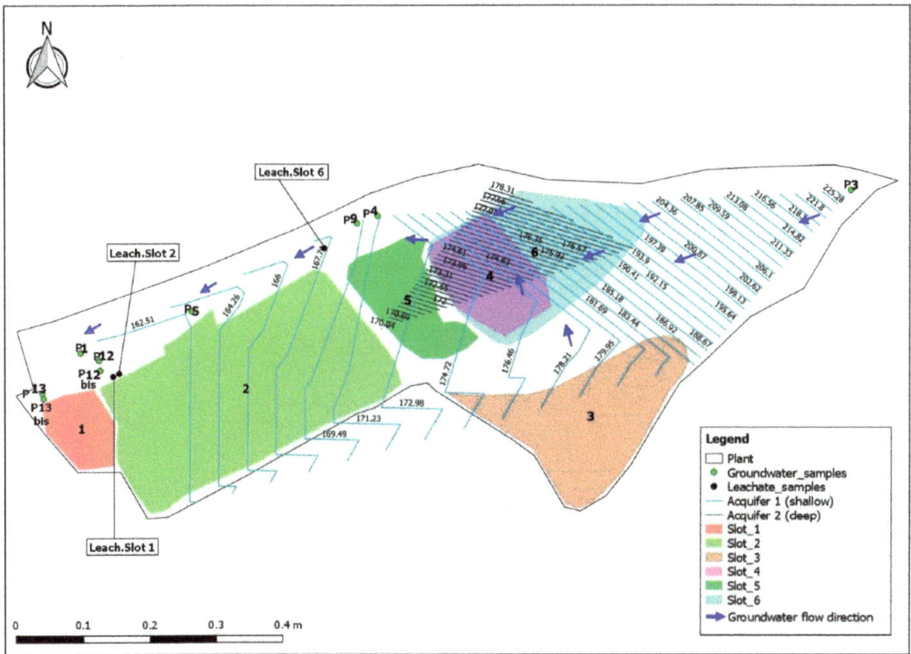

Figure 2. Water table map of Case History I.

The supply area of two aquifers is mainly represented by the fractured schist and granite settings of the Nurallao deposits, which dates to the Miocene epoch. In fact, both aquifers are characterized by a North-East supply, coming from the Miocene sediments of the Nurallao deposits, with groundwater flow direction from east to the west.

The landfill plant covers an area of approximately 0.4 km^2 and consists of 6 slots, set up partly for municipal solid waste (MSW) storage and partly for industrial waste one. In Table 1 it is shown the distinction between the slots of landfill plant based on the type of waste stored. Only Slot 6, for industrial wastes, is in operation, while the other slots have been closed.

Table 1. Municipal solid waste (MSW) and industrial waste.

Municipal Solid Waste (MSW)	Industrial Waste
Slot 2	Slot 1
Slot 3	Slot 5
Slot 4	Slot 6

2.2. Case History II in Umbria (Central Italy)

The study area is located in the province of Perugia in Umbria. The area is characterized by a hilly morphology, with altitude ranging between 500 and 600 m a.s.l., and it is located in the upper part of stream basin. Generally, the outcropping deposits are mainly formed by flysh, with marly, arenaceous and calcarenitic layers. By the Geological Map of Italy (scale 1:100,000.00), it is possible to identify the presence of two outcropping deposits (Figure 3):

- Upper sandy conglomerate deposits;
- Marly arenaceous formation.

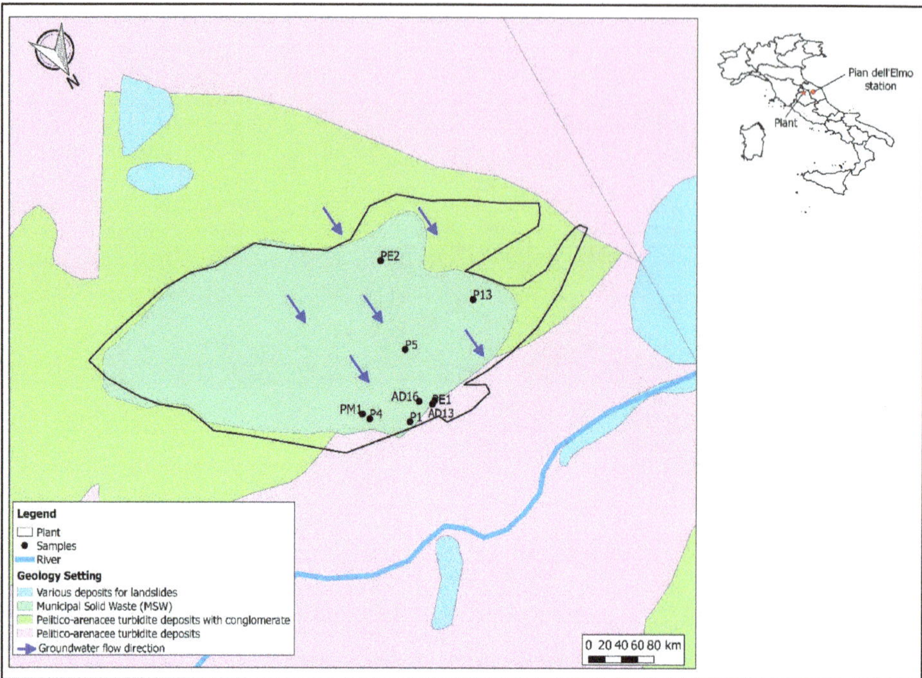

Figure 3. Geological Setting Case History II.

The marly arenaceous formation outcrops all over the area and is made of marly and arenaceous layers, alternated with clay and limestone lenses. In fact, the landfill substrate consists of flysch with compact marly and arenaceous layers.

The landfill plant, actually in operation, covers an area of approximately 0.12 km^2. Moreover, the study area is characterized by a series of marly and arenaceous layers, with low permeability, alternated with limestone lenses which, if fractured, can host suspended aquifers.

3. Material and Methods

In July 2020, eleven samples are collected for Case History I. The samples location is shown in Figure 1. Instead, for Case History II, during groundwater sampling in July 2020, nine samples are collected. Samples location is shown in Figure 3.

Table 2 shows the results of the chemical-physical parameters (pH, Eh, temperature and electrical conductivity), the Iron (Fe) concentrations (Legislative Decree 152/06 threshold equal to 200 µg/L), the Manganese (Mn) concentrations (Legislative Decree 152/06 threshold equal to 50 µg/L), the Nickel (Ni) concentrations (Legislative Decree 152/06 threshold equal to 20 µg/L) and isotopic data (δ^2H and δ^{18}O) for Case History I (Table 2a) and Case History II (Table 2b). All results refer to the July 2020 sampling survey.

Table 2. Summary of chemical-physical parameters, Fe, Mn and Ni concentrations and isotopic data for groundwater and leachate samples: (**a**) Case History I, (**b**) Case History II.

Case History I (July 2020)									
Samples	**pH**	**T**	**Eh**	**Electric Conductivity**	**Fe**	**Mn**	**Ni**	δ^2‰ H—1‰ (VSMOW)	δ^{18}‰ O—0.05‰ (VSMOW)
	-	°C	mV	µS/cm	µg/L	µg/L	µg/L		
P1	6.91	32.07	107.6	6127	18.2	197	11.5	−32	−5.86
P3	7.45	21.61	65.8	1100	8.7	1.3	1.7	−36	−6.37
P4	7.25	22.26	113.5	2228	3	1.3	1.6	−33	−5.82
P5	7.38	24.22	−104.7	1487	10.9	85.9	1.6	−35	−6.23
P12	6.48	26.66	50.1	3396	89	1271	53.9	−31	−6.13
P12BIS	6.96	28.14	−171.7	8020	8022	2941	221	−18	−6.03
P13	6.78	25.34	118	18,190	28.2	221	14.9	−29	−5.75
P13BIS	6.92	30.37	98.2	9094	5.9	292	300	−26	−5.85
Slot 1	na	na	na	na	na	na	na	−16	−2.65
Slot 2	na	na	na	na	na	na	na	−8	−6.36
Slot 6	na	na	na	na	100	26	28	−11	−1.91

(**a**)

Case History II (July 2020)									
Samples	**pH**	**T**	**Eh**	**Electric Conductivity**	**Fe**	**Mn**	**Ni**	μ^2‰ H—1‰ (VSMOW)	δ^{18}‰ O—0.05‰ (VSMOW)
	-	°C	mV	µS/cm	µg/L	µg/L	µg/L		
AD13	6.92	11.3	154	1900	74	742	45	−42	−7.0
AD16	7.54	11.1	170	5680	1150	383	105	−24	−7.5
P1	7.8	14.6	147	1660	5	0.5	9.5	−46	−7.6
P4	7.89	14.2	177	1730	5	0.1	16.6	−43	−7.5
P5	8.11	13.8	103	1020	5	5.3	1.4	−50	−7.9
P13	7.73	14.2	181	1290	5	0.3	2.4	−49	−8.0
PM1	7.69	14	39	810	5	309	14.6	−42	−6.9
PE1	8.11	na	na	15,000	5690	297	207	−21	−7.9
PE2	8.64	na	na	20,500	28,300	54	307	−4	−6.1

(**b**)

For Case History I, the Table 2a shows eight groundwater samples (identified by "P") and three samples by leachate tanks (identified by "Leach_Slot"). The samples with the same number, but "bis", are in the same point, but they track the shallowest aquifer. Instead, for Case History II, the Table 2b shows five groundwater samples (identified by "P" and "PM"), two samples by leachate tanks (identified by "PE") and two drainage water samples (identified by "AD"). In fact, to drain enough the landfill infiltration water, a series of sub-horizontal drains have been drilled on the embankment downstream of the landfill.

Concentrations of minor elements (Fe, Mn and Ni) were measured using an ICP-MS by a certified (Accredia) Italian laboratory. The analytical accuracy of these methods is equal to 5%. Ultrapure water (Millipore, Milli-Q, 16 MΩ cm) was used in preparing blanks, standard solutions, and sample dilutions.

The δ^2H and δ^{18}O contents of groundwater and leachate samples were analyzed by the Isotope Geochemistry Laboratory of the University of Parma (Italy) using the IRMS (isotope-ratio mass spectrometry) continuous flow-equilibration method with CO_2.

Isotopic abundance ratios are expressed as parts per million of their deviations, as given by the Vienna Standard Mean Ocean Water (VSMOW) (Equation (1)) [38]:

$$\delta = (R_{sample}/R_{SMOW} - 1) \times 10^3 \tag{1}$$

where:

- R_{sample} is the abundance ratio of the isotopic species, respectively $^2H/^1H$ for hydrogen and $^{18}O/^{16}O$ for oxygen, for each sample;
- R_{SMOW} is the isotopic ratio of the standard (Standard Mean Ocean Water), accepted for the isotopes in water [39].

The analytical error is 0.05‰ for $\delta^{18}O$ and 1‰ for δ^2H.

4. Results and Discussion

Several studies [14,29–31] have highlighted how methanogenesis processes can affect leachate enrichment in δ^2H. As a matter of fact, the methanogenic bacteria, during the methane production, use first the "lighter" isotope of hydrogen (^1H), therefore leaving enriched the "heavier" isotope of hydrogen in leachate (^2H) [8,22,23,32].

This paper is referred to two case histories in Italy, dealing with groundwater interaction with leachate from municipal solid waste landfills. The aim is to show how the ^2H enrichment is an index of the groundwater contamination, due to interaction with leachate in some piezometers.

Isotope composition data, reported in Table 2, have been graphed with the main isotope diagrams. For a suitable assessment, the isotope composition data for both history cases are compared with the main meteoric lines:

- Global Meteoric Water Line (GMWL) [39], described by Equation (2):

$$\delta^2H = 8\,\delta^{18}O\text{‰} + 10 \tag{2}$$

- Mediterranean Meteoric Water Line (MMWL) [40,41], defined as follow (Equation (3)):

$$\delta^2H = 8\,\delta^{18}O\text{‰} + 22 \tag{3}$$

The deviation from meteoric lines, GMWL (Equation (2)) and MMWL (Equation (3)), shows alteration phenomena due to processes occurred in the soil: mixing with marine waters of ancient or recent origin or contamination fluids leaching from soil to groundwater. In particular, the variation by meteoric lines highlights mixing phenomena of groundwater with leachate, coming from landfills of municipal solid waste consisting of a significant organic part.

4.1. Case History I in Sardinia (Southern Italy)

Figure 4 shows deuterium δ^2H and oxygen $\delta^{18}O$ isotope composition for groundwater sampled under Sardinia landfill site (Table 2a). The Figure 4 shows also the GMWL (Equation (2)), MMWL (Equation (3)) and Local Meteoric Water Line (LMWL) for southern Italy [42], described by Equation (4):

$$\delta^2H = 6.94\,\delta^{18}O\text{‰}\,(\pm0.45) + 6.41\,(\pm2.65)\text{‰} \tag{4}$$

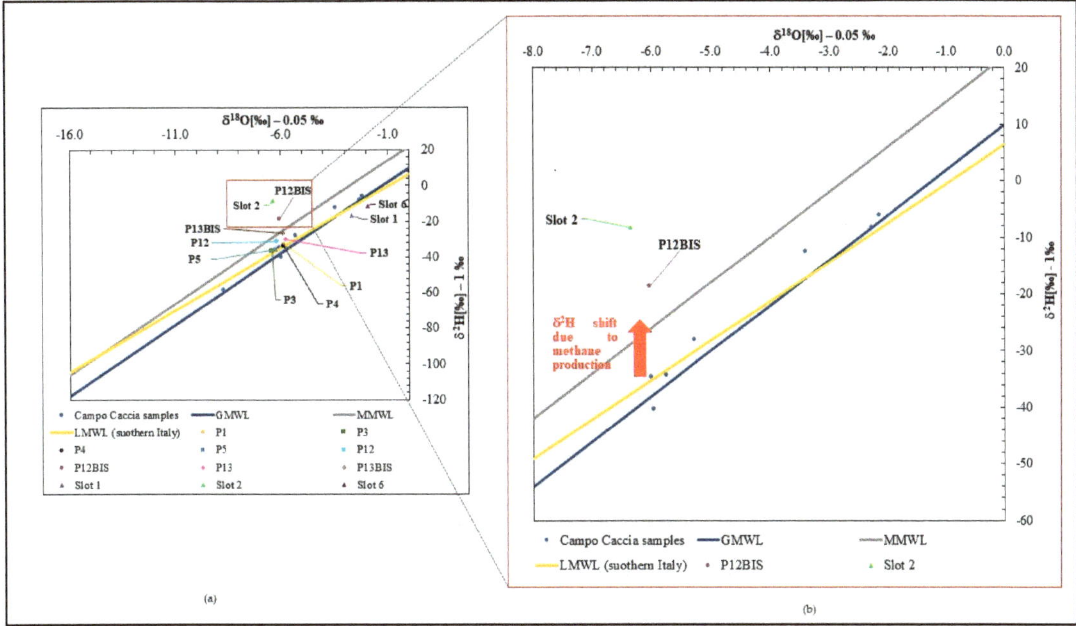

Figure 4. δ^2H vs. δ^{18}O values for Case History I: all samples (**a**) and Slot 2 & P12bis (**b**).

In addition, the Capo Caccia gauge station, located north-west of the plant (Figure 1), is identified and the isotope data δ^{18}O e δ^2H, there sampled, and analyzed by the International Atomic Energy Agency (IAEA), is reported in Figure 4. This figure shows that samples coming from Capo Caccia station are very close to the LMWL, proposed by Giustini et al., 2016 [42] for southern Italy and identified by Equation (4), which, as a consequence of it, is a good reference of isotope composition for precipitations in this area.

The isotope diagram in Figure 4a represents that the groundwater sampled in P1, P3, P4, P5, P12, P13 and P13bis is very close to meteoric lines, showing they are groundwater completely belonging to the natural hydrological cycle. In contrast, samples coming from leachate collecting tanks, located downstream of Slot 1 and Slot 6, used for the storage of industrial waste, are below the range of the reference meteoric lines. Therefore, it is possible to say that the variation of the isotope composition for deuterium δ^2H and oxygen δ^{18}O is not influenced by contamination of the leachate, as they come from industrial waste storage. On the contrary, the sample coming from the leachate collecting tank downstream Slot 2 presents anomalous values, equal to $-8‰$ for δ^2H and $-6.36‰$ for δ^{18}O, positioning itself in the isotope diagram, beyond the reference meteoric lines (Figure 4b). Slot 2 has been used for the municipal solid waste storage, which were enriched by organic component (Table 1). In particular, the isotope enrichment in deuterium δ^2H content ($-8‰$) for leachate well downstream of Slot 2, seems to be linked with methanogenesis phenomena. As a matter of fact, in municipal solid waste landfills, where the organic part in wastes is significant, the methanogenic bacteria use, preferentially, the "lighter" isotope hydrogen (^{1}H) due to the methane production. Therefore, the remaining hydrogen is enriched in deuterium (^{2}H), the "heavier" isotope [8,22]. The P12bis groundwater sample presents, as well, anomalous value for δ^2H ($-18‰$), and it is placed in the diagram close to the sample taken downstream of Slot 2 (Figure 4b). These results show contamination phenomena for groundwater sampled in P12bis point due to the mixing with leachate from Slot 2 and sampled in the collecting tank located downstream of it. The ^{2}H enrichment for P12bis point, due to interaction with the leachate by collection tank downstream of lot 2, does not cause an equal ^{18}O enrichment ($-6.03‰$). This is due to methanogenesis phenomena, during which bacteria use the "lighter" isotope hydrogen (^{1}H), than the remaining the "heavier" isotope (^{2}H), which doesn't involve oxygen [8,22].

Isotope composition results, which show anomalies for the P12bis sample, are confirmed by the concentrations of some analytes, such as iron, manganese and nickel, whose spatial trends are represented in Figure 5.

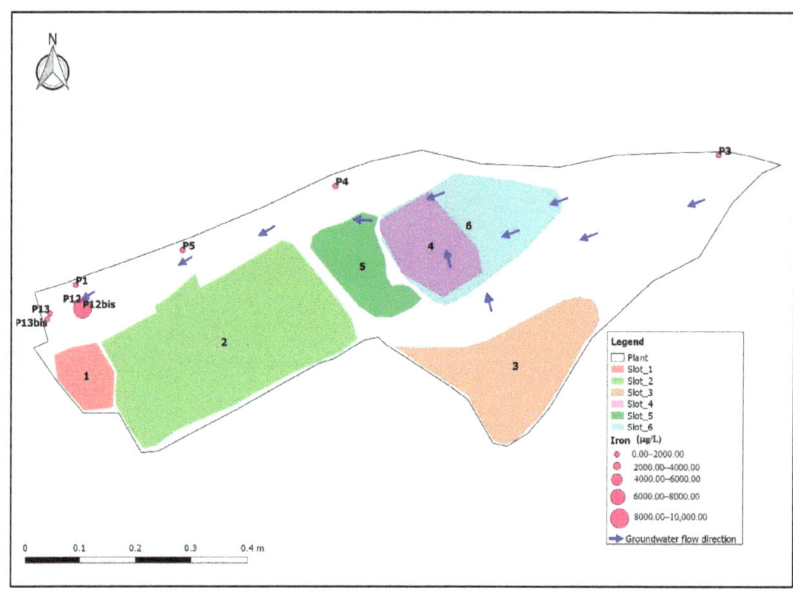

(a)

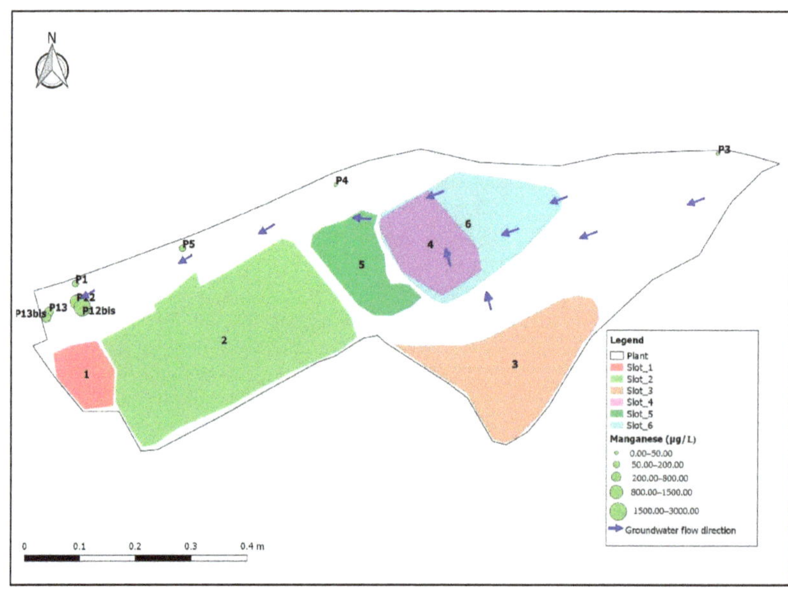

(b)

Figure 5. *Cont.*

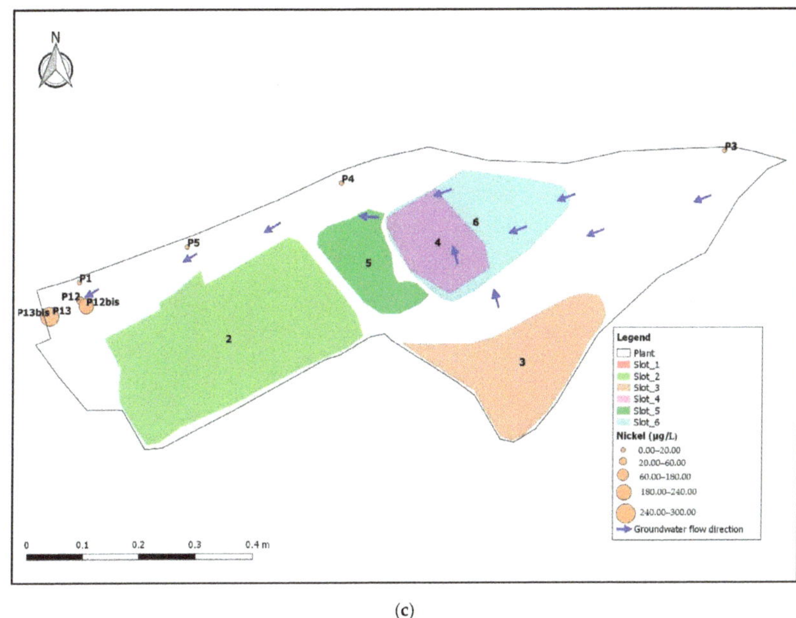

(c)

Figure 5. Concentrations spatial trend for the Case History I: (**a**) Iron, (**b**) Manganese, (**c**) Nickel.

Figure 5 shows high values from upstream to downstream of the plant, especially for the P12bis piezometer. In particular, there are concentrations equal to 8022 µg/L for Iron (Figure 5a), 2941 µg/L for Manganese (Figure 5b) and 221 µg/L for Nickel (Figure 5c), significantly higher than the threshold by Legislative Decree 152/06. Figure 5 shows very high concentrations, sometimes higher than the Legislative Decree 152/06 threshold for the Iron (200 µg/L), Manganese (50 µg/L) and Nickel (20 µg/L), also in P12, P13 and P13b is groundwater samples. This behavior is not detected by the isotope diagram in Figure 5, which is however a good indicator to verify possible interactions between groundwater and leachate.

The Pearson correlation matrix (Table 3) is used to determine the relationship between isotopic composition (δ^2H and δ^{18}O) and iron, manganese and nickel concentrations. The correlation is considered significant if greater than 0.5 [4].

Table 3. Pearson correlation matrix for Case History I (in red fair correlation, in orange moderate correlation and in green high correlation).

	pH	T	Eh	Fe	Mn	Ni	δ^2H	δ^{18}O
pH	1							
T	−0.567	1						
Eh	−0.194	0.018	1					
Fe	−0.078	0.197	−0.753	1				
Mn	−0.393	0.286	−0.692	0.919	1			
Ni	−0.274	0.544	−0.260	0.500	0.515	1		
δ^2H	−0.428	0.504	−0.451	0.836	0.833	0.791	1	
δ^{18}O	−0.433	0.444	0.463	−0.047	−0.077	0.214	0.352	1

The Pearson matrix correlation in Table 3 does not consider the sampling points by the leachate collection tanks because the data are incomplete (Table 2a). Pearson matrix shows high correlations between δ^2H and Fe (0.836), Mn (0.833) and Ni (0.791). On the contrary, Table 3 highlights fair correlations between δ^{18}O and Fe (−0.047), Mn (−0.077) and a moderate correlation with Ni (0.214). In fact, some scatter plots are graphed in Figure 6.

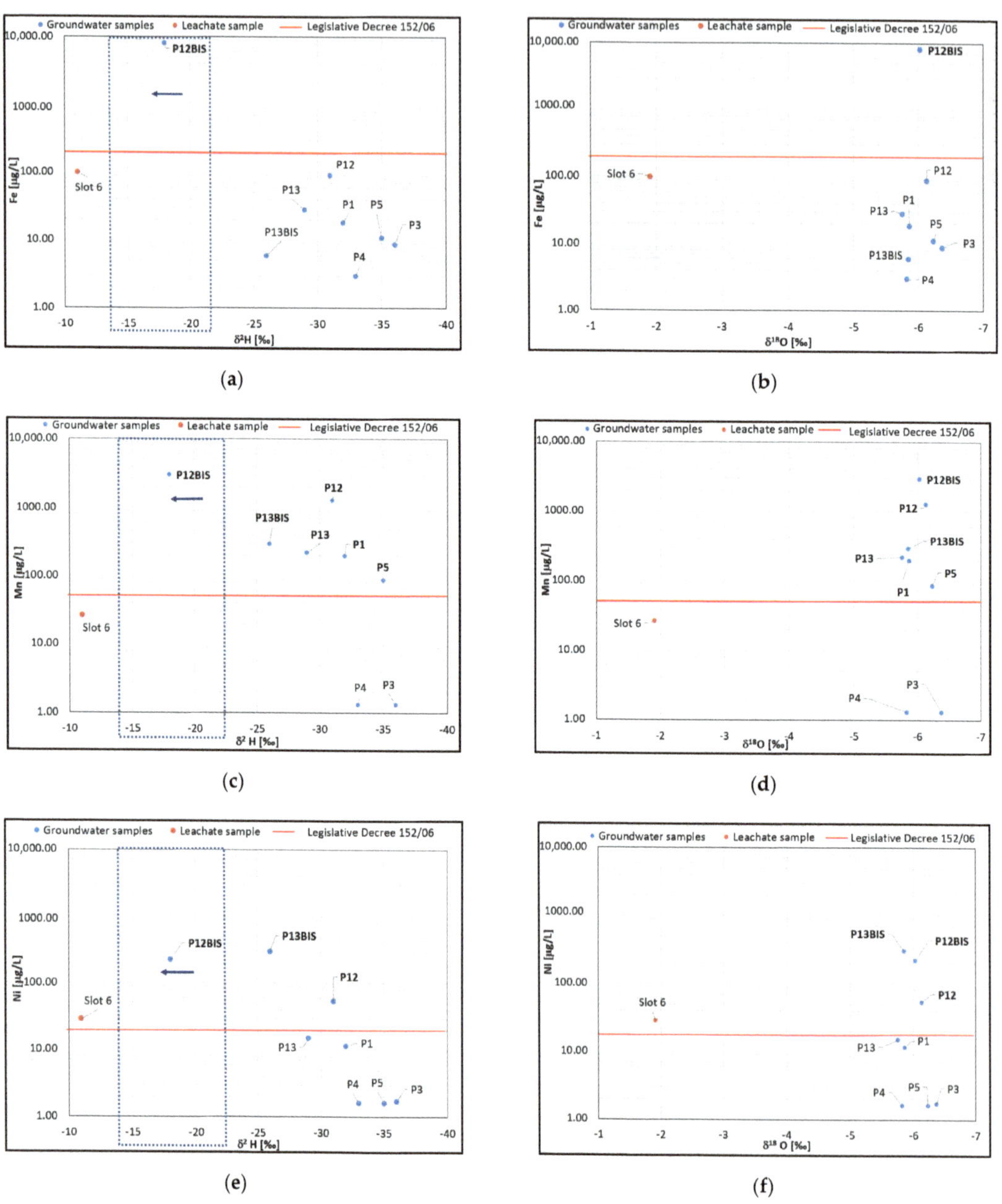

Figure 6. Scatter plot (**a**) Fe and δ^2H, (**b**) Fe and δ^{18}O, (**c**) Mn and δ^2H, (**d**) Mn and δ^{18}O, (**e**) Ni and δ^2H and (**f**) Ni and δ^{18}O.

Figure 6 supports what it is showed in isotope diagram of Figure 4, as it presents not only values over the legal threshold for Iron (Figure 6a), Manganese (Figure 6c) and Nickel (Figure 6e) in P12bis groundwater sample, but also a δ^2H enrichment. Other groundwater samples, such as P12, P13 and P13bis, have concentrations above the legal threshold, as already showed in Figure 5, without however having a δ^2H enrichment. As regards the relationships between δ^{18}O and concentrations of Iron (Figure 6b), Manganese (Figure 6d) and Nickel (Figure 6e), no enrichments are highlighted. Nevertheless, the Legislative Decree 152/06 thresholds are over for three analytes in P12, P12bis, P13 and P13bis points. The leachate sampled in Slot 6, used for the industrial waste storage, shows different behavior for the three analytes in relation to δ^2H and δ^{18}O, confirming different sources of water, with respect to the groundwater samples. Therefore, this processing confirms that the methanogenesis phenomena cause a deuterium enrichment, due to bacteria use the "lighter" isotope hydrogen (^{1}H), without generating variations for 18 oxygen isotope. Moreover, the presented comparison between δ^2H enrichment and other metals let us outline as, when there is a δ^2H enrichment and one metal high concentration values, these latter values are due to contamination process, and they can't be referred to a natural background level.

4.2. Case History II in Umbria (Central Italy)

Figure 7 represents isotope composition for deuterium δ^2H and oxygen δ^{18}O referred to the Umbria case history. According to Equation (2) for GMWL and Equation (3) for MMWL, the isotope composition (Table 2b) is graphed with the Local Meteoric Water Line (LMWL) for central Italy [42], described as follow:

$$\delta^2H = 7.46\ \delta^{18}O\,‰\ (\pm 0.32) + 8.29\ (\pm 2.33)\,‰ \tag{5}$$

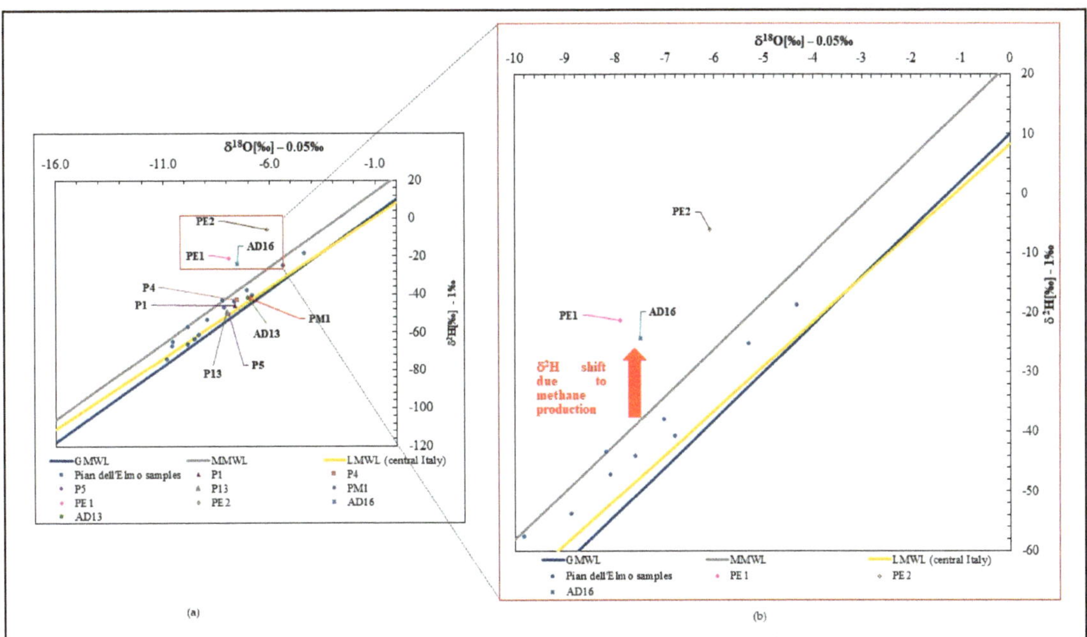

Figure 7. δ^2H vs. δ^{18}O values for Case History II: all samples (**a**) and PE1, PE2 & AD16 (**b**).

The Pian dell'Elmo pluviometric station has been identified and referred to, as it is placed at north-east to the study area (Figure 3), and its isotopic data are provided by the International Atomic Energy Agency (IAIEA).

Groundwater samples values show δ^2H and δ^{18}O isotope contents close to reference meteoric lines (Figure 7a). However, Figure 7b shows a significant deviation, with respect to the reference meteoric lines, for points representing samples taken in PE1 and PE2, which outline δ^2H isotope values respectively equal to $-21‰$ and $-4‰$. PE1 and PE2 have been used to assess the leachate levels of isotopes composition, in fact they confirm a sound enrichment in δ^2H, due to methanogenesis processes. Anomalous behavior has been found for sampling coming from AD16 point, as it presents values equal to $-24‰$ for δ^2H, showing a significant variation upwards respect to the reference meteoric lines (Figure 7b). AD16 is a sub-horizontal drain positioned at the embankment downstream of the landfill and it is near PE1, a tank used for the leachate storage (Figure 3). The δ^2H enrichment for AD16 point, with values equal to $-24‰$, seems to be linked to mixing with leachate coming from solid waste landfill, which can disperse leachate, enriched in the "heavier" isotope [8,22], due to the methanogenesis processes. The results, therefore, prove an interaction between groundwater and leachate in the AD16 drain. As shown for Case History I, the δ^2H enrichment for the AD16 groundwater sample, due to the bacteria during methanogenic processes, does not also induce a $\delta^{18}O$ enrichment, with value equal to $-7.5‰$.

The concentrations spatial trend of some analytes was also represented for plant in Perugia: Iron (Figure 8a), Manganese (Figure 8b) and Nickel (Figure 8c).

Figure 8 shows anomalous concentrations for AD16 groundwater sample, higher than Legislative Decree 152/06 threshold values, whose concentrations are equal to 1150 µg/L for Iron, 383 µg/L for Manganese and 105 µg/L for Nickel. These high concentration values confirm results of the δ^2H isotope composition for AD16 groundwater sample, thus highlighting a groundwater contamination phenomenon. Actually, Figure 8 shows elevated values, sometimes higher than the Legislative Decree 152/06 threshold for three analytes, also in AD13 and PM1 groundwater samples. This behavior is not detected by the isotope diagram in Figure 7, which is however a good indicator to verify possible interactions between groundwater and leachate.

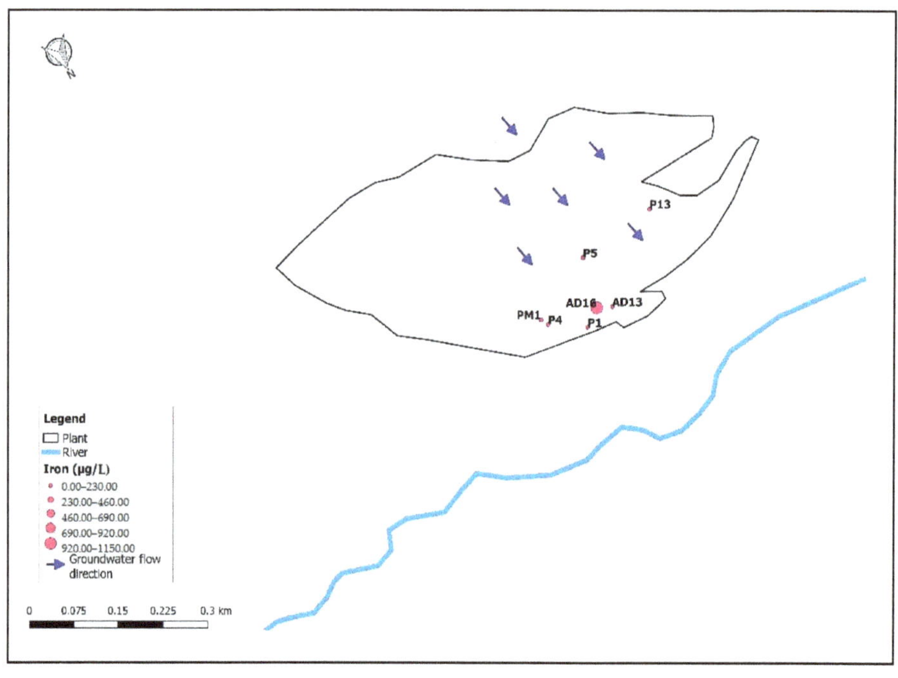

(**a**)

Figure 8. *Cont.*

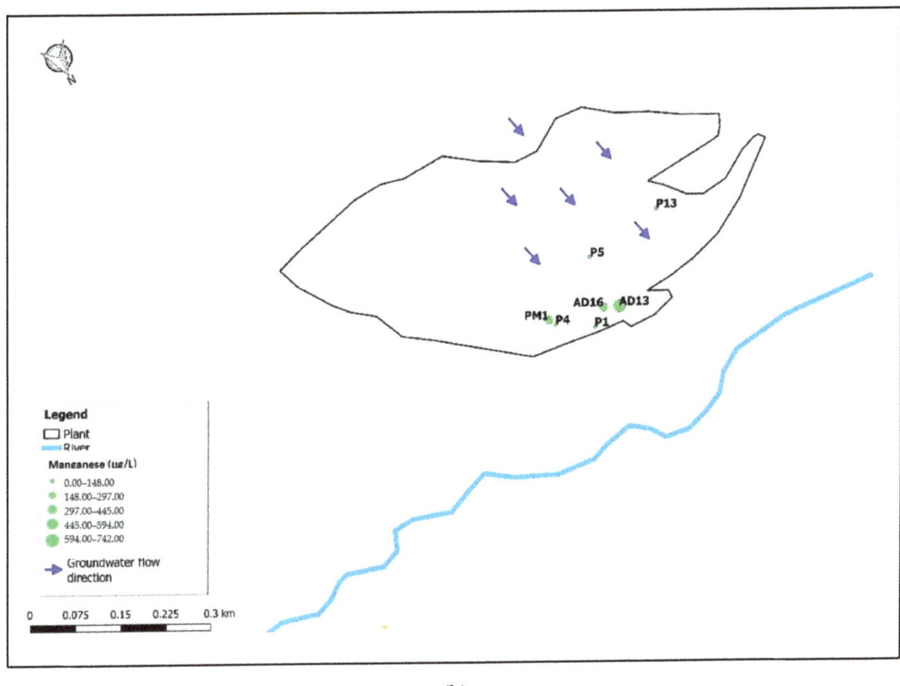

(**b**)

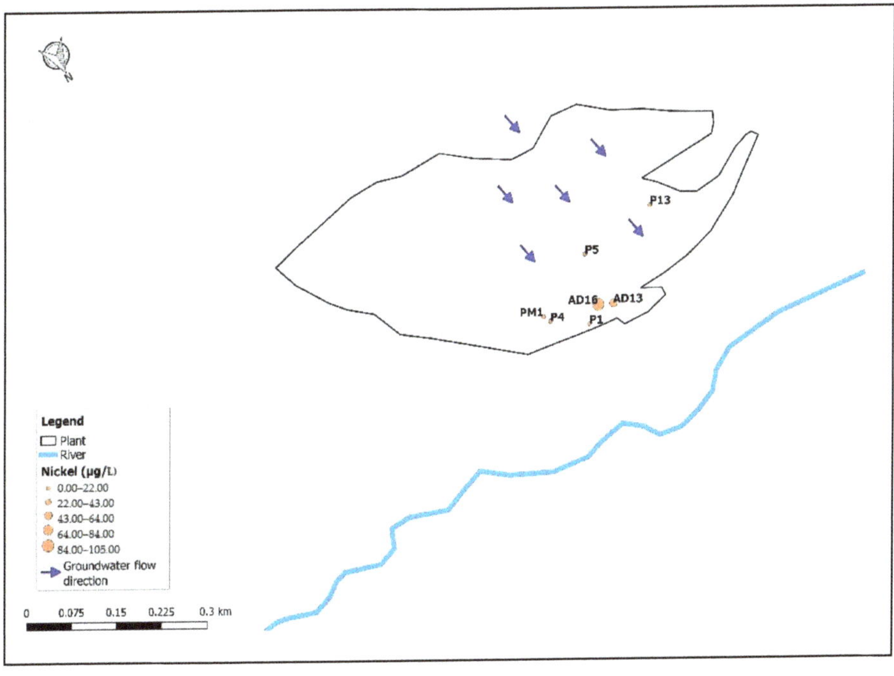

(**c**)

Figure 8. Concentrations spatial trend for the Case History II: (**a**) Iron, (**b**) Manganese, (**c**) Nickel.

As the Case History I, the Pearson correlation matrix (Table 4) is used to determine the relationship between isotopic composition (δ^2H and δ^{18}O) and Iron, Manganese and Nickel concentrations.

Table 4. Pearson correlation matrix for Case History II (in red fair correlation, in orange moderate correlation and in green high correlation).

	pH	T	Eh	Fe	Mn	Ni	δ^2H	δ^{18}O
pH	1							
T	0.741	1						
Eh	−0.184	−0.230	1					
Fe	−0.206	−0.709	0.279	1				
Mn	−0.921	−0.841	−0.087	0.324	1			
Ni	−0.495	−0.870	0.288	0.939	0.596	1		
δ^2H	−0.359	−0.732	0.180	0.936	0.484	0.964	1	
δ^{18}O	−0.625	−0.374	−0.470	0.016	0.747	0.280	0.330	1

The Pearson matrix correlation in does not consider the sampling points PE1 and PE2, by the leachate collection tanks, because the data are incomplete (Table 2b). Pearson matrix shows high correlations between δ^2H and Fe (0.936) and Mn (0.964) and a moderate correlation between with Ni (0.484). On the contrary, Table 4 indicates fair correlations between δ^{18}O and Fe (0.016), Mn (0.280) and a high correlation with Ni (0.747).

Figure 9, again, confirms what showed in the isotope diagram of Figure 7, as it shows not only values over the legal threshold for Iron (Figure 9a), Manganese (Figure 9c) and Nickel (Figure 9e) in AD16 groundwater sample, but also a δ^2H enrichment. In fact, the AD16 groundwater sample shows trends closer to the sample by PE1, a tank used for the leachate storage. In fact, the leachate samples coming from points PE1 and PE2 show different behavior with respect to the groundwater samples. Moreover, the other groundwater samples, although they have concentrations over Legislative Decree 152/06 thresholds for the three analytes, do not show the same deuterium enrichment. In regards to the relationships between δ^{18}O and Iron (Figure 9b), Manganese (Figure 9d) and Nickel (Figure 9e), no enrichments are highlighted. Therefore, this processing confirms that the methanogenesis phenomena cause a deuterium enrichment, due to bacteria use the "lighter" isotope hydrogen (^{1}H), without generating variations for 18 oxygen isotope.

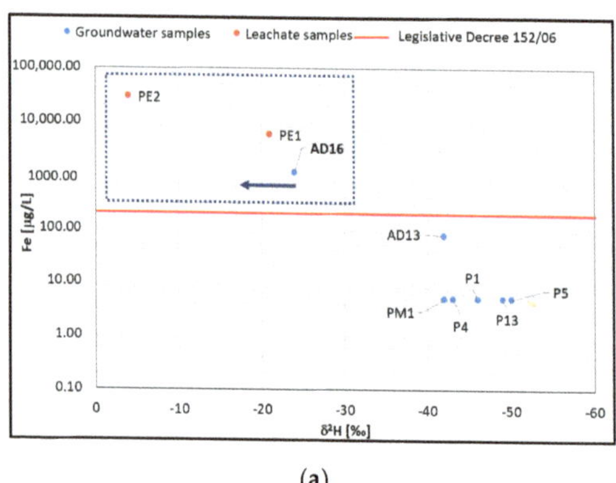

(a)

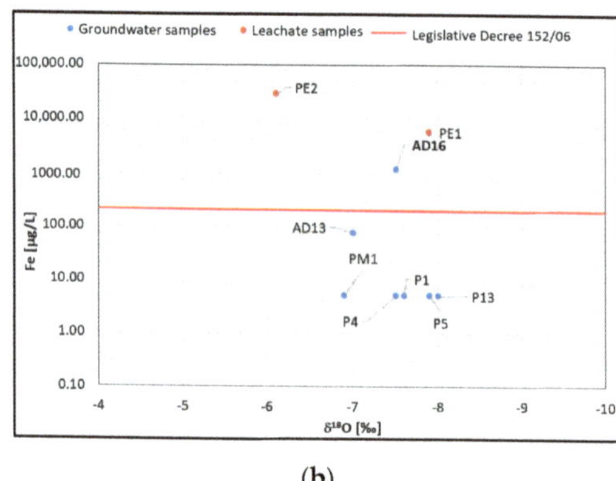

(b)

Figure 9. *Cont.*

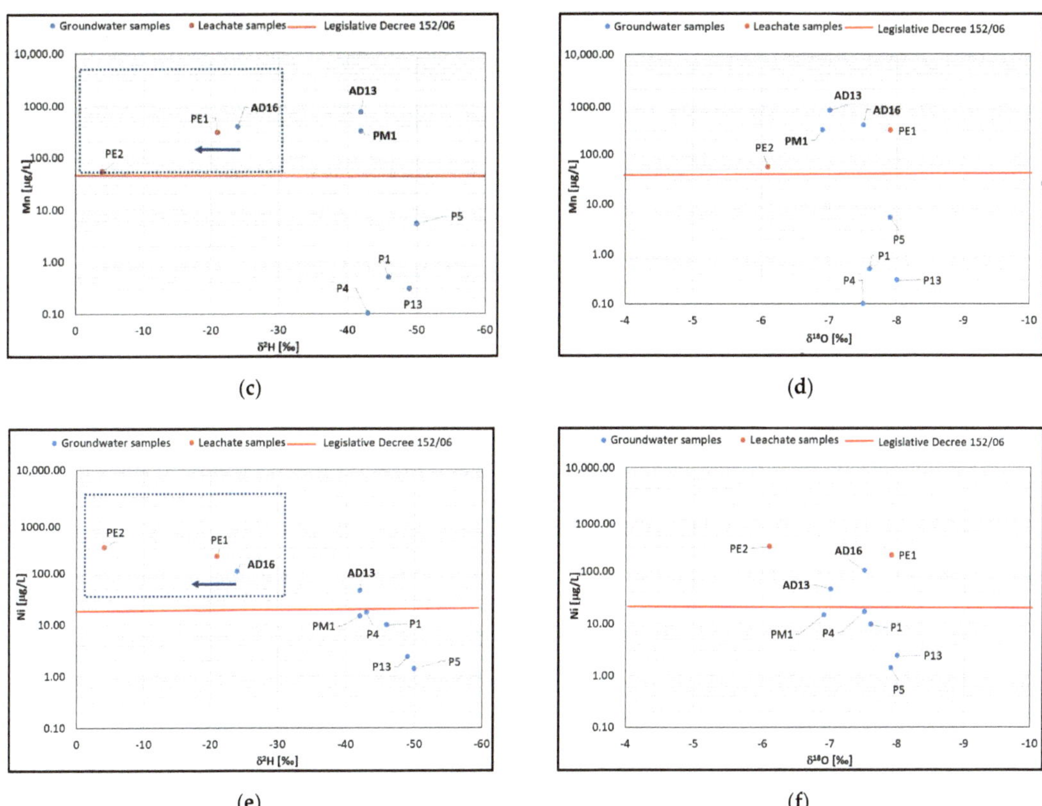

Figure 9. Scatter plot (**a**) Fe and δ^2H, (**b**) Fe and δ^{18}O, (**c**) Mn and δ^2H, (**d**) Mn and δ^{18}O, (**e**) Ni and δ^2H and (**f**) Ni and δ^{18}O.

Further regarding to this case study, the presented comparison between δ^2H enrichment and other metals, outlines that, when there is a δ^2H enrichment and one metal high concentration values, these latter values are due to contamination process, and they can't be referred to a natural background level. The same outline is not present for ^{18}O, as this element one is not involved in such a process.

5. Conclusions

This paper aims to assess the effectiveness of deuterium and oxygen isotopes application as environmental traces for contamination phenomena, due to the leachate interaction with groundwater.

Two Italian case history are showed: Case History I is referred to a plant in Cagliari province of Sardinia region, and Case History II deals with a plant in Perugia province of, Umbria Region. In both study areas, there are in operation landfills used for the storage of municipal solid waste (MSW), organic component of municipal solid wastes was not negligible. Results of the two presented Italian case histories show that isotope composition of groundwater samples is significantly influenced by interaction phenomena by leachate mixing, due to a δ^2H enrichment. The case histories prove how δ^2H enrichment determines a significant shift of the groundwater samples from the reference meteoric lines: Global Meteoric Water Line (GMWL) [39], Mediterranean Meteoric Water Line (MMWL) [40,41] and Local Meteoric Water Line (LMWL) [42].

The Case History I highlighted δ^2H anomalous values for the P12bis groundwater sample ($-18‰$), located downstream of Slot 2, used for the municipal solid waste storage.

The δ^2H enrichment confirms an interaction of the leachate from Slot 2 with groundwater, phenomenon therefore an index of contamination processes. This behavior is also confirmed by the concentration values of some analytes, such as Iron (8022 µg/L), Manganese (2941 µg/L) and Nickel (221 µg/L), validating the hypothesis of pollution for the P12bis groundwater sample. On the other side, Case History II anomalous values, referred to deuterium isotope have been highlighted in AD16 groundwater sample (−24‰). The AD16 point appears to be located near tank 1 (PE1), used for leachate collecting. The δ^2H enrichment indicates a pollution phenomenon caused by interaction between leachate and groundwater. This phenomenon also is confirmed by the concentrations of some analytes, such as Iron (1150 µg/L), Manganese (383 µg/L) and Nickel (105 µg/L), with values over Legislative Decree 152/06 threshold. On the contrary, there are no significant variations for the 18-oxygen isotope content. In the Case History I, the point P12bis has a value of δ^{18}O equal to −6.03‰ very close to the result of leachate Point 2 (−6.36‰). In the same way, for Case History II, the groundwater sampling point AD16 has a δ^{18}O value equal to −7.5‰ very close to the result of point PE1 (−7.9‰). However, the ^{2}H enrichment does not lead to a following δ^{18}O variation, confirming that the methanogenesis phenomena do not influence the oxygen isotope content.

Furthermore, for both case histories, a statistical approach was applied to study correlations between δ^{18}O and δ^2H isotopes and concentrations of Iron, Manganese and Nickel. The Pearson matrix, performed for both case histories, confirms a high relationship between δ^2H and the concentrations of Iron, Manganese and Nickel. For Case History II, scatter plots showed a similar trend between the groundwater sample AD16 and PE1. On the contrary, regarding relationships between δ^{18}O and the Iron, Manganese and Nickel concentrations, no enrichments are highlighted, as δ^{18}O is not involved in such a process. On the other side, it has come out that the δ^2H isotope excess, occurring at the meantime with high concentration values of metals, is a tracer that these latter ones can't become from a natural background level, but they are due to contamination processes.

Therefore, the results of these case histories confirm that the δ^2H isotope enrichment is a valid tracer to identify contamination processes between leachate, coming from municipal solid waste landfills and groundwater.

Author Contributions: Conceptualization, G.S. and F.A.; methodology, G.S., M.B. and F.A.; validation, G.S. and M.B.; investigation, F.A.; data curation, F.A.; writing—original draft preparation, F.A.; writing—review and editing, G.S.; supervision, G.S. and M.B.; project administration, G.S. All authors have read and agreed to the published version of the manuscript.

Funding: This research received no external funding.

Institutional Review Board Statement: Not applicable.

Informed Consent Statement: Not applicable.

Data Availability Statement: Isotopic data about Capo Caccia and Pian dell'Elmo stations are provided by the International Atomic Energy Agency (IAEA). All other data are provided in the text.

Conflicts of Interest: The authors declare no conflict of interest.

References

1. Lee, K.S.; Ko, K.S.; Kim, E.Y. Application of stable isotopes and dissolved ions for monitoring landfill leachate contamination. *Environ. Geochem. Health* **2020**, *42*, 1387–1399. [CrossRef]
2. Mukherjee, S.; Mukhopadhyay, S.; Hashim, M.A.; Gupta, B. Sen Contemporary environmental issues of landfill leachate: Assessment and remedies. *Crit. Rev. Environ. Sci. Technol.* **2015**, *45*, 472–590. [CrossRef]
3. Qin, M.; Molitor, H.; Brazil, B.; Novak, J.T.; He, Z. Recovery of nitrogen and water from landfill leachate by a microbial electrolysis cell-forward osmosis system. *Bioresour. Technol.* **2016**, *200*, 485–492. [CrossRef]
4. Adeolu, A.O.; Ada, O.V.; Gbenga, A.A.; Adebayo, O.A. Assessment of groundwater contamination by leachate near a municipal solid waste landfill. *Afr. J. Environ. Sci. Technol.* **2011**, *5*, 933–940.
5. Liu, Z.P.; Wu, W.H.; Shi, P.; Guo, J.S.; Cheng, J. Characterization of dissolved organic matter in landfill leachate during the combined treatment process of air stripping, Fenton, SBR and coagulation. *Waste Manag.* **2015**, *41*, 111–118. [CrossRef] [PubMed]

6. Foo, K.Y.; Hameed, B.H. An overview of landfill leachate treatment via activated carbon adsorption process. *J. Hazard. Mater.* **2009**, *171*, 54–60. [CrossRef] [PubMed]

7. de Medeiros Engelmann, P.; dos Santos, V.H.J.M.; Barbieri, C.B.; Augustin, A.H.; Ketzer, J.M.M.; Rodrigues, L.F. Environmental monitoring of a landfill area through the application of carbon stable isotopes, chemical parameters and multivariate analysis. *Waste Manag.* **2018**, *76*, 591–605. [CrossRef] [PubMed]

8. Castañeda, S.S.; Sucgang, R.J.; Almoneda, R.V.; Mendoza, N.D.S.; David, C.P.C. Environmental isotopes and major ions for tracing leachate contamination from a municipal landfill in Metro Manila, Philippines. *J. Environ. Radioact.* **2012**, *110*, 30–37. [CrossRef]

9. Moravia, W.G.; Amaral, M.C.S.; Lange, L.C. Evaluation of landfill leachate treatment by advanced oxidative process by Fenton's reagent combined with membrane separation system. *Waste Manag.* **2013**, *33*, 89–101. [CrossRef]

10. Puig, S.; Serra, M.; Coma, M.; Cabré, M.; Dolors Balaguer, M.; Colprim, J. Microbial fuel cell application in landfill leachate treatment. *J. Hazard. Mater.* **2011**, *185*, 763–767. [CrossRef]

11. Varjani, S.; Gnansounou, E.; Baskar, G.; Pant, D. *Waste Bioremediation*; Springer: Singapore, 2018; pp. 233–247.

12. Preziosi, E.; Frollini, E.; Zoppini, A.; Ghergo, S.; Melita, M.; Parrone, D.; Rossi, D.; Amalfitano, S. Disentangling natural and anthropogenic impacts on groundwater by hydrogeochemical, isotopic and microbiological data: Hints from a municipal solid waste landfill. *Waste Manag.* **2019**, *84*, 245–255. [CrossRef] [PubMed]

13. Naveen, B.P.; Mahapatra, D.M.; Sitharam, T.G.; Sivapullaiah, P.V.; Ramachandra, T.V. Physico-chemical and biological characterization of urban municipal landfill leachate. *Environ. Pollut.* **2017**, *220*, 1–12. [CrossRef] [PubMed]

14. Porowska, D. Identification of groundwater contamination zone around a reclaimed landfill using carbon isotopes. *Water Sci. Technol.* **2017**, *75*, 328–339. [CrossRef] [PubMed]

15. Wimmer, B.; Hrad, M.; Huber-Humer, M.; Watzinger, A.; Wyhlidal, S.; Reichenauer, T.G. Stable isotope signatures for characterising the biological stability of landfilled municipal solid waste. *Waste Manag.* **2013**, *33*, 2083–2090. [CrossRef]

16. Nigro, A.; Sappa, G.; Barbieri, M. Application of boron and tritium isotopes for tracing landfill contamination in groundwater. *J. Geochem. Explor.* **2017**, *172*, 101–108. [CrossRef]

17. Wachniew, P. Environmental tracers as a tool in groundwater vulnerability assessment. *Acque Sotter.-Ital. J. Groundw.* **2015**, 19–25. [CrossRef]

18. Elliot, T. Environmental tracers. *Water* **2014**, *6*, 3264–3269. [CrossRef]

19. Coplen, T.B.; Herczeg, A.L.; Barnes, C. Isotope Engineering—Using Stable Isotopes of the Water Molecule to Solve Practical Problems. Available online: https://link.springer.com/chapter/10.1007/978-1-4615-4557-6_3 (accessed on 31 March 2021).

20. Jasechko, S. Global Isotope Hydrogeology—Review. *Rev. Geophys.* **2019**, *57*, 835–965. [CrossRef]

21. Gat, J.R. Oxygen and hydrogen isotopes in the hydrologic cycle. *Annu. Rev. Earth Planet. Sci.* **1996**, *24*, 225–262. [CrossRef]

22. Pujiindiyati, E.R. Application of deuterium and oxygen-18 to trace leachate movement in Bantar Gebang sanitary landfill. *Atom Indones.* **2011**, *37*, 76–82. [CrossRef]

23. Tazioli, A. Landfill investigation using tritium and isotopes as pollution tracers. *Acquae Mundi* **2011**, *18*, 83–92.

24. Hackley, L.C.; Liu, C.I.; Coleman, D.D. Environmental Isotope Characteristics of Landfill Leachates and Gases. *Groundwater* **1996**, *34*, 827–836. [CrossRef]

25. Kerfoot, H.B.; Baker, J.A.; Burt, D.M. The use of isotopes to identify landfill gas effects on groundwater. *J. Environ. Monit.* **2003**, *5*, 896–901. [CrossRef]

26. Venkata Mohan, S.R.; Devi, M.P.; Reddy, M.; Chandrasekhar, K.; Juwarkar, A.; Sarma, P.N. Bioremediation of petroleum sludge under anaerobic microenvironment: Influence of biostimulation and bioaugmentation. *Environ. Eng. Manag. J.* **2011**, *10*, 1609–1616. [CrossRef]

27. Kumar, P.; Pant, D.C.; Mehariya, S.; Sharma, R.; Kansal, A.; Kalia, V.C. Ecobiotechnological Strategy to Enhance Efficiency of Bioconversion of Wastes into Hydrogen and Methane. *Indian J. Microbiol.* **2014**, *54*, 262–267. [CrossRef] [PubMed]

28. Kalia, V.C. Microbial applications. *Microb. Appl.* **2017**, *2*, 1–336.

29. Teh, Y.A.; Silver, W.L.; Conrad, M.E.; Borglin, S.E.; Carlson, C.M. Carbon isotope fractionation by methane-oxidizing bacteria in tropical rain forest soils. *J. Geophys. Res. Biogeosci.* **2006**, *111*, 1–8. [CrossRef]

30. Grossman, E.L.; Cifuentes, L.A.; Cozzarelli, I.M. Anaerobic methane oxidation in a landfill-leachate plume. *Environ. Sci. Technol.* **2002**, *36*, 2436–2442. [CrossRef] [PubMed]

31. North, J.C.; Frew, R.D.; Van Hale, R. Can stable isotopes be used to monitor landfill leachate impact on surface waters? *J. Geochem. Explor.* **2006**, *88*, 49–53. [CrossRef]

32. Bates, B.L.; McIntosh, J.C.; Lohse, K.A.; Brooks, P.D. Influence of groundwater flowpaths, residence times and nutrients on the extent of microbial methanogenesis in coal beds: Powder River Basin, USA. *Chem. Geol.* **2011**, *284*, 45–61. [CrossRef]

33. Morasch, B.; Richnow, H.H.; Schink, B.; Vieth, A.; Meckenstock, R.U. Carbon and hydrogen stable isotope fractionation during aerobic bacterial degradation of aromatic hydrocarbons. *Appl. Environ. Microbiol.* **2002**, *68*, 5191–5194. [CrossRef]

34. Mohammadzadeh, H.; Clark, I. Degradation pathways of dissolved carbon in landfill leachate traced with compound-specific 13C analysis of DOC. *Isotopes Environ. Health Stud.* **2008**, *44*, 267–294. [CrossRef]

35. Kjeldsen, P.; Barlaz, M.A.; Rooker, A.P.; Baun, A.; Ledin, A.; Christensen, T.H. Present and long-term composition of MSW landfill leachate: A review. *Crit. Rev. Environ. Sci. Technol.* **2002**, *32*, 297–336. [CrossRef]

36. Rodrigo-Ilarri, J.; Rodrigo-Clavero, M.E. Mathematical modeling of the biogas production in msw landfills. Impact of the implementation of organic matter and food waste selective collection systems. *Atmosphere* **2020**, *11*, 1306. [CrossRef]

37. Rank, D.; Papesch, W.; Rajner, V. Envinronmental Isotopes Study at a Research Landfill (Breitenau, Lower Austria). Available online: https://inis.iaea.org/collection/NCLCollectionStore/_Public/27/061/27061798.pdf?r=1 (accessed on 31 March 2021).
38. Nisi, B.; Raco, B.; Dotsika, E. Groundwater contamination studies by environmental isotopes: A review. *Handb. Environ. Chem.* **2016**, *40*, 115–150.
39. Craig, H. Isotopic variations in meteoric waters. *Science* **1961**, *133*, 1702–1703. [CrossRef] [PubMed]
40. Gat, J.R.; Carmi, I. Effect of climate changes on the precipitation patterns and isotopic composition of water in a climate transition zone: Case of the Eastern Mediterranean Sea area. *Influ. Clim. Chnage Clim. Var. Hydrolic Regime Water Resour.* **1987**, 513–524.
41. Bajjali, W. Spatial variability of environmental isotope and chemical content of precipitation in Jordan and evidence of slight change in climate. *Appl. Water Sci.* **2012**, *2*, 271–283. [CrossRef]
42. Giustini, F.; Brilli, M.; Patera, A. Mapping oxygen stable isotopes of precipitation in Italy. *J. Hydrol. Reg. Stud.* **2016**, *8*, 162–181. [CrossRef]

Article

Evaluating the Impact of Explosive Volcanic Eruptions on a Groundwater-Fed Water Supply System: An Exploratory Study in Ponta Delgada, São Miguel (Azores, Portugal)

Francisco Valente [1], José Virgílio Cruz [2,*], Adriano Pimentel [2,3], Rui Coutinho [2], César Andrade [2,3], Jorge Nemésio [4] and Selma Cordeiro [4]

[1] AmbiValente—Engenharia e Serviços, Unipessoal, Lda, 9504-535 Ponta Delgada, Portugal; ambivalente@outlook.pt
[2] IVAR—Instituto de Investigação em Vulcanologia e Avaliação de Riscos, Universidade dos Açores, 9500-321 Ponta Delgada, Portugal; adriano.hg.pimentel@azores.gov.pt (A.P.); rui.ms.coutinho@uac.pt (R.C.); cesar.cc.andrade@azores.gov.pt (C.A.)
[3] CIVISA—Centro de Informação e Vigilância Sismovulcânica dos Açores, Universidade dos Açores, 9500-321 Ponta Delgada, Portugal
[4] SMAS—Serviços Municipalizados de Água e Saneamento de Ponta Delgada, 9504-507 Ponta Delgada, Portugal; jorgenemesio@smaspdl.pt (J.N.); selmacordeiro@smaspdl.pt (S.C.)
* Correspondence: jose.vm.cruz@uac.pt

Abstract: Tephra fall is among the set of hazardous phenomena associated with volcanic activity that can impact water resources and services. The aim of this paper is to characterize the potential impacts of tephra fall on the groundwater-fed water supply system of Ponta Delgada (São Miguel, Azores) by comparing two scenarios of explosive eruptions. Vulnerability matrices were used to compute indexes, by multiplying the thickness of tephra fall deposits, corresponding to increasing hazard levels, by descriptors of the water supply system, representing the elements at risk. In a worst-case scenario, tephra covers a large area inland, severely constraining the abstraction of water to public supply, as 84.8% of the springs are affected. In 12 springs the expected hazard level is of perturbation (5157 m^3/day; 12.1% of the daily abstraction) or damage (16,094 m^3/day; 37.8%). The same trend is observed considering the storage capacity, as 75.4% of the reservoirs may be somehow affected. Moreover, 72.4% of the 68,392 inhabitants served by the water supply system will be in a zone where the damage level will be achieved. Results point out to the need of preparedness measures to mitigate consequences of a volcanic crisis over the water supply.

Keywords: water supply system; groundwater; volcanism; eruptive scenarios; vulnerability; Azores

Citation: Valente, F.; Cruz, J.V.; Pimentel, A.; Coutinho, R.; Andrade, C.; Nemésio, J.; Cordeiro, S. Evaluating the Impact of Explosive Volcanic Eruptions on a Groundwater-Fed Water Supply System: An Exploratory Study in Ponta Delgada, São Miguel (Azores, Portugal). *Water* **2022**, *14*, 1022. https://doi.org/10.3390/w14071022

Academic Editors: Helder I. Chaminé, Maria José Afonso and Maurizio Barbieri

Received: 11 February 2022
Accepted: 19 March 2022
Published: 23 March 2022

Publisher's Note: MDPI stays neutral with regard to jurisdictional claims in published maps and institutional affiliations.

1. Introduction

The impacts of volcanic activity on society and the environment are widely recognized, both by the scientific community and by the public. Nevertheless, and despite the growing awareness, the vulnerability of societies is increasing due to several reasons that lead to the rising settlement in active volcanic areas [1]. A large set of hazardous phenomena is directly and indirectly associated with volcanic activity, such as in the former group lava flows, pyroclastic density currents, tephra fall, gas and acid particle emissions, and in the latter group earthquakes, landslides, tsunamis, ground deformation, lahars and debris flows and acid rain [2]. Regarding tephra fall, the effects on water resources are the most widespread, when compared to other elements at risk, and besides the destruction of water services infrastructure, water quality may also be affected by volatiles or ash particles and may lead to indirect impacts on agricultural activity and food production. Moreover, among other hazardous volcanic phenomena that may impact water supply, such as pyroclastic density currents or lahars, the available reports suggest that tephra fall explains most of the constraints and damages recorded [3].

Water supply systems may be constrained through tephra fall due to four main processes, namely physical damages, disruption of water treatment, water quality deterioration and water shortages [3–5], and several case studies have been published on this subject [6,7]. Physical damages may occur in the water sources or on treatment and distribution facilities, and tephra fall or tephra-water slurries may cause abrasion of moving parts or metal corrosion [5,8]. Water treatment capacity may also be constrained due to blockage of water intake, filters and pipes [5], but in the present study the effects over storage are indirectly analyzed considering the potential damages in reservoirs to which chlorination stations are coupled. The deposition of tephra in the water source areas may deteriorate water quality, both in terms of overall chemical content and turbidity [5,8,9]. Water shortages are also a consequence of tephra fall, resulting from water supply disruption or from the increasing water use for cleaning operations [3], and may lead to the reduction of human consumption or even to infectious disease outbreaks when the amount of water needed for hygiene and sanitation is also lower [4]. Agriculture activities may also be affected by the impact of tephra fall on water availability and quality [10], and studies have shown these effects over the dairy industry [11], the latter being an important economic sector in the Azores archipelago.

With an area of 744.6 km^2, São Miguel is the largest and most populated (137, 307 inhabitants) island of the Azores archipelago, located in the North Atlantic Ocean between latitudes 37°42′13″ N and 37°54′38″ N, and longitudes 25°08′03″ W and 25°51′17″ W (Figure S1—electronic supplementary material—and Figure 1A). Ponta Delgada municipality occupies the westernmost part on the island, corresponding to an area of 233 km^2 and a population of 68,392 inhabitants.

The geology of São Miguel is dominated by three active quaternary central volcanoes (Sete Cidades, Fogo and Furnas), associated with highly explosive trachytic eruptions that produced widespread tephra fall deposits across the island [12–15]. Besides the characterization of the different types of eruptions associated with these central volcanoes, already addressed in detail in the cases of Sete Cidades [16], Fogo [17] and Furnas [18], or in what regards the fissural volcanic systems on the island [19], other hazardous phenomena have been also described in the last decades, such as gas emanations both in the ground [20,21] or in water bodies [22–25], tsunamis [26], atmospheric processes, floods and mass wasting [26–31] or abnormal content of minor and trace elements [32].

Despite all the knowledge regarding hazardous processes in São Miguel, the first attempt to quantify the economic loss resulting from explosive volcanic eruptions on an important economic activity, such as the tourism industry, was only accomplished recently [33]. Before that, some studies on potential damages on buildings in São Miguel Island were carried out [34], or widespread vulnerability analysis of specific islands or volcanoes [35–37].

The objective of this paper is to characterize the potential impacts of tephra fall deposition on the water supply system of the municipality of Ponta Delgada, through the comparison of two scenarios of explosive eruptions sourced from Sete Cidades Volcano which can be taken as a proxy for future eruptive events. The effects are characterized in terms of the volume of water abstracted, the storage capacity affection and the inhabitants that may suffer water supply disruption.

The present study is the first research regarding the impact of volcanic eruptions on critical infrastructure in the Azores archipelago, and further studies should take into account other possible scenarios for explosive eruptions (in terms of the location of the eruptive centers, eruptive source parameters and wind conditions), as well as scenarios of effusive eruptions (location of the source, effusive rate and main flow path of the lava flows).

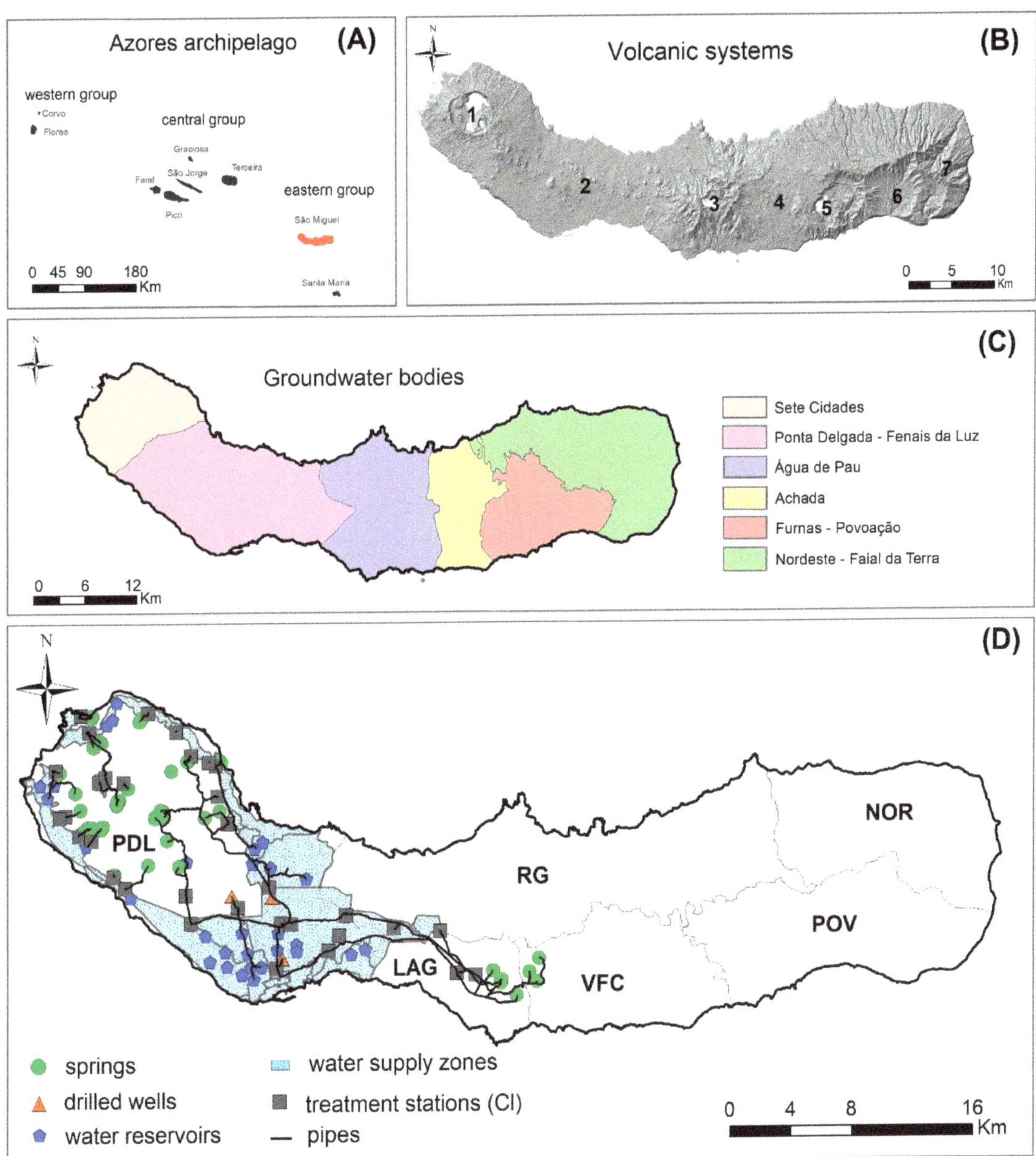

Figure 1. Study area setting: (**A**) Location of São Miguel Island in the Azores archipelago; (**B**) volcanic systems of São Miguel from west to east (1—Sete Cidades Volcano; 2—Picos Fissural Volcanic System; 3—Fogo Volcano, also known as Água de Pau Volcano; 4—Congro Fissural Volcanic System; 5—Furnas Volcano; 6—Povoação Volcano; 7—Nordeste Volcanic System; according to [38]); (**C**) groundwater bodies delimitation from [39]; (**D**) layout of the water supply system from Ponta Delgada (data from SMAS) and Municipalities (PDL—Ponta Delgada; LAG—Lagoa; RG—Ribeira Grande; VFC—Vila Franca do Campo; POV–Povoação; NOR—Nordeste).

2. Materials and Methods

2.1. Study Area

São Miguel Island is formed by seven volcanic systems (Figure 1B), namely, from the west to the east, Sete Cidades Volcano, Picos Fissural Volcanic System, Fogo Volcano (also known as Água de Pau Volcano), Congro Fissural Volcanic System, Furnas Volcano, Povoação Volcano and Nordeste Volcanic System [38]. All these systems are active, except for the Nordeste Volcanic System and Povoação Volcano, which are considered extinct [40].

The latter eruptive scenarios presented are associated with explosive eruptions in the Sete Cidades Volcano, which is the westernmost central volcano on São Miguel Island. This volcanic edifice has an area of 110 km^2, a mean subaerial diameter of approximately 12 km and reaches a maximum altitude of 845 m a.s.l. [13,41]. In the summit of the volcano a 5 km-diameter sub-circular caldera can be observed with walls as high as 350 m [41].

The growth of the Sete Cidades Volcano started approximately 200,000 years ago, and its stratigraphy comprises two main geological groups, according to reference [42]: The Inferior Group, consisting of lava flows and pyroclastic deposits, which date from more than 200,000 years ago to 36,000 years BP, and the Superior Group, comprising all volcanic products erupted in the last 36,000 years.

The last major explosive eruption of Sete Cidades occurred 16,000 years BP and was related to the last stage of formation of the summit caldera. The products of this eruption are recorded by the Santa Bárbara Formation and include trachytic tephra fall and pyroclastic density current deposits [43,44]. Over the past 5000 years at least 17 trachytic explosive eruptions took place in Sete Cidades, the last one about 500–600 years BP [42]. These recent eruptions, some of which of sub-Plinian dimensions, were sourced from the summit caldera and their products are dominated by tephra fall deposits. This high eruptive frequency of Sete Cidades makes it the most active and hazardous central volcano of São Miguel [38].

The mean annual precipitation in São Miguel amounts to 1722 mm [45], and the mean annual runoff is 686 mm, corresponding to a total discharge of 5.11 $\times$ 10^8 m^3/a [45]. According to the Azores Regional Water Plan and the River Basin Management Plan, made to comply with the EU Water-Framework Directive requirements, there are six main groundwater bodies on São Miguel (Figure 1C) [39], which are closely related to the volcanic systems on the island. A comprehensive description of the hydrogeology of São Miguel can be found in the papers [46,47].

2.2. The Water Supply System of Ponta Delgada

The water supply on São Miguel is fully assured by the six municipalities, with 100% of the 137,307 inhabitants being served. The 60 water supply zones on the island are vertically integrated and are run directly by four municipalities (Lagoa, Ribeira Grande, Vila Franca do Campo and Povoação), while the other two are run indirectly through a municipal company and a semi-autonomous service provider [48]. The mean number of inhabitants per water system is equal to 2311. On the island, a total of 157 groundwater sources are nowadays abstracted, corresponding to a volume of 2.53 $\times$ 10^7 m^3/a, contrasting to a single surface water source exploited for human supply (6.23 $\times$ 10^7 m^3/a) [39].

The Ponta Delgada water supply system serves a total of 68,392 inhabitants and corresponds to a vertically integrated system run indirectly by the Ponta Delgada municipality through a semi-autonomous water services utility company (Serviços Municipalizados de Água e Saneamento de Ponta Delgada—SMAS).

The supply system is divided into 18 zones (Figure 1D), based on the homogeneity of the water sources that feed the system. All the water is abstracted from groundwater bodies, namely through 46 springs or groups of springs and four drilled wells (despite a larger number that can be exploited), corresponding to a mean volume of about 42,545 m^3/day (2019; SMAS). The volume of water abstracted has been relatively constant since 2003, reflecting the influence of the springs presenting higher discharge rates, where seasonal variations are nonexistent [32], and in 2018 the overall abstracted volume was estimated as equal to 1.42 $\times$ 10^7 m^3/a [49].

Springs correspond to discharges from perched aquifers in the Sete Cidades and Água de Pau groundwater bodies (Figure 1C), respectively, toward the west and the east, with the latter being used also to transfer water to systems run by the Ribeira Grande and Lagoa municipalities. These springs are located at various altitudes on the flanks of the quaternary central volcanoes of Sete Cidades and Fogo, such as, for example, the so-called Chã dos Tanques (Figure 2). Hydrogeological studies have shown that in the Azores discharge is generally higher in springs from lava flow aquifers, basaltic or trachytic in nature, compared to discharges from aquifers made of pyroclastic deposits. Discharge is also often higher in winter than in summer [50], presenting an inverse linear relationship between the elevation of the discharge and the water conductivity, which suggests that groundwater discharging at the high-altitude springs have a lower residence time [51].

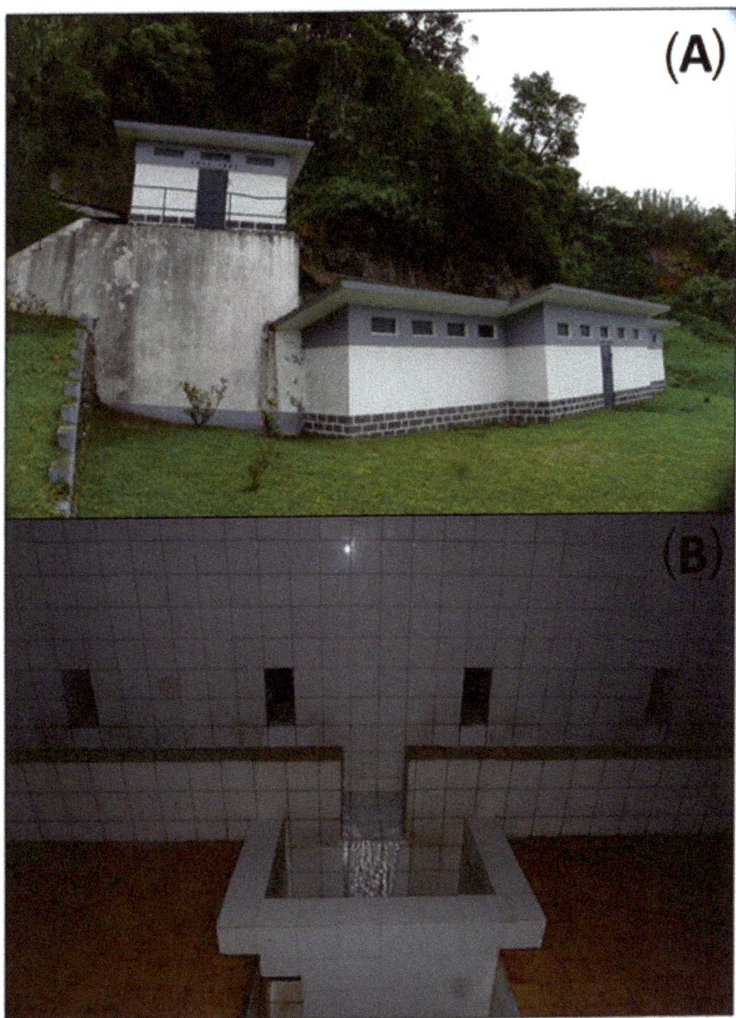

Figure 2. View of the Chã dos Tanques spring (Feteiras, Sete Cidades): (**A**) External view; (**B**) interior of the spring capture.

The Sete Cidades and Fogo central volcanoes are linked by a volcanic fissural zone (Picos Fissural Volcanic System), corresponding to a flatter area at a lower altitude where wells were drilled in the last decades, abstracting groundwater from perched or from basal aquifers. This fissural system corresponds to the so-called Ponta Delgada–Fenais da Luz groundwater body, and drilled wells present depths in the range 89 m up to 284 m, with a mean transmissivity of 3×10^{-2} m^2/s.

Water storage is ensured by using 61 reservoirs, mainly ground reservoirs, corresponding to a total capacity of 48,790 m^3 and the system also includes nearly 710 km of pipes. Coupled to reservoirs, water treatment is made in 33 treatment stations through chlorination. As the groundwater abstracted in springs is mainly of acid soft waters type, it requires only a simple correction of pH and hardness, which is typically made in the spring capture using tanks with limestone fragments in the bottom. Generally, the supplied water is of very good quality, complying to normative values and monitoring demands, despite some concerns about abnormal fluoride content in some springs [32].

Water usage in Ponta Delgada is dominated by domestic supply, which corresponds to about 3.34×10^6 m^3/a (year 2018; [52]), to which 3.39×10^5 m^3/a (commerce and services) must be added to estimate urban water services. The water supply system managed by SMAS also provided 9.03×10^5 m^3/a for social purposes and temporary supplies, about 7.53×10^5 m^3/a to other public demands, as well as 8.36×10^5 m^3/a and 5.42×10^5 m^3/a to agriculture and industry, respectively [52]. The volume of water charged by SMAS to water users in 2018 was equal to 6.54×10^6 m^3/a [52], and the leaks in the system were computed at more than 40%, which justifies the gap to the volume of water being abstracted annually [49].

2.3. Methodology

The potential impact of explosive volcanic eruptions on the water supply system of Ponta Delgada was evaluated through the application of vulnerability matrices. In the present paper mainly the physical vulnerability is approached, in the sense presented by [53], representing the susceptibility to tephra fall associated with trachytic explosive eruptions from the Sete Cidades Volcano.

The first step in the assessment of the impact of tephra fall deposition on the water supply systems was the definition of eruptive scenarios. This is a common approach used to assess volcanic hazard, as it helps to mitigate the impacts of future eruptions by anticipating the consequences that may occur [54]. For this paper, two of the eruptive scenarios of [44] were selected, which were obtained from the reconstruction of the violent sub-Plinian phase of the Santa Bárbara eruption [43,44]. The numerical simulations of tephra fall scenarios were obtained using an advection–diffusion model (for details see [55]) in the Volcanic Risk Information System (VORIS) 2.0.1 tool [56], implemented in a GIS framework, which allows for the rapid visualization of the simulation results. This methodology has been successfully applied to other cases, such as of Deception Island, Antarctica [57], El Hierro and Lanzarote, Canary Islands [54,58] and Fogo Volcano (São Miguel) [33].

Eruptive scenario A (corresponding to scenario 2 of [44]) reconstructs the known distribution and thickness of the tephra fall deposit of the Santa Bárbara eruption. The main eruptive source parameters used in scenario A were an erupted bulk volume of 0.27 km^3 and an eruption column height of 17,000 m, which are in the same range of values estimated for some of the recent (<5000 years) sub-Plinian eruptions of the Sete Cidades Volcano [42,59]. Therefore, it can be taken as a proxy for future explosive eruptions. The wind conditions used correspond to a vertical profile with wind directions ranging from SW-blowing winds at lower altitudes to WNW-blowing winds at higher altitudes.

Eruptive scenario B (corresponding to scenario 3 of [44]) simulates a hypothetical future explosive eruption with the same eruptive source parameters as in scenario A, but with a unidirectional NW-blowing wind profile. It reproduces a worst-case scenario of a violent sub-Plinian eruption of the Sete Cidades Volcano under wind conditions blowing toward Ponta Delgada city. Given the westerly prevailing winds throughout most of the year in the Azores region (see wind statistical analysis in [33,38]), scenario B is a plausible most hazardous scenario with the highest impact on critical infrastructure of Ponta Delgada municipality.

The main input parameters used for the simulations of eruptive scenarios A and B with the VORIS 2.0.1 tool are shown in Table 1 (for further details see [44]).

Table 1. Main input parameters for the two eruptive scenarios (adapted from [44]).

		Scenario A	Scenario B
Eruptive source parameters	Erupted bulk volume (km^3)	0.27	0.27
	Eruption column height (m)	17,000	17,000
Wind conditions	Altitude (m)	Direction (°)/ Intensity (m/s)	Direction (°)/ Intensity (m/s)
	2000	230/2	322/2
	4000	245/4	322/4
	8000	265/10	322/10
	13,000	280/21	322/21
	18,000	290/32	322/32

Vulnerability matrices were applied to compute indexes, as through these matrices, the thickness of tephra fall deposits, corresponding to an increasing hazard level, is multiplied by descriptors of the water supply system, representing the elements at risk. For this latter purpose, the elements at risk considered in this study were the volume of water being abstracted (Table 2), the storage capacity of the reservoirs (Table 3) and the number of inhabitants being served by the system (Table 4). The application of this semi-quantitative approach does not require ex-ante data [60] and may provide an important insight into the definition of strategies and structural measures for risk mitigation [61]. Nevertheless, as only one single volcanic hazard (tephra fall) is addressed, difficulties associated with a multi-hazard analysis are avoided, despite recognizing that in volcanic regions this approach may be of higher value as these areas are affected by a large array of hazardous phenomena [62].

Table 2. Vulnerability matrix for the volume of water being abstracted (threshold values for tephra fall thickness adapted from [3]. Tolerance status (1–4) in yellow, disturbance in orange (5–9) and damage in red (10–15).

Spring/Drilled Well		Tephra Fall Thickness (mm)		
		1–20	20–100	>100
Discharge (m^3/day)	Value	1	2	3
0–125	1	1	2	3
125–250	2	2	4	6
251–500	3	3	6	9
501–1000	4	4	8	12
>1000	5	5	10	15

Table 3. Vulnerability matrix for the storage capacity of the water reservoirs (threshold values for tephra fall thickness adapted from [3]. Tolerance status (1–4) in yellow, disturbance in orange (5–9) and damage in red (10–15).

Reservoir		Tephra Fall Thickness (mm)		
		10–100	100–500	>500
Storage Capacity (m^3)	Value	1	2	3
0–250	1	1	2	3
251–500	2	2	4	6
501–1000	3	3	6	9
1001–2000	4	4	8	12
>2000	5	5	10	15

Table 4. Vulnerability matrix for the number of inhabitants being served by the system (threshold values for tephra fall thickness adapted from [3]. Tolerance status (1–4) in yellow, disturbance in orange (5–9) and damage in red (10–15).

Water Supply Zone		Tephra Fall Thickness (mm)		
Inhabitants (no.)		1–20	20–100	>100
	Value	1	2	3
0–500	1	1	2	3
501–2000	2	2	4	6
2001–6000	3	3	6	9
6001–8000	4	4	8	12
>8000	5	5	10	15

Threshold values for tephra fall thickness used in Tables 2–4 were taken from [3]. For springs the increasing hazard levels are, respectively, equal to 1 to 20 mm, 20 to 100 mm and higher than 100 mm, while for values lower than 1 mm no damage/disruption is expected (Table 2). If along the first level (1 to 20 mm), clogging of filters and some abrasion occurs, as well as higher water turbidity, the following levels may depict an increasing deterioration of the water quality, damages to pumping equipment, infilling of tanks and in the limit the collapse of the water source roofs [3]. The same threshold values were considered for the analysis of the number of inhabitants being affected (Table 4). Indices in the matrix were classified as representative of a tolerance status (1–4), of disturbance (5–9) and of damage (10–15). Five classes were considered for the spring discharge (Table 2), taking into account data for the several sources being abstracted [32].

For reservoirs, the hazard levels used are, respectively, equal to 10 to 100 mm, 100 to 500 mm and higher than 500 mm, and for values lower than 10 mm no damage/disruption in the building is expected (Table 3). These threshold values are the same as those proposed by [3] for buildings, more appropriate for the type of reservoirs in the Ponta Delgada water supply system. For the first level (10 to 100 mm), only light roof damages are expected, with abrasion of windows and cladding; nevertheless, tephra will infiltrate the interior of the building [3]. Roof damages will increase along the following levels, from severe to total collapse, as well as the extent of damages in the buildings. The expected amount of tephra inside the reservoirs will also be much higher. Five classes were considered for the storage capacity (Table 3), considering data for the several water reservoirs that made the water supply system of Ponta Delgada [32].

Finally, for the vulnerability matrix based on the number of inhabitants being served by the system, five classes were considered regarding the number of people served in each of the 18 water supply zones into which the system is subdivided (Table 4).

As the water treatment stations are in the main reservoirs, results for the latter structures also reflect any constraints to the treatment capability following explosive volcanic eruptions.

Despite groundwater-based water supply systems perhaps being less vulnerable to tephra fall than surface water-fed systems or wastewater facilities, which are not addressed in this paper, well-head equipment is still vulnerable [5]. Moreover, the water supply system design is instrumental in what concerns impact occurrence and severity [4], and in the present case study spring-fed sources will also be impacted as suspended ash will penetrate inside the infrastructure, which may cause blockage of the water intake, and filter and pipe clogging, besides the possibility of the damages to the overall infrastructure. Tephra-water slurries may also cause clogging [3].Poor water quality is also to be expected due to the deposition of tephra inside the water sources infrastructure and in the recharge areas.

3. Results and Discussion

3.1. Impacts on Water Abstraction

The impacts of the two eruptive scenarios on the water supply system of Ponta Delgada are shown in Figures 3 and 4, respectively, for scenarios A and B. In scenario A, tephra fall deposits extend mostly to the north coast of São Miguel, while in scenario B tephra covers a much larger area inland, thus impacting at a larger scale the water supply infrastructure.

According to scenario A, 70.5% of the springs that feed the water supply are impacted (31 springs from a total of 44); nevertheless, the remaining 29.5% (13 springs) represent a major portion of the total abstraction (Figure 3A). The latter group is responsible for 64.5% of the total daily abstraction (27,446 m^3/day). In 17 springs, the hazard level is tolerable, corresponding to a volume abstracted of 1947 m^3/day (4.6% of the total daily volume), in six springs the level is of perturbance (2274 m^3/day; 5.3% of the total) and in seven springs the level is of damage (10,884 m^3/day; 25.6% of the total) (Figures 3A and 5A). The drilled wells area was not affected in this scenario.

If the tolerance level only requires an increasing frequency of maintenance operations, due to clogging of filters and equipment abrasion in the spring abstraction infrastructure, the disturbance level may imply restrictions on the water usage, due to damages to equipment or water pollution. These impacts are aggravated in the damage hazard level, where the water source infrastructure may collapse and the expected water pollution risk increases.

Scenario B reveals that only 15.2% of the springs (seven of a total of 46) are not affected at any level by tephra fall deposits, namely the ones located at Água de Pau groundwater body (Figures 1C,D and 4A). Instead, springs spread over Sete Cidades groundwater body, as well as all the drilled wells in the Ponta Delgada–Fenais da Luz groundwater body, are affected by the deposition of tephra. In 18 springs the hazard level is tolerable, corresponding to 1821 m^3/day (4.3% of the total daily volume), but in 12 springs the expected level is of perturbation (5157 m^3/day; 12.1%) or damage (16,094 m^3/day; 37.8%) (Figures 4A and 5A). The latter percentages show how scenario B has a much larger impact compared to scenario A, severely constraining the daily abstraction of water to public supply.

3.2. Impacts on Water Storage

The water storage capacity is also strongly impacted on both scenarios, but with different magnitudes. According to scenario A, in 12 reservoirs the hazard level is of tolerance, corresponding to a storage capacity of 4730 m^3 (9.7% of the total capacity), and in the other two the hazard is of perturbation (1060 m^3; 2.2%) (Figures 3B and 5B). The damage level only occurs in a single reservoir, corresponding to a storage volume of 2000 m^3 (4.1%).

If within the tolerance level only small damage in the roofs and abrasion of doors, windows and cladding of the reservoirs is to be expected, thus not constraining the storage capacity, with the perturbation level partial collapses of the infrastructure may occur, as well as serious damage in roofs and vertical structures, thus compromising water storage volume. The water storage is fully compromised at the damage level, as damages are beyond repair, as total roof collapse may occur as well as severe damage to the remaining structures.

The overall storage volume is strongly aggravated with scenario B, as 75.4% of the reservoirs may be affected, corresponding to 87.7% of the overall storage capacity of the water supply system (Figures 4B and 5B). A total of 22 reservoirs are at the tolerance level (13,360 m^3; 27.4% of the total), 18 at the perturbation level (17,140 m^3; 35.1%) and six at the damage level (12,300 m^3; 25.2%).

3.3. Impacts on Water Supply Zones

Using the same hazard levels applied for the water sources and considering the overall number of inhabitants served by each water supply zone, scenario A reveals that in four zones the level is of tolerance, with 2169 inhabitants affected (3.2% of the total). In other eight water supply zones the level is of perturbance (40,868 inhabitants; 59.6%) (Figures 3C and 5C).

Scenario B affects six zones (33% of the total) with a tolerance hazard level (5087 inhabitants; 7.4%), another six zones depict a perturbation level (13,323; 19.5%) and five present a damage hazard level (Figures 4C and 5C). The latter ones correspond to 49,513 inhabitants, 72.4% of the total population served by the Ponta Delgada water supply system.

3.4. Prepradness Measures and Forthcoming Research

The main findings suggest that the water supply system of Ponta Delgada can be severely impacted by an explosive eruption at the Sete Cidades Volcano. Moreover, despite not being analyzed in the present paper, an explosive eruption at the nearby Fogo Volcano also has the potential to disrupt the water supply system, as suggested by volcanological studies [12,33], both by directly affecting springs located in the Água de Pau groundwater body or by tephra deposition toward the west if the wind conditions are favorable. Therefore, a set of preventive measures must be adopted to mitigate the potential effects of a future explosive eruption. Several actions have been recommended to water supply managers in the literature [63], such as further protection of the water intakes and the empowerment of water monitoring, besides the increase of spare technical components.

Taken as priority criteria for the springs/wells, or similarly the reservoirs, that depict a higher hazard index for both eruptive scenarios, the water intake protection should be reinforced to prevent physical damages and avoid water turbidity increase, maintaining the interiors as cleanly as possible. Moreover, air thigh conditions will also allow a higher protection of any indoor equipment, such as electrical control panels or chlorination devices (in reservoirs).

An overall increase of the storage capacity, as well as creating conditions for water transfer between water supply zones in the system, also have to be prepared to allow a response to a volcanic crisis, or to anticipate the increasing water demand for cleaning operations in the aftermath of the eruption. The identification of alternative routes toward water intakes and reservoirs, facilitating cleaning operations in the facilities and personnel training will also be useful for a prompt response.

Monitoring of water quality is also to be reinforced during and in the aftermath of the eruption, to allow any adjustment to be made in the water treatment operations, for example, due to acidity decrease or higher water turbidity, or to control some elemental contamination from toxic elements (e.g., fluoride). Preparedness should also include analyzing tephra to characterize those toxic elements' leaching (Al, As, Cd, Cr, Cu, F, Fe, Mn, Ni, Pb, Zn) to which some protocols have been already developed [64].

The geology of the Azores depicts a large array of eruptive styles, from more effusive eruptions to highly explosive events, thus other eruptive scenarios must be fully investigated in the future to consider water supply as a major issue of civil defense in the Azores archipelago, in general, and in São Miguel Island, in particular.

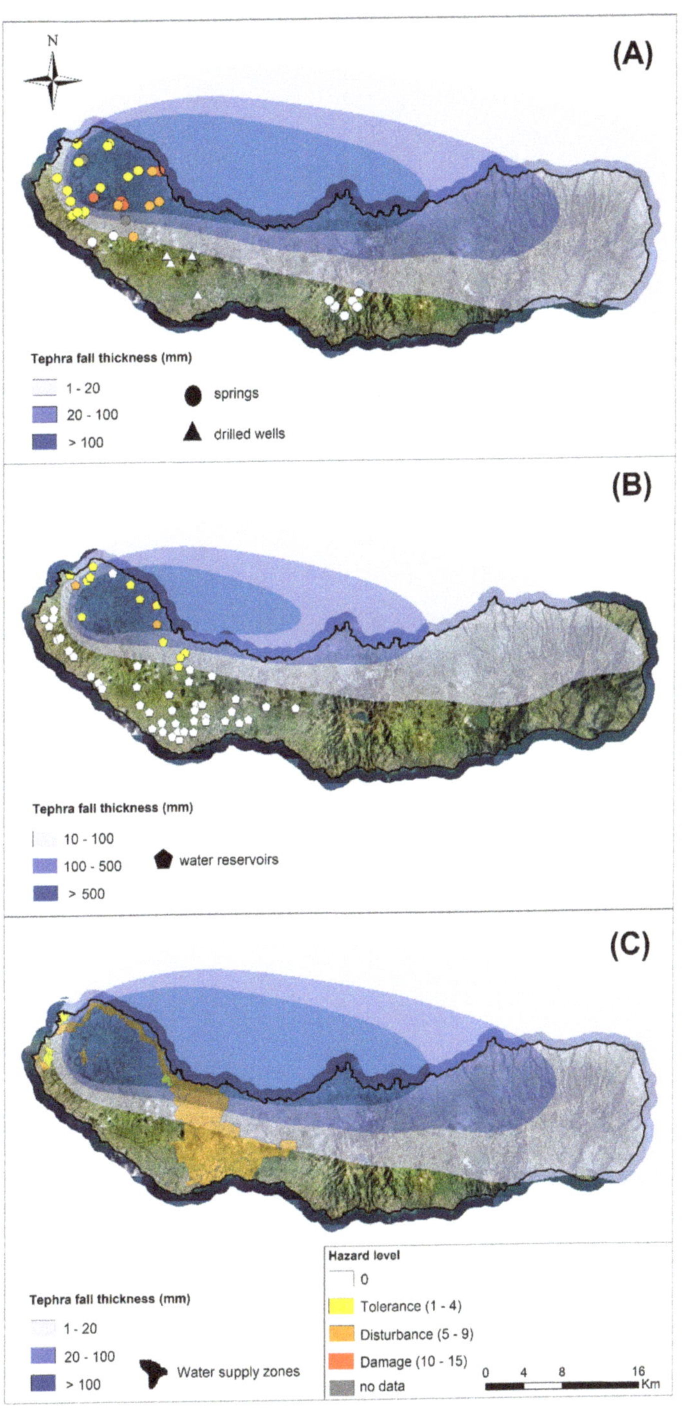

Figure 3. Cartographic representation of the impacts of eruptive scenario A, respectively, for: (**A**) Volume of water being abstracted; (**B**) storage capacity of the reservoirs; (**C**) number of inhabitants being served by the system over the water supply system of Ponta Delgada.

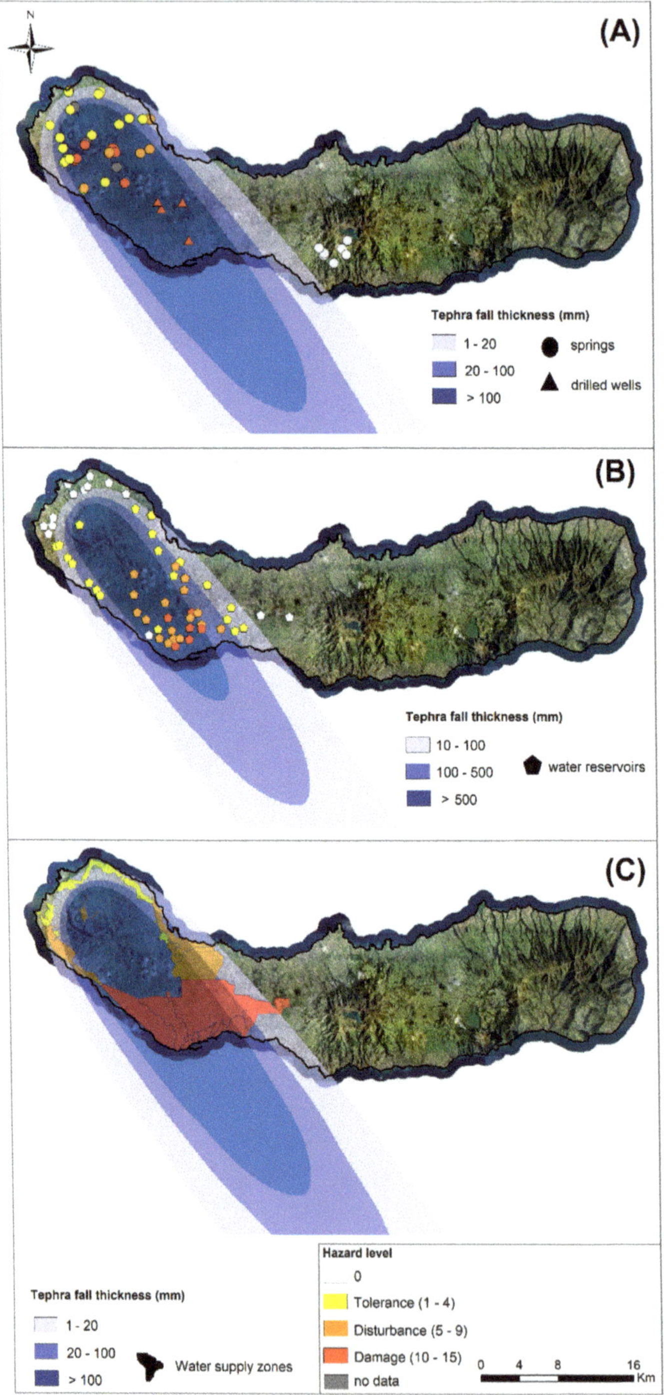

Figure 4. Cartographic representation of the impacts of eruptive scenario B, respectively, for: (**A**) Volume of water being abstracted; (**B**) storage capacity of the reservoirs; (**C**) number of inhabitants being served by the system over the water supply system of Ponta Delgada.

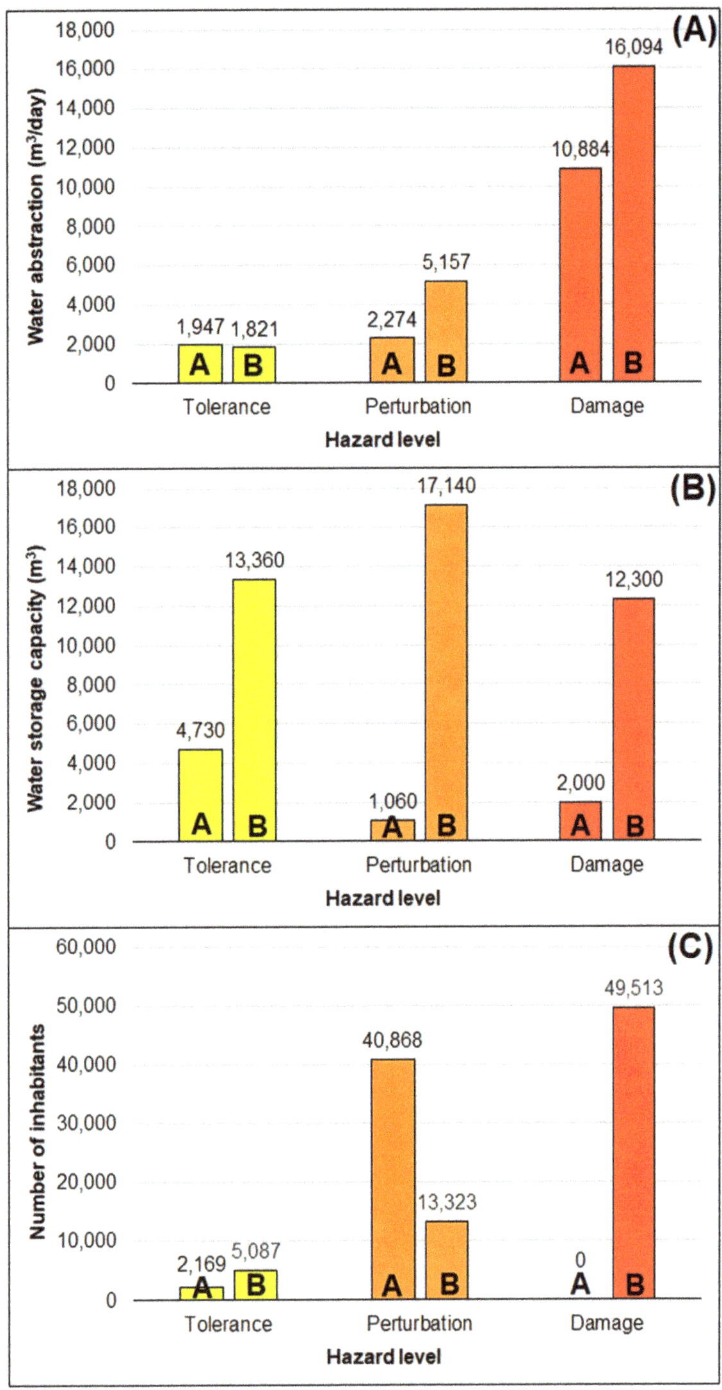

Figure 5. Comparison of the two eruptive scenarios (**A**,**B**), respectively, for: (**A**) Volume of water being abstracted; (**B**) storage capacity of the reservoirs; (**C**) number of inhabitants being served by the system over the water supply system of Ponta Delgada.

4. Conclusions

Ponta Delgada is the municipality with the largest population in the Azores and is located in the vicinity of two active central volcanoes (Sete Cidades and Fogo). Therefore, the water supply system in this area, which serves a total of 68,392 inhabitants, may be affected by volcanic eruptions in the future, with severe consequences if the supply is partially or totally interrupted. Moreover, the location of the water captures, mainly spread in the flanks of these volcanoes, as well as the widespread distribution of the urban areas along the coastal area of the municipality, poses additional difficulties in preventing or mitigating the consequences of an eruption over the water supply system.

The present study also points to the need for the municipal emergency plan to provide measures specifically focused on protecting the water supply, to maintain it during and after a volcanic crisis with a minimum of constraints. The results of this paper can serve as a benchmark for other municipalities in the Azores, where the issue of water supply during and after a volcanic event is equally crucial and stresses the need to develop such studies in the near future.

Supplementary Materials: The following supporting information can be downloaded at: https://www.mdpi.com/article/10.3390/w14071022/s1, Figure S1: Location of the Azores archipelago in the North Atlantic Ocean.

Author Contributions: Conceptualization, J.V.C. and F.V.; methodology, J.V.C., F.V., A.P., C.A., J.N. and S.C.; software, C.A.; investigation, F.V. and J.V.C.; validation, J.V.C.; writing—original draft preparation, J.V.C., F.V. and A.P.; writing—review and editing, J.V.C. and R.C.; visualization, C.A.; supervision, J.V.C. All authors have read and agreed to the published version of the manuscript.

Funding: This research received no external funding.

Institutional Review Board Statement: Not applicable.

Informed Consent Statement: Not applicable.

Data Availability Statement: The data presented in this study are available on request from the corresponding author.

Acknowledgments: The authors wish to express their gratitude to the Serviços Municipalizados de Água e Saneamento de Ponta Delgada for their valuable support to the research, namely by the data kindly made available.

Conflicts of Interest: The authors declare no conflict of interest.

References

1. Schmincke, H.-U. *Volcanism*; Springer: Berlin/Heidelberg, Germany, 2004; p. 324.
2. Francis, P.; Oppenheimer, C. *Volcanoes*, 2nd ed.; Oxford University Press: Oxford, UK, 2004; p. 521.
3. Wilson, G.; Wilson, T.M.; Deligne, N.I.; Cole, J.W. Volcanic hazard impacts to critical infrastructure: A review. *J. Volcanol. Geotherm. Res.* **2014**, *286*, 148–182. [CrossRef]
4. Stewart, C.; Wilson, T.M.; Leonard, G.S.; Johnston, D.M.; Cole, J.W. Volcanic hazards and water shortages. In *Water Shortages: Environmental Economic and Social Impacts*; Briggs, A.C., Ed.; Nova Publishing: New York, NY, USA, 2010; pp. 105–124.
5. Wilson, T.M.; Stewart, C.; Sword-Daniels, V.; Leonard, G.S.; Johnston, D.M.; Cole, J.W.; Wardman, J.; Wilson, G.; Barnard, S.T. Volcanic ash impacts on critical infrastructure. *Phys. Chem. Earth* **2012**, *45–46*, 5–23. [CrossRef]
6. Johnston, D.; Stewart, C.; Leonard, G.S.; Hoverd, J.; Thordasson, T.; Cronin, S. *Impacts of Volcanic Ash on Water Supplies in Auckland: Part I. Institute of Geological & Nuclear Sciences Science Report 2004/25*; Institute of Geological & Nuclear Sciences Limited: Lower Hutt, New Zealand, 2004.
7. Michaud-Dubuy, A.; Carazzo, G.; Kaminski, E. Volcanic hazard assessment for tephra fallout in Martinique. *J. Appl. Volcanol.* **2021**, *10*, 8. [CrossRef]
8. Stewart, C.; Johnston, D.M.; Leonard, G.; Horwell, C.J.; Thordason, T.; Cronin, S. Contamination of water supplies by volcanic ashfall: A literature review and simple impact modelling. *J. Volcanol. Geotherm. Res.* **2006**, *158*, 296–306. [CrossRef]
9. Wilson, G.; Wilson, T.M.; Deligne, N.I.; Blake, D.M.; Cole, J.W. Framework for developing volcanic fragility and vulnerability functions for critical infrastructure. *J. Appl. Volcanol.* **2017**, *6*, 14. [CrossRef]

10. Wilson, T.M.; Cole, J.W.; Stewart, C.; Cronin, S.J.; Johnston, D.M. Ash storms: Impacts of wind-remobilised volcanic ash on rural communities and agriculture following the 1991 Hudson eruption, southern Patagonia, Chile. *Bull. Volcanol.* **2010**, *73*, 223–239. [CrossRef]
11. Wild, A.J.; Wilson, T.M.; Bebbington, M.S.; Cole, J.W.; Craig, H.M. Probabilistic volcanic impact assessment and cost-benefit analysis on network infrastructure for secondary evacuation of farm livestock: A case study from the dairy industry, Taranaki, New Zealand. *J. Volcanol. Geotherm. Res.* **2019**, *387*, 106670. [CrossRef]
12. Booth, B.; Croasdale, R.; Walker, G. A quantitative study of five thousand years of volcanism on São Miguel, Azores. *Philos. Trans. R. Soc. Lond.* **1978**, *288*, 271–319.
13. Moore, R.B. Volcanic geology and eruption frequency, São Miguel, Azores. *Bull. Volcanol.* **1990**, *52*, 602–614. [CrossRef]
14. Guest, J.E.; Gaspar, J.L.; Cole, P.D.; Queiroz, G.; Duncan, A.M.; Wallenstein, N.; Ferreira, T.; Pacheco, J.M. Volcanic geology of Furnas volcano, São Miguel, Azores. *J. Volcanol. Geotherm. Res.* **1999**, *92*, 1–29. [CrossRef]
15. Pacheco, J.M.; Ferreira, T.; Queiroz, G.; Wallenstein, N.; Coutinho, R.; Cruz, J.V.; Pimentel, A.; Silva, R.; Gaspar, J.L.; Goulart, C. Notas sobre a geologia do arquipélago dos Açores. In *Geologia de Portugal*; Dias, R., Araújo, A., Terrinha, P., Kullberg, J.C., Eds.; Escolar Editora: Lisboa, Portugal, 2013; pp. 595–690. (In Portuguese)
16. Queiroz, G.; Gaspar, J.L.; Guest, J.E.; Gomes, A.; Almeida, M.H. Eruptive history and evolution of Sete Cidades Volcano, São Miguel Island, Azores. In *Volcanic Geology of São Miguel Island (Azores Archipelago)*; Gaspar, J.L., Guest, J.E., Duncan, A.M., Barriga, F., Chester, D.K., Eds.; Geological Society: London, UK, 2015; Memoirs 44; pp. 87–104.
17. Wallenstein, N.; Chester, D.; Coutinho, R.; Duncan, A.; Dibben, C. Volcanic hazard vulnerability on São Miguel Island, Azores. In *Volcanic Geology of São Miguel Island (Azores Archipelago)*; Gaspar, J.L., Guest, J.E., Duncan, A.M., Barriga, F., Chester, D.K., Eds.; Geological Society: London, UK, 2015; Memoirs 44; pp. 213–225.
18. Guest, J.E.; Pacheco, J.M.; Cole, P.D.; Duncan, A.M.; Wallenstein, N.; Queiroz, G.; Gaspar, J.L.; Ferreira, T. The volcanic history of Furnas Volcano, São Miguel, Azores. In *Volcanic Geology of São Miguel Island (Azores Archipelago)*; Gaspar, J.L., Guest, J.E., Duncan, A.M., Barriga, F., Chester, D.K., Eds.; Geological Society: London, UK, 2015; Memoirs 44; pp. 125–134.
19. Ferreira, T.; Gomes, A.; Gaspar, J.L.; Guest, J. Distribution and significance of basaltic eruptive centres: São Miguel, Azores. In *Volcanic Geology of São Miguel Island (Azores Archipelago)*; Gaspar, J.L., Guest, J.E., Duncan, A.M., Barriga, F., Chester, D.K., Eds.; Geological Society: London, UK, 2015; Memoirs 44; pp. 135–146.
20. Viveiros, F.; Ferreira, T.; Cabral Vieira, J.; Silva, C.; Gaspar, J.L. Environmental influences on soil CO_2 degassing at Furnas and Fogo volcanoes (São Miguel Island, Azores archipelago). *J. Volcanol. Geotherm. Res.* **2008**, *177*, 883–893. [CrossRef]
21. Viveiros, F.; Cardellini, C.; Ferreira, T.; Caliro, S.; Chiodini, G.; Silva, C. Soil CO_2 emissions at Furnas volcano, São Miguel Island, Azores archipelago: Volcano monitoring perspectives, geomorphologic studies, and land use planning application. *J. Geophys. Res.* **2010**, *115*, B12208. [CrossRef]
22. Cruz, J.V.; Coutinho, R.M.; Carvalho, M.R.; Oskarsson, N.; Gislason, S.R. Chemistry of waters from Furnas volcano, São Miguel, Azores: Fluxes of volcanic carbon dioxide and leached material. *J. Volcanol. Geotherm. Res.* **1999**, *92*, 151–167. [CrossRef]
23. Freire, P.; Andrade, C.; Coutinho, R.; Cruz, J.V. Spring geochemistry in an active volcanic environment (São Miguel, Azores): Source and fluxes of inorganic solutes. *Sci. Total Environ.* **2013**, *454–455*, 154–169. [CrossRef]
24. Andrade, C.; Viveiros, F.; Cruz, J.V.; Coutinho, R. Global carbon dioxide output of volcanic lakes in the Azores archipelago, Portugal. *J. Geochem. Explor.* **2021**, *229*, 106835. [CrossRef]
25. Branco, R.; Cruz, J.V.; Silva, C.; Coutinho, R.; Andrade, C.; Zanon, V. Radon (^{222}Rn) occurrence in groundwater bodies on São Miguel Island (Azores archipelago, Portugal). *Environ. Earth Sci.* **2021**, *80*, 609. [CrossRef]
26. Andrade, C.; Trigo, R.M.; Freitas, M.C.; Gallego, M.C.; Borges, P.; Ramos, A.M. Comparing historic records of storm frequency and the North Atlantic Oscillation (NAO) chronology for the Azores region. *Holocene* **2008**, *18*, 745–754. [CrossRef]
27. Valadão, P.; Gaspar, J.L.; Queiroz, G.; Ferreira, T. Landslide density map of S. Miguel Island, Azores archipelago. *Nat. Hazards Earth Syst. Sci.* **2002**, *2*, 51–56. [CrossRef]
28. Marques, R. Estudo de Movimentos de Vertente no Concelho da Povoação (ilha de São Miguel, Açores): Inventariação, Caracterização e Análise da Susceptibilidade. Ph.D. Thesis, Universidade dos Açores, Ponta Delgada, Portugal, 2013. (In Portuguese).
29. Wallenstein, N.; Chester, D.K.; Duncan, A.M. Methodological implications of volcanic hazard evaluation and risk assessment: Fogo Volcano, São Miguel, Azores. *Z. För Geomorphol. Suppl.* **2005**, *140*, 129–149.
30. Wallenstein, N.; Duncan, A.; Chester, D.; Marques, R. Fogo Volcano (São Miguel, Azores): A hazardous edifice Le volcan Fogo, un édifice générateur d'aléas indirects. *Géomorphologie Relief Processus Environ.* **2007**, *3*, 259–270. [CrossRef]
31. Pacheco, D.; Mendes, S.; Cymbron, R. Azores assessment and management of floods risks. In *Advances in Natural Hazards and Hydrological Risks: Meeting the Challenge, Proceedings of the 2nd International Workshop on Natural Hazards (NATHAZ'19), Lajes do Pico, Portugal, 9–10 May 2019*; Fernandes, F., Malheiro, A., Chaminé, H., Eds.; Springer Nature: Cham, Switzerland, 2020; pp. 133–136.
32. Cordeiro, S.; Coutinho, R.; Cruz, J.V. Fluoride content in drinking water supply in São Miguel volcanic island (Azores, Portugal). *Sci. Total Environ.* **2012**, *432*, 23–36. [CrossRef] [PubMed]
33. Medeiros, J.; Carmo, R.; Pimentel, A.; Vieira, J.C.; Queiroz, G. Assessing the impact of explosive eruptions of Fogo volcano (São Miguel, Azores) on the tourism economy. *Nat. Hazards Earth Syst. Sci.* **2021**, *21*, 417–437. [CrossRef]
34. Pomonis, A.; Spence, R.; Baxter, P. Risk assessment of residential buildings for an eruption of Furnas Volcano, São Miguel the Azores. *J. Volcanol. Geotherm. Res.* **1999**, *92*, 107–131. [CrossRef]

35. Dibben, C.; Chester, D.K. Human vulnerability in volcanic environments: The case of Furnas Volcano, São Miguel, Azores. *J. Volcanol. Geotherm. Res.* **1999**, *92*, 133–150. [CrossRef]

36. Baxter, P.J.; Baubron, J.C.; Coutinho, R. Health hazards and disaster potential of ground gas emission at Furnas Volcano, São Miguel, Azores. *J. Volcanol. Geotherm. Res.* **1999**, *92*, 95–106. [CrossRef]

37. Wallenstein, N.; Duncan, A.; Guest, J.E.; Almeida, M.H. Eruptive history of Fogo Volcano, São Miguel, Azores. In *Volcanic Geology of São Miguel Island (Azores Archipelago)*; Gaspar, J.L., Guest, J.E., Duncan, A.M., Barriga, F., Chester, D.K., Eds.; Geological Society: London, UK, 2015; Memoirs 44; pp. 105–123.

38. Gaspar, J.L.; Guest, J.E.; Queiroz, G.; Pacheco, J.M.; Pimentel, A.; Gomes, A.; Marques, R.; Felpeto, A.; Ferreira, T.; Wallenstein, N. Eruptive frequency and volcanic hazards zonation in São Miguel Island, Azores. In *Volcanic Geology of São Miguel Island (Azores Archipelago)*; Gaspar, J.L., Guest, J.E., Duncan, A.M., Barriga, F., Chester, D.K., Eds.; Geological Society: London, UK, 2015; Memoirs 44; pp. 155–166.

39. AHA-DRA. Plano de Gestão da Região Hidrográfica dos Açores—RH9. Versão para consulta pública (Azores River Basin Management Plan). In *Report AHA-SRAM*; Direção Regional do Ordenamento do Território e dos Recursos Hídricos: Ponta Delgada, Portugal, 2015. (In Portuguese)

40. Duncan, A.M.; Guest, J.E.; Wallenstein, N.; Chester, D.K. The older volcanic complexes of São Miguel, Azores: Nordeste and Povoação. In *Volcanic Geology of São Miguel Island (Azores Archipelago)*; Gaspar, J.L., Guest, J.E., Duncan, A.M., Barriga, F., Chester, D.K., Eds.; Geological Society: London, UK, 2015; Memoirs 44; pp. 147–153.

41. Queiroz, G. Vulcão das Sete Cidades (S.Miguel, Açores)—História Eruptiva e Avaliação do Hazard. Ph.D. Thesis, Universidade dos Açores, Ponta Delgada, Portugal, 1998. (In Portuguese).

42. Queiroz, G.; Pacheco, J.M.; Gaspar, J.L.; Aspinall, W.P.; Guest, J.E.; Ferreira, T. The last 5000 years of activity at Sete Cidades volcano (São Miguel Island, Azores): Implications for hazard assessment. *J. Volcanol. Geotherm. Res.* **2008**, *178*, 562–573. [CrossRef]

43. Porreca, M.; Pimentel, A.; Kueppers, U.; Izquierdo, T.; Pacheco, J.; Queiroz, G. Event stratigraphy and emplacement mechanisms of the last major caldera eruption on Sete Cidades Volcano (São Miguel, Azores): The 16 ka Santa Bárbara Formation. *Bull. Volcanol.* **2018**, *80*, 76. [CrossRef]

44. Kueppers, U.; Pimentel, A.; Ellis, B.; Forni, F.; Neukampf, J.; Pacheco, J.; Perugini, D.; Queiroz, G. Biased Volcanic Hazard Assessment Due to Incomplete Eruption Records on Ocean Islands: An Example of Sete Cidades Volcano, Azores. *Front. Earth Sci.* **2019**, *7*, 122. [CrossRef]

45. DROTRH-INAG. *Plano Regional da Água. Relatório Técnico*; DROTRH-INAG: Ponta Delgada, Portugal, 2001. (In Portuguese)

46. Cruz, J.V.; Freire, P.; Costa, A.; Fontiela, J.; Cabral, L.; Coutinho, R. Hydrogeochemical characterization of mineral waters in São Miguel Island, Azores. In *Volcanic Geology of São Miguel Island (Azores Archipelago)*; Gaspar, J.L., Guest, J.E., Duncan, A.M., Barriga, F., Chester, D.K., Eds.; Geological Society: London, UK, 2015; Memoirs 44; pp. 257–269.

47. Coutinho, R.; Fontiela, J.; Freire, P.; Cruz, J.V. Hydrogeology of São Miguel Island, Azores: A review. In *Volcanic Geology of São Miguel Island (Azores Archipelago)*; Gaspar, J.L., Guest, J.E., Duncan, A.M., Barriga, F., Chester, D.K., Eds.; Geological Society: London, UK, 2015; Memoirs 44; pp. 289–296.

48. Cruz, J.V.; Melo, M.; Medeiros, D.; Costa, S.; Cymbron, R.; Rocha, S.; Medeiros, C.; Valente, A.; Mendes, S.; Silva, D.; et al. Water management and planning in a small island archipelago: The Azores case study (Portugal) in the context of the Water Framework Directive. *Water Policy* **2017**, *19*, 1097–1118. [CrossRef]

49. DROTRH. Plano Regional da Água. Relatório técnico. In *Report DROTRH*; Direção Regional do Ordenamento do Território e dos Recursos Hídricos: Ponta Delgada, Portugal, 2020. (In Portuguese)

50. Freire, P.; Andrade, C.; Coutinho, R.; Cruz, J.V. Fluvial geochemistry in São Miguel Island (Azores, Portugal): Source and fluxes of inorganic solutes in an active volcanic environment. *Sci. Total Environ.* **2014**, *466–467*, 475–489. [CrossRef]

51. Cabral, L.; Andrade, C.; Coutinho, R.; Cruz, J.V. Groundwater composition in perched-water bodies in the north flank of Fogo volcano (São Miguel, Azores): Main causes and comparison with river water chemistry. *Environ. Earth Sci.* **2015**, *73*, 2779–2792. [CrossRef]

52. DROTRH. Plano de Gestão da Região Hidrográfica dos Açores—RH9. Versão para consulta pública (Azores River Basin Management Plan). In *Report DROTRH*; Direção Regional do Ordenamento do Território e dos Recursos Hídricos: Ponta Delgada, Portugal, 2021. (In Portuguese)

53. Fuchs, S.; Ornetsmüller, C.; Totschnig, R. Spatial scan statistics in vulnerability assessment: An application to mountain hazards. *Nat. Hazards* **2012**, *64*, 2129–2151. [CrossRef]

54. Becerril, L.; Martí, J.; Bartolini, S.; Geyer, A. Assessing qualitative long-term volcanic hazards at Lanzarote Island (Canary Islands). *Nat. Hazards Earth Syst. Sci.* **2017**, *17*, 1145–1157. [CrossRef]

55. Folch, A.; Felpeto, A. A coupled model for dispersal of tephra during sustained explosive eruptions. *J. Volcanol. Geoth. Res.* **2005**, *145*, 337–349. [CrossRef]

56. Felpeto, A.; Martí, J.; Ortiz, R. Automatic GIS-based system for volcanic hazard assessment. *J. Volcanol. Geotherm. Res.* **2007**, *166*, 106–116. [CrossRef]

57. Bartolini, S.; Geyer, A.; Martí, J.; Pedrazzi, D.; Aguirre-Díaz, G. Volcanic hazard on deception Island (South Shetland Islands, Antarctica). *J. Volcanol. Geotherm. Res.* **2014**, *285*, 150–168. [CrossRef]

58. Becerril, L.; Bartolini, S.; Sobradelo, R.; Martí, J.; Morales, J.M.; Galindo, I. Long-term volcanic hazard assessment on El Hierro (Canary Islands). *Nat. Hazards Earth Syst. Sci.* **2014**, *14*, 1853–1870. [CrossRef]

59. Cole, P.; Pacheco, J.M.; Gunasekera, R.; Queiroz, G.; Gonçalves, P.; Gaspar, J.L. Contrasting styles of explosive eruption at Sete Cidades, São Miguel, Azores, in the last 5000 years: Hazard implications from modeling. *J. Volcanol. Geotherm. Res.* **2008**, *178*, 574–591. [CrossRef]
60. Papathoma-Köhle, M.; Gems, B.; Sturm, M.; Fuchs, S. Matrices, curves and indicators: A review of approaches to assess physical vulnerability to debris flows. *Earth-Sci. Rev.* **2017**, *171*, 272–288. [CrossRef]
61. Papathoma-Köhle, M. Vulnerability curves versus vulnerability indicators: Application of an indicator-based methodology for debris-flow hazards. *Nat. Hazards Earth Syst. Sci.* **2016**, *16*, 1771–1790. [CrossRef]
62. Kappes, M.S.; Papathoma-Köhle, M.; Keiler, M. Assessing physical vulnerability for multi-hazards using an indicator-based methodology. *Appl. Geogr.* **2012**, *32*, 577–590. [CrossRef]
63. Wilson, T.M.; Stewart, C.; Wardman, J.B.; Wilson, G.; Johnston, D.M.; Hill, D.; Hampton, S.J.; Villemure, M.; McBride, S.; Leonard, G.; et al. Volcanic ashfall preparedness poster series: A collaborative process for reducing the vulnerability of critical infrastructure. *J. Appl. Volcanol.* **2014**, *3*, 10. [CrossRef]
64. Stewart, C.; Damby, D.E.; Tomašek, I.; Horwell, C.J.; Plumlee, G.S.; Armienta, M.A.; Hinojosa, M.G.; Appleby, M.; Delmelle, P.; Cronin, S.; et al. Assessment of leachable elements in volcanic ashfall: A review and evaluation of a standardized protocol for ash hazard characterization. *J. Volcanol. Geotherm. Res.* **2020**, *392*, 106756. [CrossRef]

Article

Urban Groundwater Processes and Anthropogenic Interactions (Porto Region, NW Portugal)

Maria José Afonso [1,2,*], **Liliana Freitas** [1,3], **José Manuel Marques** [4], **Paula M. Carreira** [5], **Alcides J.S.C. Pereira** [3], **Fernando Rocha** [2] and **Helder I. Chaminé** [1,2]

[1]   Laboratory of Cartography and Applied Geology (LABCARGA), Department of Geotechnical Engineering, School of Engineering (ISEP), Polytechnic of Porto, 4200-072 Porto, Portugal; lfsfr@isep.ipp.pt (L.F.); hic@isep.ipp.pt (H.I.C.)

[2]   GeoBioTec Centre (Georesources, Geotechnics, Geomaterials research group, University of Aveiro, 3810-193 Aveiro, Portugal; tavares.rocha@ua.pt

[3]   CITEUC, Department of Earth Sciences, Faculty of Sciences, University of Coimbra, 3030-790 Coimbra, Portugal; apereira@dct.uc.pt

[4]   Centro de Recursos Naturais e Ambiente (CERENA), Instituto Superior Técnico, University of Lisbon, 1049-001 Lisbon, Portugal; jose.marques@tecnico.ulisboa.pt

[5]   Centro de Ciências e Tecnologias Nucleares (C2TN), Instituto Superior Técnico, University of Lisbon, 2695-066 Bobadela, Portugal; carreira@ctn.tecnico.ulisboa.pt

*   Correspondence: mja@isep.ipp.pt; Tel.: +351-228-340-500

Received: 28 July 2020; Accepted: 3 October 2020; Published: 9 October 2020

Abstract: Groundwater in fissured rocks is one of the most important reserves of available fresh water, and urbanization applies an extremely complex pressure which puts this natural resource at risk. Two-thirds of Portugal is composed of fissured aquifers. In this context, the Porto urban region is the second biggest metropolitan area in mainland Portugal. In this study, a multidisciplinary approach was developed, using hydrogeological GIS-based mapping and modeling, combining hydrogeochemical, isotopic, and hydrodynamical data. In addition, an urban infiltration potential index (IPI-Urban) was outlined with the combination of several thematic layers. Hydrogeochemical signatures are mainly Cl-Na to Cl-SO$_4$-Na, being dependent on the geographic proximity of this region to the ocean, and on anthropogenic and agricultural contamination processes, namely fertilizers, sewage, as well as animal and human wastes. Isotopic signatures characterize a meteoric origin for groundwater, with shallow flow paths and short residence times. Pumping tests revealed a semi- to confined system, with low long-term well capacities (<1 L/s), low transmissivities (<4 m^2/day), and low storage coefficients (<10^{-2}). The IPI-Urban index showed a low groundwater infiltration potential, which was enhanced by urban hydraulic and sanitation features. This study assessed the major hydrogeological processes and their dynamics, therefore, contributing to a better knowledge of sustainable urban groundwater systems in fractured media.

Keywords: urban groundwater; hydrogeochemistry; hydrodynamics; IPI-Urban; NW Portugal

---

1. Introduction

Urbanization is the foremost global phenomenon of our time, and groundwater from springs and wells has been a vital source of urban water supply since the first settlements [1]. Urban hydrogeology is a key scientific field, and an understanding of groundwater in urban areas has increased greatly in the last decades. Attention was given to the relationship between urban development and groundwater resources beginning in the 1950s and 1960s of the 20th century, when accelerated growth after the Second World War began to cause significant diversity of hydrological problems. Most of these problems were related to urban runoff and floods, therefore, urban hydrology established itself in an affirmative way [2,3]. Thus, urban groundwater rapidly emerged as a necessary and urgent branch of study. Particularly, after the 1990s of the 20th century, several scientific publications on issues addressing the subject of hydrogeology in urban areas are found, namely [2–22].

The impact of urban development on geological processes, in general, and groundwater systems, in particular, has been highlighted for a long time (e.g., [12,16,23–36]). It is remarkable that only one century ago, 20% of the world population lived in urban areas, whereas, presently, 54% of the world population lives in urban environments [37]. Anthropogenic activity is the major geomorphic agent affecting the Earth's land surface, while urbanization and agriculture are, perhaps, the major processes currently affecting the land (e.g., [7,14,20,23,38–41] and references therein). The need for provision of safe water, sanitation, and drainage systems are key elements which are vital to the understanding and management of groundwater resources in urban environments. Groundwater problems associated with urbanization and urban sprawl include both quantity and quality issues, namely (e.g., [2,42]), subsidence, covering of shallow systems, increased recharge from leaky water ascribed to sewage systems and irrigation return flow, changes in subsurface secondary porosity, permeability from utility systems and other constructions, as well as the effects of imported water resources, saline intrusion, and contamination from point and nonpoint sources. In this context of global anthropogenic activity, groundwater is a key source of urban supply worldwide, and aquifer storage represents a key resource for achieving water-supply security under climate change that will contribute to aggravate these pressures, with decreasing rainfall, as well as longer and more frequent and extended drought periods (e.g., [1,43,44]). Additionally, the impact of the complex system of man-made infrastructures (e.g., sewer, storm sewers, pipes, trenches, tunnels, and other buried structures) and impervious surfaces and or cover areas must be highlighted as anthropogenic disturbances in groundwater media (e.g., [11,42,45–47]).

Integrated multidisciplinary approaches are often required to address the scientific issues involving groundwater resources. In this way, environmental isotopes such as ^{2}H, ^{18}O, ^{13}C, and ^{3}H, can be used as fingerprints to assess some of these groundwater problems and to provide information needed for the rational management of these water bodies (e.g., [48–50]). Furthermore, the delineation of groundwater potential zones (GWPZ's) using remote sensing, geographic information systems (GIS), and geovisualization techniques are effective tools for groundwater exploration and groundwater recharge (e.g., [13,17,18,35,36,51,52]). This approach encompasses the integration of several factors, organized from different geodatabases and a multiparametric approach, which influence the occurrence of groundwater, namely lithology, tectonic lineaments, land use, slope, drainage, precipitation, and urban hydraulics. Nearly half of the world depends on groundwater as the main source of drinking water [53]. However, groundwater is an underexploited resource in many urban areas, due to its invisibility, inadequate management, scientific uncertainty, and political strategies that favor the use of surface waters (e.g., [54–57]).

The main objectives of this study in the Porto urban region (NW Portugal) were:

1. To develop an integrated geoenvironmental assessment of groundwater resources in urban environments using geotechnological capabilities, particularly GIS mapping and geovisualization techniques, and also extensive field and laboratory work;
2. To evaluate groundwater quality and groundwater flow paths, by combining hydrogeochemical and environmental isotopic data and to identify the leading processes responsible for groundwater disrepair;
3. To assess groundwater quantity, by means of pumping test data;
4. To delineate groundwater infiltration potential zones at a regional scale, using the innovative infiltration potential index for urban areas (IPI-Urban) that integrates several layers of information properly weighted and overlaid in a GIS platform; a tool which should improve our understanding of complex urban groundwater recharge processes in future investigations;
5. To refine the regional hydrogeological conceptual model, merging all the data, to improve the understanding of urban groundwater systems in the Porto urban region.

2. Study Area: Land Cover and Hydrogeological Background

The Porto urban region is a densely populated region which includes other cities such as Vila Nova de Gaia, Matosinhos, Maia, and Vila do Conde, in the Porto Metropolitan Area, with about 1.7 million inhabitants [58], (Figure 1). According to the land use map for mainland Portugal [59], most of the investigated region (93.3%) is occupied by the following three classes: urban and industrial areas (41.4%), agricultural areas (28.5%), and forest areas (23.4%). Urban areas are concentrated primarily in the municipality of Porto and in the surrounding municipalities. Agricultural spaces are located mainly in the northern part, especially in the municipality of Vila do Conde. Forest areas are mainly found in the NE, especially to the east of Vila do Conde and in the municipality of Trofa.

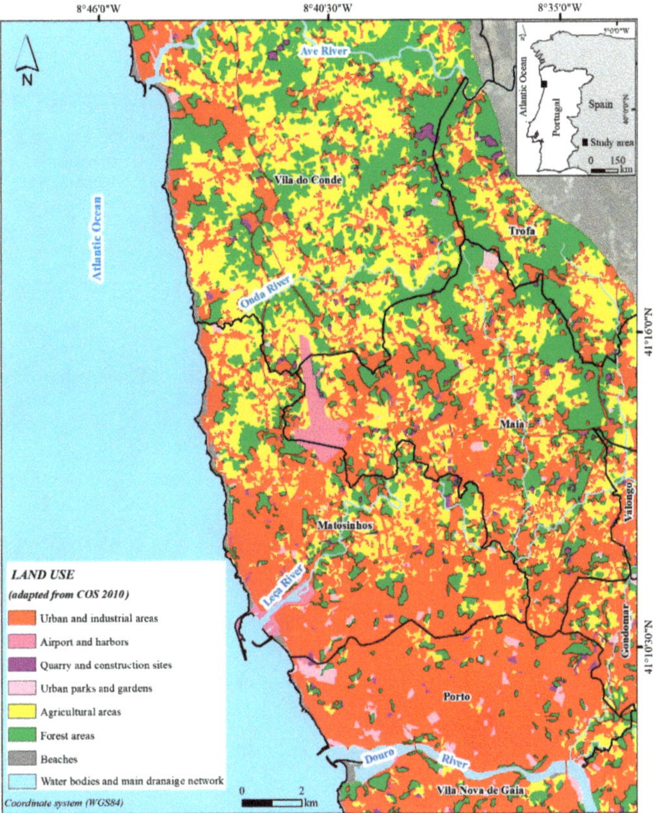

Figure 1. Land use of the urban coastal region between the Municipalities of Vila do Conde and Vila Nova de Gaia (adapted from [59]).

Considering agriculture and farming, the most important food supplies produced in this region are potatoes, vegetables, fruits, and maize, as well as cattle raising, namely bovines. According to [20], the two principal types of fertilization are organic which uses mostly manure, as well as sewage and inorganic. The inorganic fertilizers are mostly composed of nitrogen, phosphorous, potassium, and nitrogen which is the most prevalent. A sprinkler system is the leading irrigation system.

This region is drained by the following four major streams: the Ave River, in the north; rivers Onda and Leça in the centre; and Douro River, in the south. Located close to the Atlantic Ocean, this region has a temperate climate with dry and mild summers (Köppen climate classification Csb), with a mean annual precipitation around 1200 mm and a mean annual air temperature of 14 °C. There is a water deficit from June to September, mainly in July and August [60–62]. The availability of groundwater resources is reflected by these climatic conditions, along with geologic and morphotectonic features. The regional geological framework comprises a crystalline fissured basement of highly deformed and overthrust Late Proterozoic/Palaeozoic metasedimentary and granitic rocks. The post-Miocene sedimentary deposits frequently cover the bedrock [63–65] and references therein. The morphotectonic background includes a littoral platform characterized by a regular planation surface dipping gently to the west, ending around 100 m a.s.l [66]. The Ave, Leça, and Douro rivers have incised valleys that cross the flatness of this morphological surface, suggesting a regional tectonic control. The main regional hydrogeologic units [12] in this area are (Figure 2) the following: (i) sedimentary cover, particularly beach and dune sands, alluvia, sandy silts, and clays; (ii) metasedimentary rocks, namely micaschists, metagraywackes, and paragneisses; and (iii) granitic rocks, mainly two-mica granite, medium to coarse grained and biotitic granite, medium to fine grained.

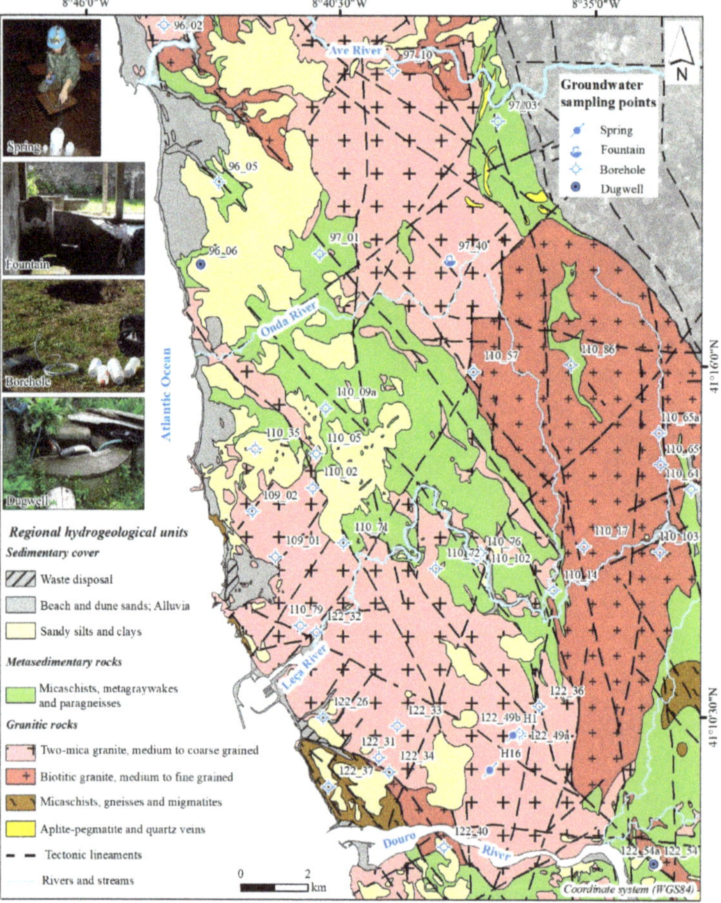

Figure 2. Regional hydrogeological setting of the Porto urban region (NW Portugal). Geological background updated from [63,65,69,70]. Hydrogeological framework adapted from [61,71,72]. Hydrogeological inventory and groundwater sampling points from [61].

3. Materials and Methods

The study area was assessed using various tools, as proposed by [67,68], namely geological and hydrogeological mapping, hydrogeochemical and isotopic techniques, and hydrodynamic characterization. Fieldwork campaigns were first carried out to identify major geologic features accountable for groundwater circulation pathways and to assess litho-structural heterogeneities. The field and laboratory data were evaluated in a GIS environment (ArcGis|ESRI, Redlands, CA, USA, and OCAD for Cartography Inc., Baar, Switzerland).

An urban hydrogeological field inventory was developed, which included more than 300 water points (springs, water mines, fountains, boreholes, and dug wells). Three fieldwork campaigns were carried out in May 2008, November 2008–January 2009, and May–June 2009. A total of 40 water points were selected for groundwater sampling and for chemical analyses consisting of 34 boreholes (mean depth of 117 m), 3 dug wells (mean depth of 11 m), 2 springs, and 1 fountain (connected to a water mine) (see Figure 2), most of them used for industrial and agricultural water supply. The inclusion of 2 dug wells and 2 springs in the fieldwork campaigns basically developed in the period 2005–2007 and were justified as follows: (i) to have dug wells represented in the hydrogeological unit "micaschists, metagraywackes and paragneisses"; (ii) to have springs represented (the 2 springs selected were two important springs of the historic Paranhos and Salgueiros water galleries in Porto City, which date back to the 16th Century).

Among the 40 samples, isotopic determinations of $\delta^{18}O$, δ^2H, and ^3H were performed in 14 boreholes, 2 dugwells, and 2 springs in one fieldwork campaign. All the inventory and sampling sites were georeferenced with a high-accuracy GPS (Trimble® GeoExplorer, Sunnyvale, CA, USA). In situ measurements included water temperature (°C), pH, electrical conductivity (μ Scm^{-1}), dissolved oxygen (mg L^{-1}), and redox potential (mV) using a multiparametric portable equipment (Hanna Instruments, HI9828). The hydrogeochemical analyses included major and minor element concentrations and were acquired at the "Instituto Nacional de Saúde Doutor Ricardo Jorge" (Porto, Portugal). Environmental isotopes (^{18}O, ^2H, and ^3H) were measured at the "Instituto Tecnológico e Nuclear", meanwhile renamed "Campus Tecnológico e Nuclear/Instituto Superior Técnico (CTN/IST)" (Bobadela, Portugal). The $\delta^{18}O$ and δ^2H results were reported to the Vienna-Standard Mean Ocean Water (V-SMOW), and determinations were carried out using a mass spectrometer SIRA 10 VG-ISOGAS, applying the analytical methods described in [73,74], respectively. The tritium content was determined using the electrolytic enrichment and liquid scintillation counting method described by [75,76], using a PACKARD TRI-CARB 2000 CA/LL (see [77,78]). AquaChem 5.1 software was used for the hydrochemical interpretation.

The hydrodynamic interpretation of the groundwater systems was carried out based on pumping tests which were gathered from hydrogeological and geotechnical scientific and technical reports. Pumping data were analyzed using the Theis method, the Cooper–Jacob method, and the Theis recovery method [79–82], and the transmissivity and storage coefficient were evaluated. Moreover, the Logan approximation [83] was used to estimate transmissivity. The values of transmissivity, storage coefficient, and long-term well capacity were compiled from the hydrogeological reports and also contributed to the discussion.

The delineation of groundwater infiltration potential zones was assessed by the IPI-Urban index (details in [13,17–19]). The IPI-Urban is a weighted sum of the following eight factors, calculated using the analytical hierarchy process (AHP) [84]: land use, hydrogeological units, tectonic lineament density, drainage density, slope, and three other factors related to urban hydraulic and sanitation, sewer network density, stormwater network density and water supply network density.

4. Results and Discussion

4.1. Hydrogeochemical Approach

The physical and chemical parameters of groundwater samples collected during the fieldwork campaigns are summarized in the supplementary materials (Table S1). These groundwaters are slightly acidic with a median pH value of 6.2, and electrical conductivity (EC) values ranging from 95 to 1035 μS/cm, with a median value of 382 μS/cm, therefore, characterized as medium mineralized waters. These values are in good agreement with other data reported for this area [20,61,62]. The dissolved oxygen (DO) values in these groundwaters are in the range of 0.30–4.14 mg/L, with a median value of 1.75 mg/L. According to [85], these values are in accordance with those expected for most groundwaters (DO concentrations <5 mg/L, frequently <2 mg/L). The redox potential (Eh) measurements revealed that most of the studied groundwaters were in the range +0.1–+0.3 V, with a median of +0.15 V, which were values characteristic of most groundwaters [86] and within the water stability domain.

The major anions analyzed included bicarbonate (HCO_3^-), sulphate (SO_4^{2-}), chloride (Cl^-), and nitrate (NO_3^-). The groundwaters showed the following global rating in terms of median concentrations (mg/L): HCO_3^- (45.0) > Cl^- (41.2) > NO_3^- (40.0) > SO_4^{2-} (35.9). HCO_3^- has the highest standard deviation, while SO_4^{2-} has the lowest, and 31% of the water samples have concentrations of NO_3^- > 50 mg/L. Considering the European guidelines for the use of groundwater for human consumption and the Portuguese legislation for irrigation activities [87], nearly one-third of these water samples exceed the parametric value for nitrate (50 mg/L). This trend corroborates the results presented by [20,61,62]. Moreover, [20] classified a peri-urban area of Vila do Conde, between Ave and Onda rivers, with a potential high to moderate risk of nitrate contamination, based on the SINTACS and IPNOA indexes.

The major cations studied were sodium (Na^+), calcium (Ca^{2+}), potassium (K^+), and magnesium (Mg^{2+}), and the global rating was the following in terms of median concentrations (mg/L): Na^+ (37.0) > Ca^{2+} (13.0) > Mg^{2+} (7.9) > K^+ (4.0). The highest and the lowest standard deviation belong to Na^+ and Mg^{2+}, respectively. Among the other cations, most of the samples have concentrations of aluminium (Al^{3+}) and iron (Fe^{2+}) below 50 μg/L and concentrations of ammonia (NH_4) below 0.1 mg/L. Concerning the environmental isotopic approach, the median values for $\delta^{18}O$, δ^2H, and ^3H are, respectively, −4.42‰, −26.5‰ and 1.0 TU.

Considering the major anions and cations, we conclude that the hydrogeochemical signatures of these groundwaters are predominantly Cl-Na to Cl-SO$_4$-Na, however, the following facies may occur, Cl-SO$_4$-Na-Mg, Cl-SO$_4$-Na-Ca, and Cl-SO$_4$-Mg-Na (Figure 3a). Nevertheless, in the three fieldwork campaigns, five groundwater samples from boreholes drilled in diverse geological environments revealed a distinct hydrogeochemical facies in the bicarbonate domain: 110_17, HCO_3-Ca (biotitic granite, medium to fine grained); 110_64, HCO_3-Na to HCO_3-Na-Mg (mica schists, metagraywackes, and paragneisses); 110_65, HCO_3-Na (biotitic granite, medium to fine grained); 122_32, HCO_3-Na-Ca (two-mica granite, medium to coarse grained); 122_34, HCO_3-Ca-Mg (two-mica granite, medium to coarse grained). Moreover, two other borehole groundwater samples had HCO_3-Cl-Na facies, i.e., 110_02 (geological contact between two-mica granite, medium to coarse grained and gneisses) and 110_05 (geological contact between two-mica granite, medium to coarse grained and migmatites). [11,60–62,71,88] achieved similar hydrogeochemical signatures for the same hydrogeological units.

In order to better understand the hydrochemical signatures of these waters and the main processes controlling water mineralization, several correlations were established. Figure 3b–d indicates the following:

1. The strongest linear correlations were obtained in the water samples belonging to the shallow groundwater systems, represented by dug wells, fountains and springs.
2. Cl and Mg seem to have the same source for the shallow and deep groundwater systems, while the relation Cl and Na seems to point to a unique source only for the shallow systems;

    however, the common source of Cl and Mg is not marine, and the sea spray seems to be partially responsible for the Cl and Na; moreover, the proximity of several data to the Mg/Na seawater line outlines a partially common marine origin for these parameters.

3. The isotopic signatures of the groundwaters characterize a meteoric origin, since most of the water samples are positioned very close to the Global Meteoric Water Line (GMWL), defined by [89] and later improved by [90–92], and similar to the isotopic composition of the precipitation water samples, from the Portuguese Isotopes in Precipitation Network [93]. From the isotopic point of view, the following two main groundwater clusters have been identified: Group I stands for groundwaters collected from dug wells and boreholes, which presents a more enriched isotopic composition, similar to the Porto precipitation [93], corresponding to normal deviations related with seasonal variations of $\delta^{18}O$ and δ^2H on precipitation; Group II is composed of the spring samples and presents more depleted $\delta^{18}O$ and δ^2H values, which may be attributed to the fact that they could be ascribed to random precipitation events, resulting into a direct infiltration of meteoric waters along the fractured granitic rocks;

4. The meteoric origin of the shallow groundwaters seems to be reinforced by the SO_4-Ca facies, concerning the relationship with the precipitation data; therefore, the partial origin of SO_4 should be atmospheric pollution, enriched in SO_2 gases.

5. The common source of Cl and Mg must be anthropogenic, related to organic fertilizers, including sewage and livestock residues (liquid and semiliquid manure), and animal waste (e.g., bovines), in areas with a higher agricultural and/or livestock production activity, especially in the Vila do Conde, Trofa, and Northern Maia municipalities. Moreover, the good relationship between NO_3 and Cl for the shallow groundwater systems shows their partially common source; in fact, NO_3 is also an important constituent of fertilizers, either organic or synthetic, sewage, and animal and human wastes (e.g., [94]). The studied groundwaters do not present such a trend, indicating different sources for Cl and NO_3. Several studies developed in this region have reached similar conclusions (e.g., [20,62,95–97]).

6. The relations among Cl and Mg, and Cl and NO_3, may be ascribed to the urban and industrial sectors, particularly in the Porto, Matosinhos, and Vila Nova de Gaia municipalities, due to numerous groundwater potential contamination activities and their high density in some areas (cf. [17] for Porto and Vila Nova de Gaia urban areas), i.e., wastewater leakages, cesspools, and solid waste tanks contamination, hydrocarbons present from vehicle fuels and industrial processes, such as solvents and degreasing agents, namely trichloroethylene, which is one of the most common.

7. HCO_3 and Ca have a reasonable correlation among the borehole water samples, and the correlation between $HCO_3 + NO_3$ and Na + Ca is good for the same water samples. This trend seems to indicate that these parameters have partially the same origin that probably should be ascribed to water–rock interaction, namely the hydrolysis of plagioclases presented in granitic rocks.

    Therefore, the mineralization of these groundwaters seems to be controlled by the following three main sources/processes: (i) meteoric, including sea spray, contributing with the Cl^- and Na^+ ions, and atmospheric pollution, particularly responsible by the presence of SO_4^{2-}; (ii) anthropogenic contamination, particularly introducing Cl^-, Mg^{2+}, and NO_3^-; and (iii) water–rock interaction, adding HCO_3^-, Na^+, and Ca^{2+}.

    Since the altitude differential of the springs sampling sites is rather small, in the study region, (spring waters are the most representative of local precipitation), it was difficult to estimate the altitude of recharge areas ascribed to the groundwaters from boreholes (usually recharged far from the discharge). In fact, the correlation between the altitude of the spring sites and $\delta^{18}O$ values was very low ($r = -0.14$).

    Tritium concentrations, in the range 0.0–4.6 TU, are consistent with the precipitation ^3H record measured at the Porto meteorological station, with a weight annual mean of 4.5 TU (see [100]). Therefore, for coastal areas, according to [49], the ^3H values related to the studied groundwaters could correspond to a mixture of recent waters, with short and fast underground flow paths (<5–10 years) and relatively close to the surface, and submodern waters where recharge took place before the thermonuclear events in 1952.

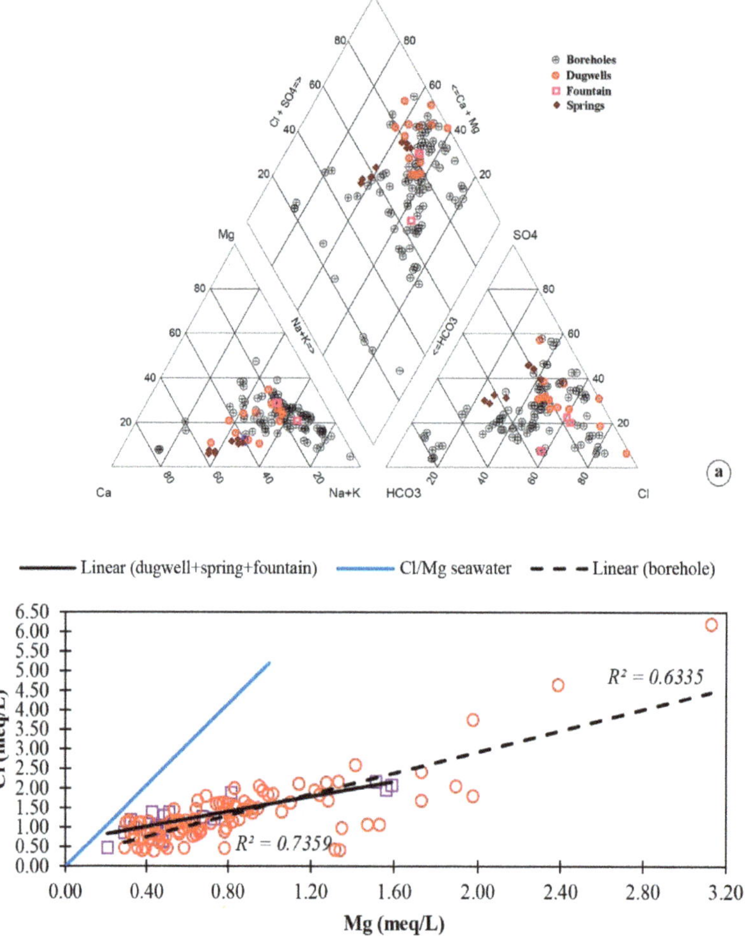

Figure 3. *Cont.*

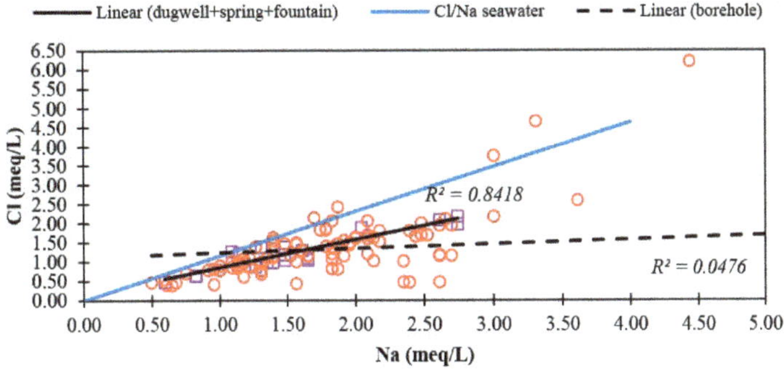

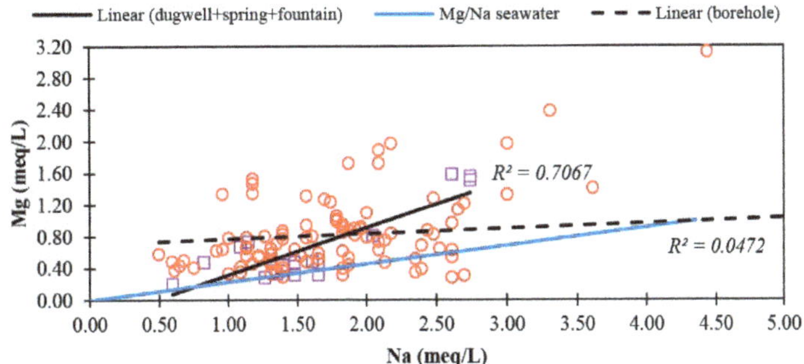

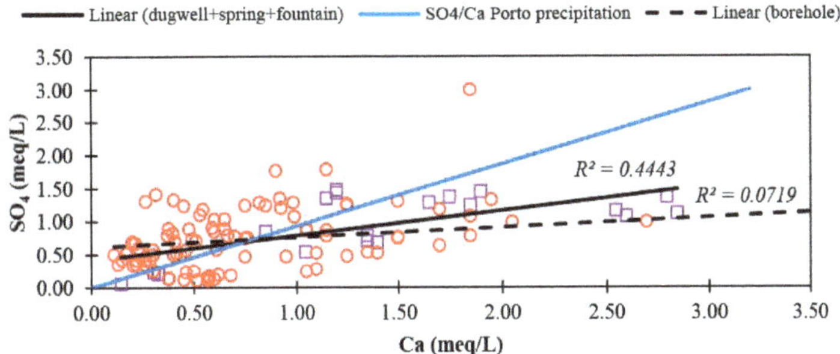

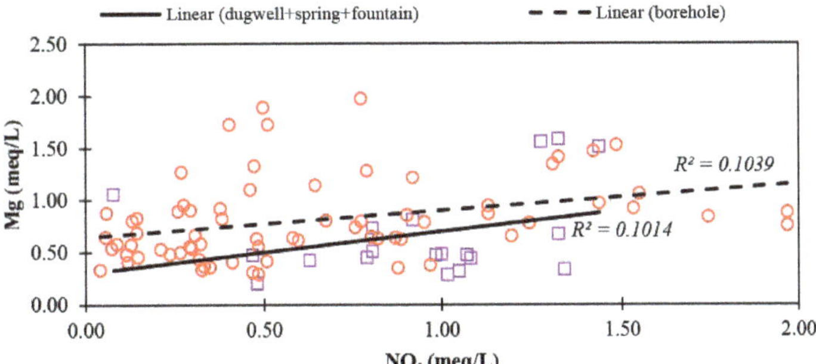

Figure 3. *Cont.*

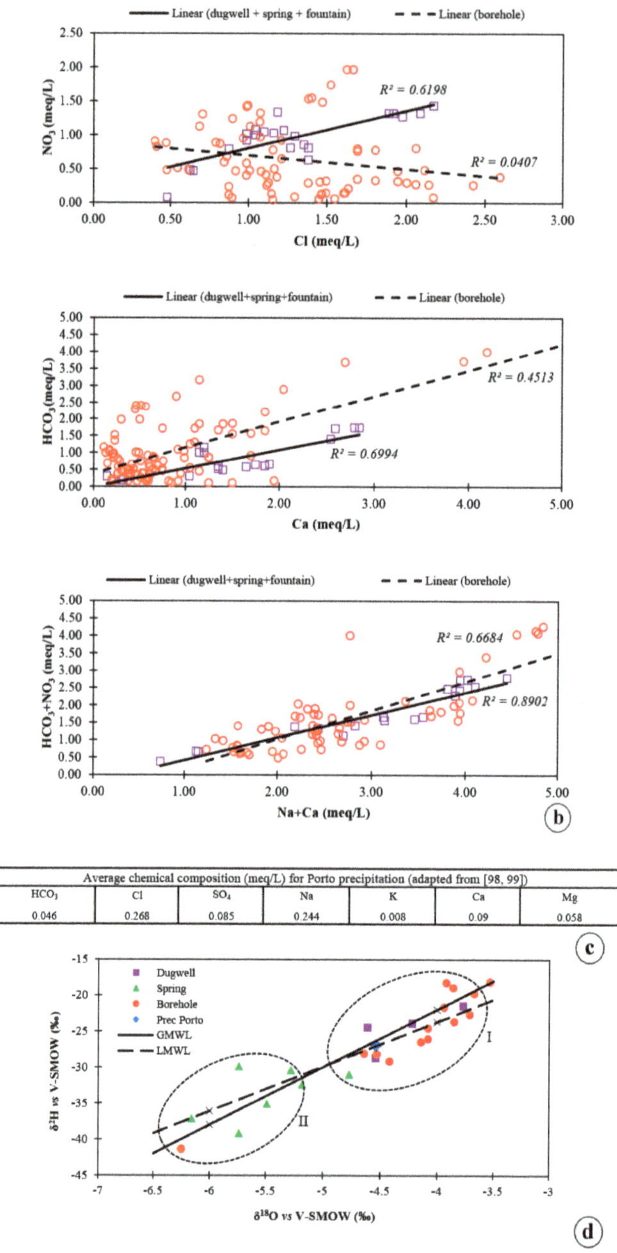

Average chemical composition (meq/L) for Porto precipitation (adapted from [98, 99])

HCO₃	Cl	SO₄	Na	K	Ca	Mg
0.046	0.268	0.085	0.244	0.008	0.09	0.058

(c)

Figure 3. Hydrochemical and isotopic relationships for groundwater samples collected during the three fieldwork campaigns. (**a**) Piper diagram; (**b**) Scatter diagrams for several hydrochemical parameters, R-squared value (R^2) is indicated for each studied linear relationship, and seawater relations are those reported by [85]; (**c**) The average chemical composition of Porto precipitation, reported by [98] and [99]; (**d**) $\delta^{18}O$ vs. δ^2H signatures of the studied groundwater samples. The global meteoric water line (GMWL) ($\delta^2H = 8\delta^{18}O + 10$, [89]), the local (Porto meteorological station) meteoric water line (LMWL) ($\delta^2H = (6.20 \pm 0.33)$ and $\delta^{18}O + (1.12 \pm 1.16)$, r = 0.93, [93]), and the long-term mean isotopic composition of Porto precipitation (Prec. Porto) ($\delta^2H = -26.9‰$ and $\delta^{18}O = 4.54$ [93]) were plotted as reference.

4.2. Hydrodynamical Assessment

For this evaluation, the study area was divided into the following two key sectors (Figure 4a), each sector dominated by a hydrogeological unit: (i) Sector 1, in the geological contact of two-mica granite, medium to coarse grained and metasedimentary rocks, and also in the biotitic granite, medium to fine grained, with 16 boreholes; (ii) Sector 2, in the two-mica granite, medium to coarse grained, with 18 boreholes and five piezometers.

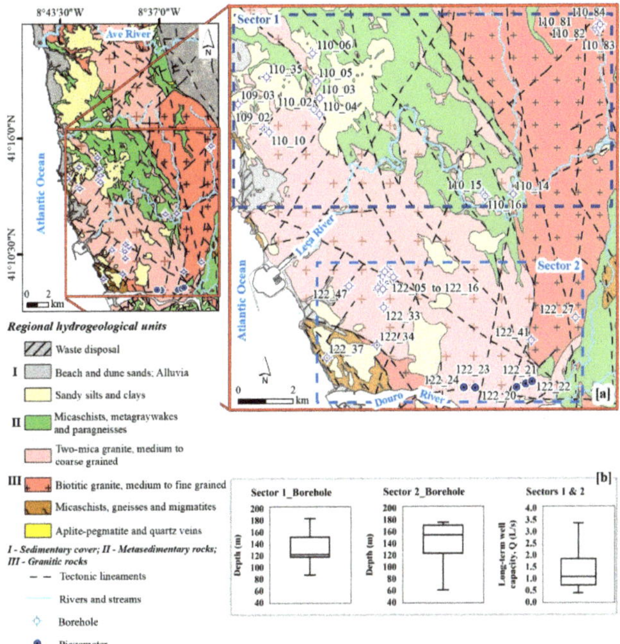

Figure 4. Inventory of boreholes and piezometers used for the hydrodynamical assessment. (**a**) Study sites (Sectors 1 and 2) are shown on the right side of the figure; (**b**) Depth of boreholes and long-term well capacity for Sectors 1 and 2.

The median depth of the boreholes is 121 m in Sector 1, and higher in Sector 2, around 154 m (Figure 4b). For piezometers, the depth range is 15–34 m. The median value for the thickness of the weathering profile is null for both sectors, and it varies in the following ranges: 0.0–17.0 m in Sector 1, and 0.0–26 m in Sector 2. Concerning long-term well capacity, the minima, maxima, and median values are very similar in both sectors, i.e., 0.4 L/s, 3.5 L/s, and 1.1 L/s, respectively (Figure 4b). There was no correlation found between depth of boreholes and long-term well capacity, or between depth of screens and long-term well capacity.

For the assessment of transmissivity and storage coefficient values, we selected two boreholes and one piezometer, i.e., boreholes 110_02 and 122_09 in Sectors 1 and 2, respectively, and piezometer 122_23 in Sector 2 (Figure 5). Concerning borehole 110_02, the transmissivity values obtained during the pumping period were 21.8 m^2/day and 34.4 m^2/day, for the borehole and the observation well 110_03, respectively, while, for the recovery period, it was 24.3 m^2/day. Additionally, the storage coefficient value was 3.1×10^{-4}. Regarding borehole 122_09, the transmissivity values obtained during the pumping period were 2.8 m^2/day and 6.7 m^2/day for the borehole and the observation well 122_07, respectively, while, for the recovery period, it was 2.1 m^2/day. Moreover, the storage coefficient value was 2.5×10^{-5}. For piezometer 122_23, the transmissivity values obtained during the pumping period were 51.3 m^2/day and 55 m^2/day. Furthermore, the storage coefficient values were 3.1×10^{-2} and 3.9×10^{-3}.

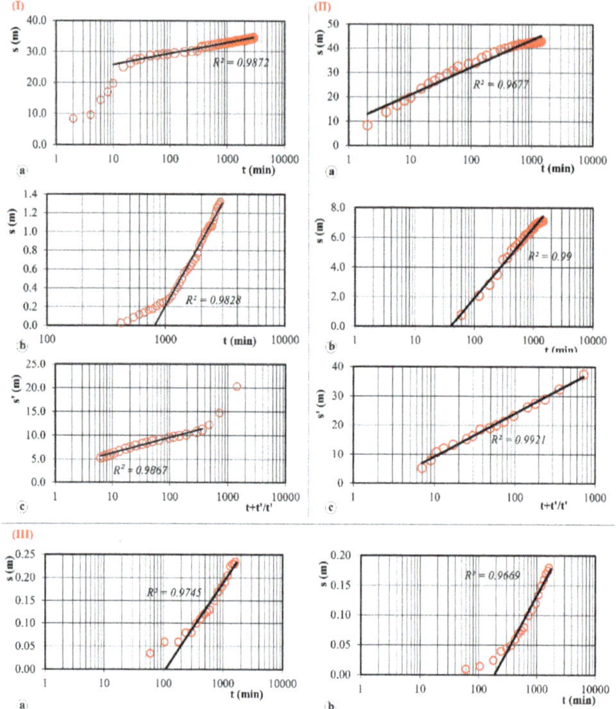

Figure 5. Interpretation of pumping tests. (**I**) 110_02, Cooper–Jacob method: (**a**) Pumping period data for 110_02; (**b**) Pumping period data for the observation well 110_03, located 370 m from 110_02; (**c**) Theis recovery method; (**II**) 122_09, Cooper–Jacob method: (**a**) Pumping period data for 122_09; (**b**) Pumping period data for the observation well 122_07, located 128 m from 122_09; (**c**) Theis recovery method. (**III**) 122_23, Cooper–Jacob method for two observation wells: (**a**) Located 16 m (**b**) Located 63 m.

Due to the lack of well-documented aquifer conditions and complete pumping tests, a Logan approximation was achieved using specific capacity values for 69% of the boreholes in two sectors (Figure 6). According to [101], the Logan method can give a reasonable first estimate of transmissivity, provided that near equilibrium conditions are achieved when the maximum drawdown is measured. Therefore, Figure 6a shows good correlations among the Logan method and the other methods, particularly the Theis and the Cooper–Jacob methods. Additionally, transmissivity values are quite variable in both sectors, in the range of 0.1–26 m^2/day. Nevertheless, the median values of transmissivity are very similar for the different methods and even for the two sectors, 1.8–4.0 m^2/day (Sector 1) and 2.5–4.2 m^2/day (Sector 2), (Figure 6b).

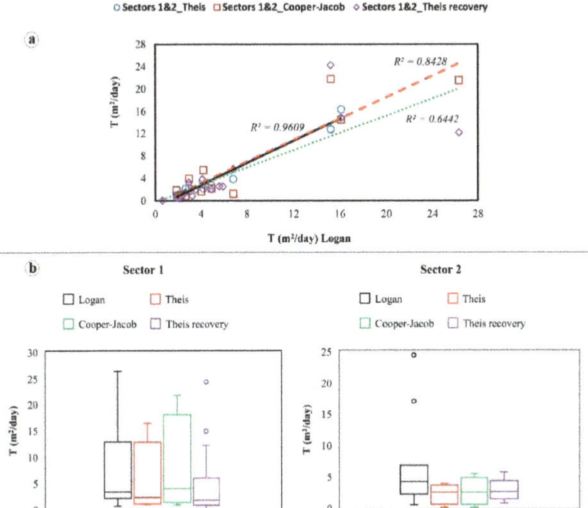

Figure 6. Transmissivity values for Sectors 1 and 2. (**a**) Logan vs. other methods; (**b**) Box-and-whisker diagrams for Sectors 1 and 2.

Considering the storage coefficient, values are also variable, in the range of 2.5×10^{-5}–4×10^{-2}, which characterizes confined to semi-confined groundwater conditions.

Very good correlations ($R^2 > 0.85$) were found between long-term well capacities and transmissivities in both sectors. The highest values of long-term well capacity and transmissivity could be explained by the local presence of a very productive fracture, or to a direct connection with a nearby surface water body or a porous overlying aquifer.

All these conclusions confirm other investigations, performed by several authors, in fractured-rock aquifers in the Porto region and Portugal (e.g., [60,61,72,102,103] and references therein).

The hydrodynamic behavior of these aquifer systems meets the criteria mentioned in the literature for fractured-rock aquifers over the last four decades (e.g., [104–113]).

4.3. Urban Infiltration Potential Index (IPI-Urban)

The spatial distribution and explanation of six thematic layers (Figure 7) involved in the computation of the IPI-Urban (Figure 8), i.e., tectonic lineaments density, slope, drainage density, sewage network density, stormwater network density, and water supply system density are presented in this section.

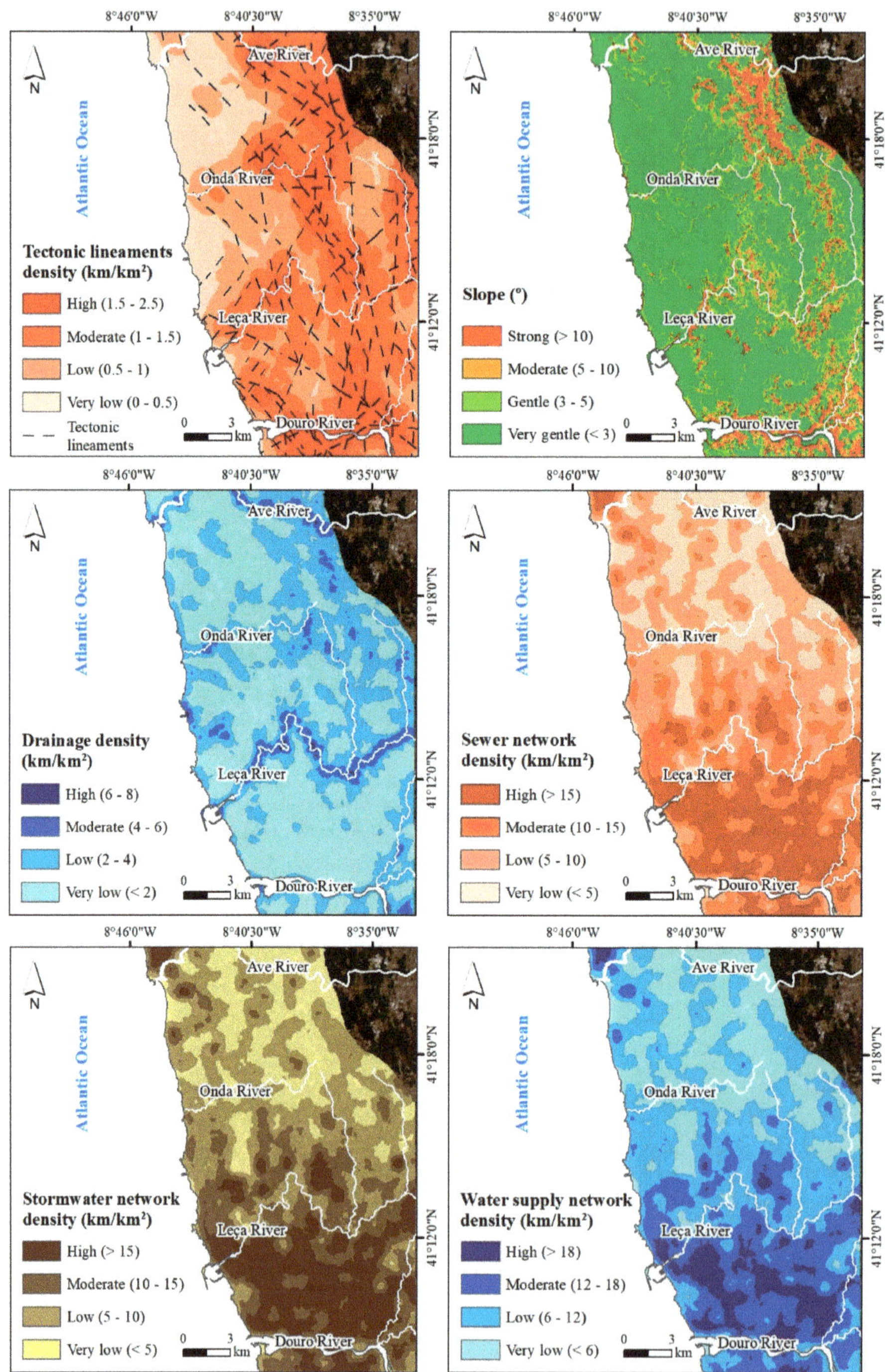

Figure 7. Tectonic lineaments density, slope, drainage density, sewage network density, stormwater network density, and water supply system density in the Porto urban region (NW Portugal).

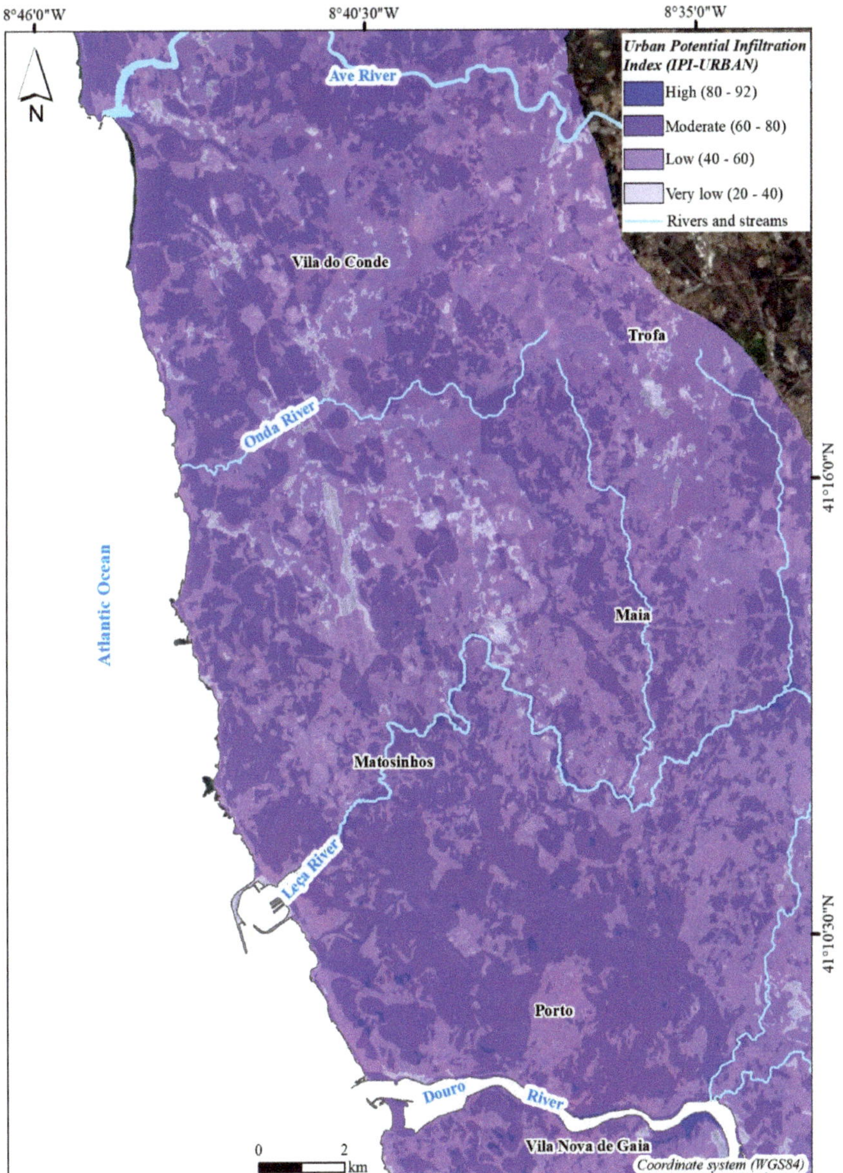

Figure 8. Urban infiltration potential index (IPI-Urban) in the Porto urban region (NW Portugal).

Land use and hydrogeological units' thematic layers were previously introduced in Section 2. Nevertheless, briefly, land use plays a crucial role in this region, and the following two classes dominate: urban and industrial areas (ca. 41%), and agricultural and forest areas (ca. 52%). This suggests that surfaces of low to very low permeability cover more than one-third of the region and must be responsible for the reduction in infiltration and, consequently, direct groundwater recharge. Hydrogeological units also play a key role in groundwater occurrence. Granitic rocks are the most representative lithology, namely the two-mica and biotitic granites (ca. 56%).

Concerning tectonic lineaments, most of the region (ca. 65%) has moderate to low densities (0.5–1.5 km/km^2). The highest densities (>1.5 km/km^2, ca. 20%) occur disseminated in the region, and the lowest densities (<0.5 km/km^2, ca. 15%) are mostly concentrated in the west-northwestern (W-NW) area, where the sedimentary cover has its most representative domain.

As for slope, much of the region (ca. 81%) has very gentle to gentle gradients (<5°). Moderate and strong gradients (>5°) arise in almost 19% of the region and are mostly connected with the Ave, Leça and Douro river valleys.

The region has mostly (ca. 92%) a very low to low drainage density (<4 km/km^2). The highest densities (4–8 km/km^2) are closely related with the four major rivers and streams (Ave, Onda, Leça, and Douro rivers).

Regarding sewer and stormwater networks, the largest part of the region (ca. 58%) has low to moderate densities (5–15 km/km^2). The highest densities (>15 km/km^2) are concentrated (ca. 21%) in the most representative municipalities, namely Porto and Matosinhos.

Ultimately, the dominant water supply system classes (ca. 66%) are low to very Low (<12 km/km^2). Most of the remaining parts of the region (ca. 25%) has moderate densities (12–18 km/km^2) which are concentrated in Porto, Matosinhos, and Maia municipalities.

Regarding the IPI-Urban index, the overall scenario shows that in 96% of the urban region, the dominant classes are low (ca. 54%) and moderate (ca. 42%). The low class seems to be mostly connected with the two-mica and biotitic granites in areas where these hydrogeological units coexist with an agricultural and forest cover. However, the moderate class appears to be associated firstly with the sedimentary cover. Moreover, this class occurs in areas covered by an urban and industrial fabric. Although this may seem incoherent, the higher densities of the urban hydraulic and sanitation features (sewer, stormwater, and water supply networks) are found in these areas. Therefore, on a regional scale, the major role of these features is to enhance the natural low infiltration potential. The very low class occurs in approximately 4% of the region and is linked to the "Micaschist, metagreywacke and paragneiss" and "Micaschist, gneiss and migmatite" hydrogeological units, especially where these units are covered by urban and industrial areas, as well as the airport and harbor areas. The high class represents less than 1% of the region and corresponds to very confined areas, mainly between the Leça and Douro rivers and to the south of the Douro river.

5. Hydrogeological Conceptual Ground Models

The building of a conceptual hydrogeological model is presented as the first step in the entire process of modeling Earth systems. A multicriteria methodology supported in an accurate GIS mapping allows a perspective from geosciences for the paradigm of smart cities, namely through multidisciplinary, interdisciplinary, and transdisciplinary approaches (e.g., [35,36]).

In the study of crystalline aquifers, hydrogeomorphology has a fundamental role in the refinement of conceptual hydrogeological models, which aim at the sustainability of groundwater resources [17,18]. In this kind of system, there are many factors, some of them very complex, that control the infiltration and circulation of groundwater at the local/regional scale.

Figure 9 shows the hydrogeological conceptual models for the Porto urban region (Figure 9I), and the Porto urban area (Figure 9II). This conceptualization included the identification of different kinds of aquifer systems in an urban and pre-urban scenario. The main criteria considered were the hydrogeological units features, the morphostructure, the land use, the hydrogeochemical, isotopic and hydrodynamical characteristics, the infiltration potential, and the recharge processes (Table 1).

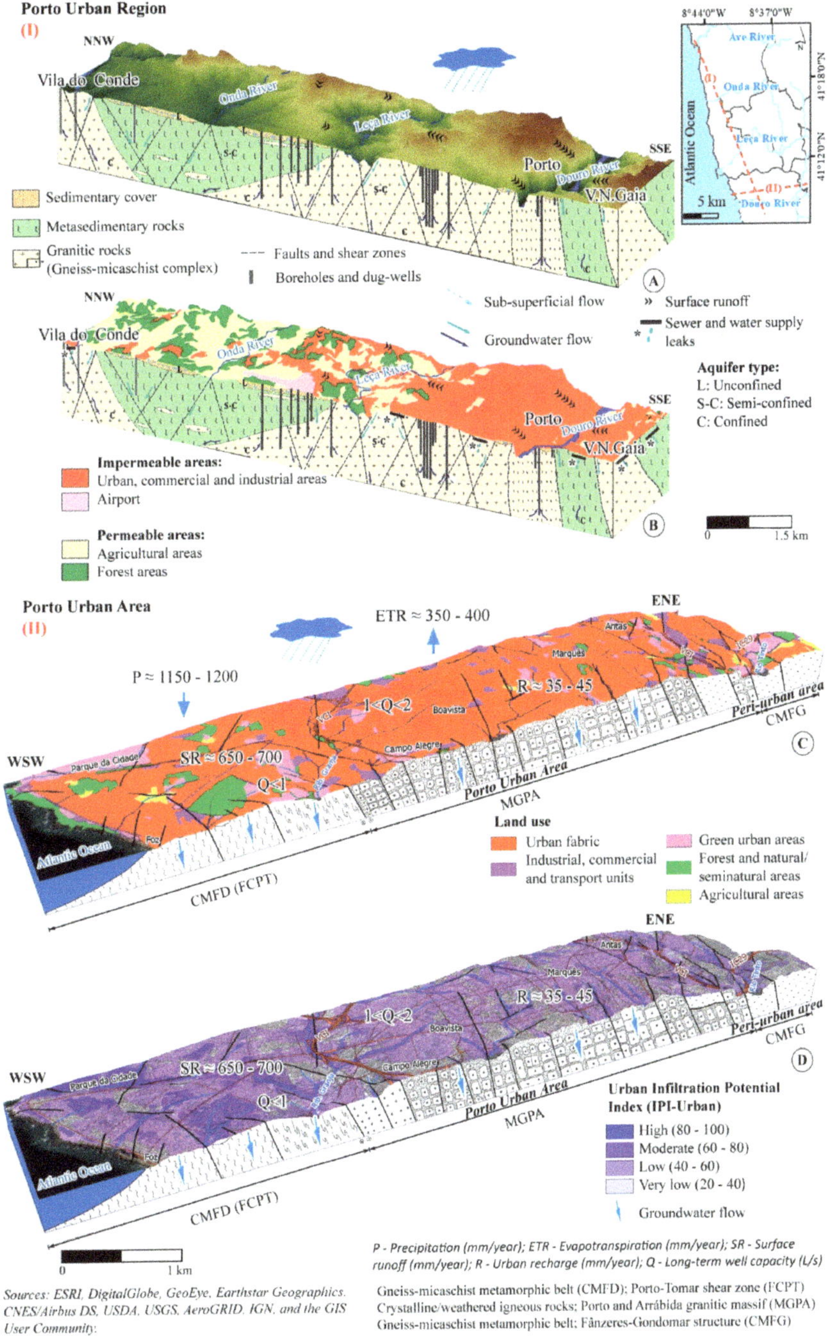

Figure 9. Urban hydrogeological conceptual models (revised and updated from [61,114]. (**I**) Porto urban region: (**A**) Pre-urban scenario; (**B**) Urban scenario). (**II**) Porto urban area: (**C**) Land use context; (**D**) IPI-Urban framework.

Table 1. Regional hydrogeological units (cf. Figure 2) and related features in the Porto urban region (summarized and updated from [17,18,20,61,62,71,72,103,115]).

	Hydrogeological Groups			Sedimentary Cover		Metasedimentary Rocks	Granitic Rocks			
	Regional Hydrogeological Units (RHU)			Beach and Dune Sands; Alluvia	Sandy Silts and Clays	Micaschists, Metagraywackes and Paragneisses	Two-mica Granite, Medium to Coarse Grained	Biotitic Granite, Medium to Fine Grained	Micaschists, Gneisses and Migmatites	Aplite-pegmatite and Quartz Veins
	Thickness (m)			<12	10–20	not applicable				
HYDROGEOLOGICAL FEATURES	Weathering profile	low (m)		not applicable		10–20			5–10	<5
		high (m)					20–40	20–40		
		silty and/or clayey				X		X	X	X
		sandy					X	X	X	X
	Connectivity to the drainage system	with		X	X					
		possible				X	X	X	X	X
	Type of media flow	porous		X	X					
		fissured				X	X	X	X	X
	Hydrochemical facies			Cl-Na to NO$_3$-Na		Cl-Na to Cl-SO$_4$-Na				
	Environmental isotopes	δ^{18}O (‰)		not determined		−6.5 to −3.5				not determined
		δ^2H (‰)				−40 to −20				
		^{3}H (TU)				< 5				not determined
	Hydrodynamic parameters	long-term well capacity, Q (L/s)	very low (Q ≤ 1)		X	X	X	X	X	
			Low (1 < Q <2)	X						X
		Transmissivity (T, m^2/day)		15–20	< 1	1–3	0.5–2			1–3
		Storage coefficient (S)		10^{-1}–10^{-2}	10^{-2}–10^{-3}	10^{-3}–10^{-5}				
	Aquifer confinement			unconfined	aquitard	semi-confined to confined				
	More suitable exploitation structures	dug-wells, galleries, and springs		X	X					X
		boreholes				X	X	X	X	X
	Direct groundwater recharge (%)			25–30	20–25	10–15	5–10			15–20
	Urban infiltration potential index (IPI-Urban)	high		X						
		moderate		X	X		X	X		
		low			X	X	X	X	X	X
		very low				X			X	

Regarding the Porto urban region, the following two perspectives are presented: a pre-urban context (Figure 9IA) and an urban context, with the main levels of land use (Figure 9IB), which is relevant to evaluate the impact on infiltration and direct recharge of these aquifer systems. For the Porto urban area, two viewpoints are shown, i.e., the main land use classes (Figure 9IIC) and the main IPI-Urban classes (Figure 9IID). For both views, several hydroclimatic and hydrodynamical parameters were added.

The analysis of both models allowed us to characterize the following three aquifer systems, i.e., superficial, intermediate, and deep, which are interconnected, sometimes in a discontinuous manner:

a. A superficial unit corresponds essentially to the sedimentary cover and the weathered/fractured zones of metasedimentary and granitic rocks that constitute a porous medium with hydraulic connection to the drainage system. The water table is close to the surface (<5 m). The more suitable exploitation structures are dug wells and springs associated, or not, to galleries, and the long-term well capacities are low (1 < Q < 2 L/s). Sedimentary cover can reach thicknesses of almost 30 m. In crystalline rocks, the weathering thickness is variable, and can reach values of 20–40 m locally (e.g., [116]), affecting transmissivity, which is generally low (<5 m^2/day), and storage coefficient. The sedimentary cover constitutes an unconfined aquifer, while, in crystalline rocks, it corresponds to a semi-confined one. The hydrochemical facies is mostly Cl-Na. The infiltration potential is moderate to high and the groundwater recharge is direct, through infiltration of precipitation.

b. Intermediate aquifers constitute a fractured media, which may have a hydraulic connection to the drainage system. The more suitable exploitation structures are boreholes and the long-term well capacities are mostly very low (Q < 1 L/s). Transmissivity values are low (<5 m^2/day) and these aquifers are semi-confined to confined. Groundwater has a short and shallow circuit with a Cl-Na to Cl-SO4-Na hydrochemical facies. The infiltration potential is moderate to low and the groundwater recharge occurs by leakage of the overlain levels or directly from the surface, namely by geological structures (e.g., geological contacts, faults, and veins), with favorable geo-hydraulic characteristics for the groundwater flow (e.g., major deep, open and not filled fractures, and intersections between tectonic lineaments).

c. Deep aquifers correspond to unweathered and massive crystalline bedrock, with closed fractures. They constitute a fissured media, where groundwater flow tends to have a weak regime with very low transmissivities and the hydraulic characteristics are confined.

6. Conclusions

Expanding our understanding of the behavior of fractured bedrock systems in urban environments will continue to be a research challenge requiring multidisciplinary approaches such as the one presented in this work.

The Porto urban region is the second largest metropolitan area in Portugal mainland, with a high population density, living in a land covered by urban and industrial areas (ca. 41%), mostly in the south part of the region, and agricultural and forest areas (ca. 52%), in the north.

The climatic conditions, the morphotectonic and geological background, mainly comprised of crystalline formations, namely granites and metasedimentary rocks, are responsible for the groundwater resources availability.

These groundwaters have median values for pH, electrical conductivity, dissolved oxygen, and redox potential of 6.2, 382 µS/cm, 1.75 mg/L, and +0.15 V, respectively. Considering major anions, the median concentrations for bicarbonate, sulphate, chloride, and nitrate are 45.0 mg/L, 35.9 mg/L, 41.2 mg/L, and 40.0 mg/L, respectively. In addition, for the major cations the median concentrations for sodium, calcium, potassium, and magnesium are 37.0 mg/L, 13.0 mg/L, 4.0 mg/L, and 7.9 mg/L, respectively. The hydrogeochemical facies are mainly Cl-Na to Cl-SO$_4$-Na. Moreover, the median values for δ^{18}O, δ^{2}H, and ^{3}H are −4.42‰, −26.5‰, and 1.0 TU, respectively. All these characteristics are mostly dependent on the proximity of the Porto urban region to the Atlantic Ocean, the anthropogenic and agricultural contamination processes, and, with a lesser impact, the water–rock interaction, since groundwater has fast, shallow flow paths, and short residence times.

This fractured environment is the key to several hydrodynamic characteristics, which characterize confined to semi-confined groundwater conditions, namely low to very low long-term well capacities, with a median value of 1 L/s; low transmissivities, with median values in the range 2–4 m^2/day; and, generally, low storage coefficients, in the range 10^{-5}–10^{-2}.

The infiltration potential and the groundwater recharge are typically low in this fissured background. This low potential largely arises in zones where granitic rocks and agricultural/forest areas coexist. However, a moderate infiltration potential also occurs, naturally associated with the sedimentary cover, and artificially to the urban hydraulic and sanitation features.

The main conclusions summarized above contribute to outline recommendations for policymakers for the environmental protection, reasoning on the planning with nature and to the sustainable water resources management for the urban areas. In fact, integrated groundwater resources management is vital to evaluate water use, water availability, and to design balanced solutions in a changing society, environment, and climate.

Last but not least, in the current context of climate change, water shortage and groundwater depletion, along with a pandemic scenario, the authors support this impressive message by [117], "There is a need for the incorporation of groundwater in national IWRM plans and promotion of groundwater resource management in the programs of river basin agencies. Planned conjunctive management of groundwater and surface-water resources in many cases represents the best prospect for improving water-supply security for urban and irrigation use, and for sustainable resources."

Supplementary Materials: The following are available online at http://www.mdpi.com/2073-4441/12/10/2797/s1, Table S1: Physical, chemical, and isotopic parameters for groundwater samples collected during the fieldwork campaigns.

Author Contributions: Conceptualization, M.J.A. and H.I.C.; methodology, M.J.A., L.F., J.M.M., P.M.C. and H.I.C; software, L.F. and M.J.A.; investigation, M.J.A., L.F., J.M.M., P.M.C., F.R., A.J.S.C.P. and H.I.C.; visualization, M.J.A., L.F. and H.I.C.; writing—original draft preparation, M.J.A., L.F., H.I.C, J.M.M., P.M.C., F.R., and A.J.S.C.P.; writing-review and editing, M.J.A., L.F., and H.I.C. All authors have read and agreed to the published version of the manuscript.

Funding: This work was partially financed by FEDER-EU COMPETE Funds and the Portuguese Foundation for the Science and Technology, FCT (UID/GEO/04035/2020, UID/Multi/00611/2020, and GroundUrban project POCI/CTE-GEX/59081/2004), and by the Labcarga|ISEP re-equipment program (IPP-ISEP|PAD'2007/08). The research was also funded by a doctoral scholarship from the Portuguese Foundation for Science and Technology (FCT) to L. Freitas (SFRH/BD/117927/2016). P.M. Carreira acknowledges the FCT support through the FCT-UIDB/04349/2020 project and J.M. Marques recognizes the FCT support through the UID/ECI/04028/2020 project.

Acknowledgments: This work was partially financed by FEDER-EU COMPETE Funds and the Portuguese Foundation for the Science and Technology, FCT (UID/GEO/04035/2020, UID/Multi/00611/2020 and GroundUrban project: POCI/CTE-GEX/59081/2004), and by the Labcarga|ISEP re-equipment program (IPP-ISEP|PAD'2007/08). The research was also funded by a doctoral scholarship from the Portuguese Foundation for Science and Technology (FCT) to L. Freitas (SFRH/BD/117927/2016) and, currently, by Project ReNATURE (Centro-01-0145-FEDER-000007-PT). P.M. Carreira acknowledge the FCT support through the FCT-UIDB/04349/2020 project and J.M. Marques recognize the FCT support through the UID/ECI/04028/2020 project. We are grateful to the anonymous reviewers for the comments that helped to improve the focus of the manuscript.

Conflicts of Interest: The authors declare no conflict of interest.

References

1. Foster, S.D.; Tyson, G. *Resilient Cities & Groundwater*; Strategic Overview Series; International Association of Hydrogeologists: Goring/Reading, UK, 2015.
2. Howard, K.W.F.; Israfilov, R.G. *Current Problems of Hydrogeology in Urban Areas, Urban. Agglomerates, and Industrial Centres*; Kluwer: Dordrecht, The Netherlands, 2002.

3. Howard, K.W.F. (Ed.) Urban. Groundwater. In Proceedings of the Meeting the Challenge: Selected Papers from the 32nd International Geological Congress (IGC), Florence, Italy, 20–28 August 2004; Taylor & Francis: London, UK, 2007.

4. Massing, H.; Packman, J.; Zuidema, F.C. *Hydrological Processes and Water Management in Urban Areas*; IAHS Publication No. 198; IAHS: Wallingford, UK, 1990; p. 362.

5. Johnson, S.P. *The Earth Summit: The United Nations Conference on Environment and Development (UNCED)*; Graham & Trotman/Martinus Nijhoff: London, UK, 1993.

6. Wilkinson, W.B. (Ed.) *Groundwater Problems in Urban Areas*; T. Telford: London, UK, 1994; p. 453.

7. Chilton, J. (Ed.) Groundwater in the Urban Environment: Problems, Processes and Management. In Proceedings of the 27th Congress, International Association of Hydrogeologists, Nottingham, UK, 21–27 September 1997.

8. Chilton, J. *Groundwater in the Urban Environment: Selected City Profiles*; A. A. Balkema: Rottterdam, The Netherlands, 1999; p. 342.

9. Howard, K.W.F. *Groundwater for Socio-Economic Development—The Role of Science*; UNESCO IHP-VI Series on Groundwater, 9, Published as CD.; UNESCO: Paris, France, 2004; ISBN 92-9220-029-1.

10. Tellam, J.H.; Rivett, M.O.; Israfilov, R.G. *Urban. Groundwater Management and Sustainability*; NATO Science Series, IV Earth and Environmental Sciences; Springer: Dordrecht, The Netherlands, 2006; Volume 74, p. 491.

11. Afonso, M.J.; Chaminé, H.I.; Marques, J.M.; Carreira, P.M.; Guimarães, L.; Guilhermino, L.; Gomes, A.; Fonseca, P.E.; Pires, A.; Rocha, F. Environmental issues in urban groundwater systems: A multidisciplinary study of the Paranhos and Salgueiros spring waters, Porto (NW Portugal). *Environ. Earth Sci.* **2010**, *61*, 379–392. [CrossRef]

12. Afonso, M.J.; Freitas, L.; Pereira, A.J.S.C.; Neves, L.J.P.F.; Guimarães, L.; Guilhermino, L.; Mayer, B.; Rocha, F.; Marques, J.M.; Chaminé, H.I. Environmental groundwater vulnerability assessment in urban water mines (Porto, NW Portugal). *Water* **2016**, *8*, 499. [CrossRef]

13. Afonso, M.J.; Freitas, L.; Chaminé, H.I. Groundwater recharge in urban areas (Porto, NW Portugal): The role of GIS hydrogeology mapping. *Sustain. Water Resour. Manag.* **2019**, *5*, 203–216. [CrossRef]

14. Sharp, J.M.; Hibbs, B.J. Special Issue on Hydrogeological Impacts of Urbanization. *Environ. Eng. Geosci.* **2012**, *18*, 111.

15. Schirmer, M.; Leschik, S.; Musolff, A. Current research in urban hydrogeology—A review. *Adv. Water Resour.* **2013**, *51*, 280–291. [CrossRef]

16. Freitas, L.; Afonso, M.J.; Devy-Vareta, N.; Marques, J.M.; Gomes, A.; Chaminé, H.I. Coupling hydrotoponymy and GIS cartography: A case study of hydrohistorical issues in urban groundwater systems, Porto NW Portugal. *Geogr. Res.* **2014**, *52*, 182–197. [CrossRef]

17. Freitas, L.; Afonso, M.J.; Pereira, A.J.S.C.; Deleru-Matos, C.; Chaminé, H.I. Assessment of sustainability of groundwater in urban areas (Porto, NW Portugal): A GIS mapping approach to evaluate vulnerability, infiltration and recharge. *Environ. Earth Sci.* **2019**, *78*, 140. [CrossRef]

18. Freitas, L.; Chaminé, H.I.; Pereira, A.J.S.C. Coupling groundwater GIS mapping and geovisualisation techniques in urban hydrogeomorphology: Focus on methodology. *SN Appl. Sci.* **2019**, *1*, 490. [CrossRef]

19. Freitas, L.; Chaminé, H.I.; Afonso, M.J.; Meerkhan, H.; Abreu, T.; Trigo, J.F.; Pereira, A.J.S.C. Integrative groundwater studies in a small-scale urban area: Case study from the municipality of Penafiel (NW Portugal). *Geosciences* **2020**, *10*, 54. [CrossRef]

20. Barroso, M.F.; Ramalhosa, M.J.; Olhero, A.; Antão, M.C.; Pina, M.F.; Guimarães, L.; Teixeira, J.; Afonso, M.J.; Delerue-Matos, C.; Chaminé, H.I. Assessment of groundwater contamination in an agricultural peri-urban area (NW Portugal): An integrated approach. *Environ. Earth Sci.* **2015**, *73*, 2881–2894. [CrossRef]

21. Hibbs, B.J. Groundwater in urban areas. *J. Contemp. Water Res. Educ.* **2016**, *159*, 143. [CrossRef]

22. Guimarães, L.; Guilhermino, L.; Afonso, M.J.; Marques, J.M.; Chaminé, H.I. Assessment of urban groundwater: Towards integrated hydrogeological and effects-based monitoring. *Sustain. Water Resour. Manag.* **2019**, *5*, 217–233. [CrossRef]

23. Sherlock, R.L. *Man as a Geological Agent*; H.F. & G. Witherby: London, UK, 1922; p. 372.

24. Legget, R.F. *Cities and Geology*; McGraw-Hill: New York, NY, USA, 1973; p. 579.

25. Dunne, T.; Leopold, L.B. *Water in Environmental Planning*; W.H. Freeman Co.: San Francisco, CA, USA, 1978; p. 818.

26. Utgard, R.O.; McKenzie, G.D.; Foley, D. *Geology in the Urban. Environment*; Burgess Publishing Company: Minneapolis, MI, USA, 1978; p. 355.
27. Leveson, D. *Geology and the Urban. Environment*; Oxford University Press: Oxford, UK, 1980; p. 386.
28. Zaporozec, A. (Ed.) Cities and water. *Geo J.* **1985**, *11*, 203–283.
29. McCall, G.J.; Demulder, E.; Marker, B.R. *Urban. Geoscience*; AGID Special Publication Series; Taylor & Francis: Rotterdam, The Netherlands, 1996; Volume 20, p. 279.
30. Gehrels, H.; Peters, N.E.; Hoehn, E.; Jensen, K.; Leibundgut, C.; Griffioen, J.; Webb, B.; Zaadnoordijk, W.J. *Impact of Human Activity on Groundwater Dynamics*; International Association of Hydrological Sciences, Publication no. 269; IAHS Press: Wallingford, UK, 2001; p. 369.
31. Pokrajac, D. (Ed.) Groundwater in urban areas. *Urban. Water* **2001**, *3*, 171–237. [CrossRef]
32. Bocanegra, E.; Hernández, M.; Usunoff, E. *Groundwater and Human Development. International Association of Hydrogeologists Selected Papers*; Taylor & Francis: London, UK, 2005; Volume 6, p. 262.
33. Ehlen, J.; Haneberg, W.C.; Larson, R.A. *Humans as Geologic Agents. Reviews in Engineering Geology*; The Geological Society of America: Boulder, CO, USA, 2005; Volume 16, p. 158.
34. Culshaw, M.G.; Reeves, H.J.; Jefferson, I.; Spink, T.W. *Engineering Geology for Tomorrow's Cities*; Engineering Geology Special Publications; Geological Society: London, UK, 2009; Volume 22, p. 315.
35. Chaminé, H.I.; Afonso, M.J.; Freitas, L. From historical hydrogeological inventory through GIS mapping to problem solving in urban groundwater systems. *Eur. Geol. J.* **2014**, *38*, 33–39.
36. Chaminé, H.I.; Teixeira, J.; Freitas, L.; Pires, A.; Silva, R.S.; Pinho, T.; Mon- teiro, R.; Costa, A.L.; Abreu, T.; Trigo, J.F. From engineering geosciences mapping towards sustainable urban planning. *Eur. Geol. J.* **2016**, *41*, 16–25.
37. UN-Habitat [United Nations Human Settlements Programme]. *Urbanization and Development: Emerging Futures—World Cities Report 2016*; United Nations Human Settlements Programme, World Urban Forum edition: Nairobi, Kenya, 2016; p. 264.
38. Underwood, J.R. Anthropic rocks as a fourth basic class. *Environ. Eng. Geosci.* **2001**, *7*, 104–110. [CrossRef]
39. Baker, L.A. (Ed.) *The Water Environment of Cities*; Springer Science & Business Media: London, UK, 2009; p. 307.
40. Sharp, J.M. The impacts of urbanization on groundwater systems and recharge: Aqua Mundi. *Aqua Mundi* **2010**, *1*, 51–56. [CrossRef]
41. Gogu, C.R.; Campbell, D.; de Beer, J. The Urban Subsurface—From Geoscience and Engineering to Spatial Planning and Management. *Procedia Eng.* **2017**, *209*, 224. [CrossRef]
42. Hibbs, B.J.; Sharp, J.M. Hydrogeological impacts of urbanization. *Environ. Eng. Geosci.* **2012**, *18*, 3–24. [CrossRef]
43. Taylor, R.G.; Scanlon, B.; Döll, P.; Rodell, M.; van Beek, R.; Wada, Y.; Longuevergne, L.; Leblanc, M.; Famiglietti, J.S.; Edmunds, M.; et al. Ground water and climate change. *Nat. Clim. Chang.* **2013**, *3*, 322–329. [CrossRef]
44. Foster, S.D.; Tyson, G. *Global Change & Groundwater*; Strategic Overview Series; International Association of Hydrogeologists: Goring/Reading, UK, 2016.
45. Wiles, T.J.; Sharp, J.M. The secondary permeability of impervious cover. *Environ. Eng. Geosci.* **2008**, *14*, 251–265. [CrossRef]
46. Pujades, E.; De Simone, S.; Carrera, J.; Vázquez-Suñé, E.; Jurado, A. Settlements around pumping wells: Analysis of influential factors and a simple calculation procedure. *J. Hydrol.* **2017**, *548*, 225–236. [CrossRef]
47. Lyu, H.M.; Shen, S.-L.; Yang, J.; Yin, Z.-Y. Inundation analysis of metro systems with the storm water management model incorporated into a geographical information system: A case study in Shanghai. *Hydrol. Earth Syst. Sci.* **2019**, *23*, 4293–4307. [CrossRef]
48. IAEA [International Atomic Energy Agency]. Stable Isotope Hydrology. In *Deuterium and Oxygen-18 in the Water Cycle*; Technical Reports Series 210; IAEA: Vienna, Austria, 1981.
49. Clark, I.D.; Fritz, P. *Environmental Isotopes in Hydrogeology*; CRC Press: Boca Raton, FL, USA; Lewis Publishers: Boca Raton, FL, USA, 1997; p. 328.
50. Hoefs, J. *Stable Isotope Geochemistry*; Completely Revised. Updated and Enlarged Edition; Springer: Berlin, Germany, 1997.
51. Şener, E.; Davraz, A.; Ozcelik, M. An integration of GIS and remote sensing in groundwater investigations: A case study in Burdur, Turkey. *Hydrogeol. J.* **2005**, *13*, 826–834. [CrossRef]

52. Thapa, R.; Gupta, S.; Gupta, A.; Reddy, D.V.; Kaur, H. Use of geospatial technology for delineating groundwater potential zones with an emphasis on water-table analysis in Dwarka River basin, Birbhum, India. *Hydrogeol. J.* **2018**, *26*, 899–922. [CrossRef]

53. Morris, B.L.; Litvak, R.G.; Ahmed, K.M. Urban groundwater protection and management: Lessons from developing cities in Bangladesh and Kyrghyztan. In *Current Problems of Hydrogeology in Urban Areas, Urban Agglomerates and Industrial Centres*; NATO Science Series, IV Earth and Environmental Sciences; Howard, K.W.F., Israfilov, R.G., Eds.; Kluwer Academic Publishers: Dordrecht, The Netherlands, 2002; Volume 8, pp. 77–102.

54. Sharp, J.M. Ground-water supply issues in urban and urbanizing areas. In *Groundwater in the Urban Environment: Problems, Process and Management*; Chilton, J., Ed.; A. A. Balkema: Rotterdam, The Netherlands, 1997; pp. 67–74.

55. Barrett, M.H.; Howard, A.G. Urban groundwater and sanitation: Developed and developing countries. In *Current Problems of Hydrogeology in Urban Areas, Urban Agglomerates and Industrial Centres*; NATO Science Series, IV Earth and Environmental Sciences; Howard, K.W.F., Israfilov, R.G., Eds.; Kluwer Academic Publishers: Dordrecht, The Netherlands, 2002; Volume 8, pp. 39–56.

56. Foster, S.D.; Hirata, R.; Howard, K.W.F. Groundwater use in developing cities: Policy issues arising from current trends. *Hydrogeol. J.* **2011**, *19*, 271–274. [CrossRef]

57. Elshall, A.S.; Arik, A.D.; El-Kadi, A.I.; Pierce, S.; Ye, M.; Burnett, K.M.; Wada, C.A.; Bremer, L.L.; Chun, G. Groundwater sustainability: A review of the interactions between science and policy. *Environ. Res. Lett* **2020**, *15*, 093004. [CrossRef]

58. INE—Instituto Nacional de Estatística. Statistical Information about Portuguese Population: Porto City. 2011. Available online: http://www.ine.pt/ (accessed on 30 January 2019).

59. Caetano, M.; Igreja, C.; Marcelino, F.; Costa, H. Estatísticas e Dinâmicas Territoriais Multiescala de Portugal Continental 1995–2007–2010 com Base na Carta de Uso e Ocupação do Solo (COS). Relatório Técnico. Direção-Geral do Território (DGT). 2017. Available online: http://www.dgterritorio.pt/ (accessed on 28 February 2019).

60. Afonso, M.J. Hidrogeologia de rochas graníticas da região do Porto (NW de Portugal). *Cad. Lab. Xeol. Laxe* **2003**, *28*, 173–192.

61. Afonso, M.J. Hidrogeologia e Hidrogeoquímica da Região Litoral Urbana do Porto, entre Vila do Conde e Vila Nova de Gaia (NW de Portugal): Implicações Geoambientais. Ph.D. Thesis, Instituto Superior Técnico da Universidade Técnica de Lisboa, Lisboa, Portugal, 2011.

62. Afonso, M.J.; Chaminé, H.I.; Carvalho, J.M.; Marques, J.M.; Gomes, A.; Araújo, M.A.; Fonseca, P.E.; Teixeira, J.; Marques da Silva, M.A.; Rocha, F.T. Urban groundwater resources: A case study of Porto City in northwest Portugal. In *Urban Groundwater: Meeting the Challenge. International Association of Hydrogeologists Selected Papers*; Howard, K.W.F., Ed.; Taylor & Francis Group: London, UK, 2007; Volume 8, pp. 271–287.

63. Pereira, E.; Ribeiro, A.; Carvalho, G.S.; Noronha, F.; Ferreira, N.; Monteiro, J.H. *Carta Geológica de Portugal, escala 1/200000. Folha 1*; Serviços Geológicos de Portugal: Lisboa, Portugal, 1989.

64. Chaminé, H.I.; Gama Pereira, L.C.; Fonseca, P.E.; Noronha, F.; Lemos de Sousa, M.J. Tectonoestratigrafia da faixa de cisalhamento de Porto–Albergaria-a-Velha–Coimbra–Tomar, entre as Zonas Centro-Ibérica e de Ossa-Morena (Maciço Ibérico, W de Portugal). *Cad. Lab. Xeol. Laxe* **2003**, *28*, 37–78.

65. Chaminé, H.I.; Afonso, M.J.; Robalo, P.M.; Rodrigues, P.; Cortez, C.; Mon- teiro Santos, F.A.; Plancha, J.P.; Fonseca, P.E.; Gomes, A.; Devy-Vareta, N.F.; et al. Urban speleology applied to groundwater and geo-engineering studies: Underground topographic surveying of the ancient Arca D'Água galleries catchworks (Porto, NW Portugal). *Int. J. Speleol.* **2010**, *39*, 1–14. [CrossRef]

66. Araújo, M.A.; Gomes, A.; Chaminé, H.I.; Fonseca, P.E.; Gama Pereira, L.C.; Pinto de Jesus, A. Geomorfologia e geologia regional do sector de Porto-Espinho (W de Portugal): Implicações morfoestruturais na cobertura sedimentar Cenozóica. *Cad. Lab. Xeol. Laxe* **2003**, *28*, 79–105.

67. Struckmeier, W.F.; Margat, J. *Hydrogeological Maps: A Guide and a Standard Legend*; International Association of Hydrogeologists: Hannover, Germany, 1995; p. 177.

68. Assaad, F.A.; LaMoreaux, P.E.; Hughes, T.H.; Wangfang, Z.; Jordan, H. *Field Methods for Geologists and Hydrogeologist*; Springer: Berlin, Germany, 2004; p. 420.

69. Carríngton da Costa, J.; Teixeira, C. *Carta Geológica de Portugal na escala de 1/50000. Notícia Explicativa da Folha 9-C (Porto)*; Serviços Geológicos de Portugal: Lisboa, Portugal, 1957; p. 38.

70. Teixeira, C.; Medeiros, A.C. *Carta geológica de Portugal na escala 1:50000. Notícia explicativa da folha 9A-Póvoa de Varzim*; Serviços Geológicos de Portugal: Lisboa, Portugal, 1965.

71. Pedrosa, M.Y. *Notícia explicativa da Carta Hidrogeológica de Portugal, à escala 1/200000. Folha 1*; Instituto Geológico e Mineiro: Lisboa, Portugal, 1999; p. 70.

72. Carvalho, J.M. Prospecção e Pesquisa de Recursos Hídricos Subterrâneos no Maciço Antigo Português: Linhas Metodológicas. Ph.D. Thesis, Universidade de Aveiro, Aveiro, Portugal, 2006; p. 292.

73. Epstein, S.; Mayeda, T. Variations of 18O content of waters from natural sources. *Geochim. Cosmochim. Acta* **1953**, *4*, 213–224. [CrossRef]

74. Friedman, I. Deuterium content of natural waters and other substances. *Geochim. Cosmochim. Acta* **1953**, *4*, 89–103. [CrossRef]

75. IAEA [International Atomic Energy Agency]. *Procedure and Technique Critique for Tritium Enrichment by Electrolysis at IAEA Laboratory*; Technical Procedure No. 19; IAEA: Vienna, Austria, 1976.

76. Lucas, L.L.; Unterweger, M.P. Comprehensive review and critical evaluation of the half-life of tritium. *J. Res. Natl. Inst. Stand. Technol.* **2000**, *105*, 541–549. [CrossRef]

77. Carreira, P.M.; Marques, J.M.; Graça, R.C.; Aires-Barros, L. Radiocarbon application in dating "complex" hot and cold CO_2-rich mineral water systems: A review of case studies ascribed to the northern Portugal. *Appl. Geochem.* **2008**, *23*, 2817–2828. [CrossRef]

78. Carreira, P.M.; Marques, J.M.; Carvalho, M.R.; Capasso, G.; Grassa, F. Mantle-derived carbon in Hercynian granites Stable isotopes signatures C/He associations in the thermomineral waters N-Portugal. *J. Volcanol. Geotherm. Res.* **2010**, *189*, 49–56. [CrossRef]

79. Theis, C.V. The relation between lowering of the piezometric surface and rate and duration of discharge of a well using ground-water storage. *Trans. Am. Geophys. Union* **1935**, *16*, 519–524. [CrossRef]

80. Cooper, H.H.J.R.; Jacob, C.E. A generalized graphical method for evaluating formation constants and summarizing well-field history. *Trans. Am. Geophys. Union* **1946**, *27*, 526–534. [CrossRef]

81. Kruseman, G.P.; de Ridder, N.A. *Analysis and Evaluation of Pumping Test. Data*, 2nd ed.; International Institute for Land Reclamation and Improvement: Wageningen, The Netherlands, 1990; Volume 47, p. 377.

82. Sterrett, R.J. *Groundwater and Wells*, 3rd ed.; Johnson Screens, A Weatherford Company: New Brighton, MN, USA, 2007; p. 812.

83. Logan, J. Estimating transmissibility from routine production tests of water wells. *Ground Water* **1964**, *2*, 35–37. [CrossRef]

84. Malczewski, J.; Rinner, C. *Multicriteria Decision Analysis in Geographic Information Science*; Springer: New York, NY, USA, 2015.

85. Custodio, E.; Llamas, M.R. *Hidrología Subterránea*; Segunda Edición Corregida; Ediciones Omega: Barcelona, Spain, 2001; p. 2350.

86. Younger, P.L. *Groundwater in the Environment: An Introduction*; Blackwell Publishing: Hoboken, NJ, USA, 2007; p. 318.

87. MA—Ministério do Ambiente. *Decreto-Lei nº 236/98, de 1 de Agosto. Diário da República—I Série-A, Nº 176*; Ministério do Ambiente: Lisbon, Portugal, 1998.

88. Pedrosa, M.Y. *Carta Hidrogeológica de Portugal, à escala 1/200000. Folha 1*; Instituto Geológico e Mineiro: Lisboa, Portugal, 1998.

89. Craig, H. Isotopic variations in meteoric waters. *Science* **1961**, *133*, 1702–1703. [CrossRef] [PubMed]

90. Rozanski, K.; Araguás-Araguás, L.; Gonfiantini, R. Isotopic patterns in modern global precipitation. In *Climate Change in Continental Isotopic Records, Geoph. Monog. Series*; Swart, P.K., Ed.; AGU: Washington, DC, USA, 1993; pp. 1–36.

91. Bowen, G.J.; Wilkinson, B. Spatial distribution of $\delta^{18}O$ in meteoric precipitation. *Geology* **2002**, *30*, 315–318. [CrossRef]

92. Terzer, S.; Wassenaar, L.I.; Araguás-Araguás, L.J.; Aggarwal, P.K. Global isoscapes for $\delta^{18}O$ and δ^2H in precipitation: Improved prediction using regionalized climatic regression models. *Hydrol. Earth. Syst. Sci.* **2013**, *17*, 4713–4728. [CrossRef]

93. Carreira, P.M.; Araújo, M.F.; Nunes, D. Isotopic composition of rain and water vapour samples from Lisbon region: Characterization of monthly and daily events. In *IAEA-TECDOC-1453 Isotopic Composition of Precipitation in the Mediterranean Basin in Relation to Air Circulation Patterns and Climate*; IAEA: Vienna, Austria, 2005; pp. 141–155.

94. Wakida, F.; Lerner, D. Non-agricultural sources of groundwater nitrate: A review and case study. *Water Res.* **2005**, *39*, 3–16. [CrossRef] [PubMed]

95. Heitor, A.M.F. Contaminação das águas subterrâneas no Norte de Portugal. In *Las Aguas Subterráneas en el Noroeste de la Península Ibérica*; Textos de las Jornadas, Mesa Redonda y Comunicaciones; Samper, J., Leitão, T., Fernández, L., Ribeiro, L., Eds.; A Coruña. AIH-Grupo Español & APRH. ITGE: Madrid, Spain, 2000; pp. 295–308.

96. Pedrosa, M.Y.; Brites, J.A.; Pereira, A.P. *Carta das Fontes e do Risco de Contaminação da Região de Entre-Douro-e-Minho. Folha Sul, Escala 1/100000, Nota Explicativa*; Instituto Geológico e Mineiro: Lisboa, Portugal, 2002.

97. Correia, M.; Barroso, A.; Barroso, M.F.; Soares, D.; Oliveira, M.B.P.P.; Delerue-Matos, C. Contribution of different vegetable types to exogenous nitrate and nitrite exposure. *Food Chem.* **2010**, *120*, 960–966. [CrossRef]

98. Begonha, A.; Sequeira Braga, M.A.; Gomes da Silva, F. A acção da água da chuva na meteorização de monumentos graníticos. In *IV Congresso Nacional de Geologia. Memórias do Museu e Laboratório Mineralógico e Geológico da Faculdade de Ciências da Universidade do Porto*; Borges, F.S., Marques, M., Eds.; MLMGFCUP: Porto, Portugal, 1995; Volume 4, pp. 177–181.

99. Begonha, A. *Meteorização do Granito e Deterioração da Pedra em Monumentos e Edifícios da Cidade do Porto*; Colecção monografias, FEUP Edições: Porto, Portugal, 2001; p. 445.

100. Carreira, P.M.; Valério, P.; Nunes, D.; Araújo, M.F. Temporal and seasonal variations of stable isotopes (2H and 18O) and tritium in precipitation over Portugal. In Proceedings of the Isotopes in Environmental Studies—Aquatic Forum, Monte Carlo, Monaco, 25–29 October 2004; IAEA: Vienna, Austria, 2006; pp. 370–373.

101. Misstear, B.D.R. The value of simple equilibrium approximations for analysing pumping test data. *Hydrogeol. J.* **2001**, *9*, 125–126. [CrossRef]

102. Afonso, M.J.; Carvalho, J.M.; Marques, J.M.; Chaminé, H.I. Hydrodynamic constraints of the Porto urban area crystalline bedrock (NW Portugal, Iberian Massif): Implications on groundwater resources. In Proceedings of the 7th Hellenic Hydrogeological Conference and 2nd MEM Workshop on Fissured Rocks Hydrology, Athens, Greece, 5–6 October 2005; Stournaras, G., Pavlopoulos, K., Bellos, T.h., Eds.; The Geological Society of Greece (Hellenic Committee of Hydrogeology): Athens, Greece, 2005; Volume 2, pp. 77–81.

103. Carvalho, J.M.; Chaminé, H.I.; Afonso, M.J.; Espinha Marques, J.; Medeiros, A.; Garcia, S.; Gomes, A.; Teixeira, J.; Fonseca, P.E. Productivity and water cost in fissured-aquifers from the Iberian crystalline basement (Portugal): Hydrogeological constraints. In *Water, Mining and Environment. Book Homage to Professor Rafael Fernández Rubio*; López-Geta, J.A., et al., Eds.; Instituto Geológico y Minero de España: Madrid, Spain, 2005; pp. 193–207.

104. Larsson, I. *Groundwater in Hard Rocks. Studies and Reports in Hydrology*; UNESCO: Paris, France, 1984; p. 234.

105. Wright, E.P.; Burgess, W.G. *The Hydrogeology of Crystalline Basement in Africa*; Geological Society Special Publication, 68; GSL: London, UK, 1992; p. 264.

106. Lloyd, J.W. *Water Resources of Hard Rock Aquifers in Arid and Semi-arid Zones*; Studies and Reports in Hydrology, 58; UNESCO: Paris, France, 1999; p. 284.

107. Robins, N.S.; Misstear, B.D.R. *Groundwater in the Celtic Regions: Studies in Hard Rock and Quaternary Hydrogeology*; Geological Society of London: London, UK, 2000; p. 273.

108. Stober, I.; Bucher, K. *Hydrogeology of Crystalline Rocks*; Water Science and Technology Library; Kluwer Academic Publishers: Dordrecht, The Netherlands, 2000; Volume 34, p. 284.

109. Krásný, J.; Sharp, J.M. Hydrogeology of fractured rocks from particular fractures to regional approaches: State-of-the-art and future challenges. In *Groundwater in Fractured Rocks. International Association of Hydrogeologists Selected Papers*; Krásný, J., Sharp, J.M., Eds.; Taylor & Francis Group: London, UK, 2007; pp. 1–30.

110. Singhal, B.B.S.; Gupta, R.P. *Applied Hydrogeology of Fractured Rocks*, 2nd ed.; Springer: Dordrecht, The Netherlands, 2010; p. 408.

111. Gustafson, G. *Hydrogeology for Rock Engineers*; BeFo: Stockholm, Sweden, 2012.

112. Sharp, J.M. *Fractured Rock Hydrogeology*; CRC Press: Boca Raton, FL, USA, 2014; p. 408.

113. Ofterdinger, U.; Macdonald, A.M.; Comte, J.C.; Young, M.E. (Eds.) *Groundwater in Fractured Bedrock Environments: Managing Catchment and Subsurface Resources*; Geological Society: London, UK, 2019.

114. Freitas, L. Avaliação Integrada de Recursos Hídricos em Áreas Urbanas: Aplicações Para a Sustentabilidade e o Ordenamento Territorial. Ph.D. Thesis, Faculdade de Ciências e Tecnologia da Universidade de Coimbra, Coimbra, Portugal, 2019; p. 425.
115. Carvalho, J.M.; Espinha Marques, J.; Afonso, M.J.; Chaminé, H.I. Prospecção e pesquisa de recursos hidrominerais e de água de nascente no Maciço Antigo Português. *Boletim e Minas* **2007**, *42*, 161–196.
116. Begonha, A.; Sequeira Braga, M.A. Weathering of the Oporto granite: Geotechnical and physical properties. *Catena* **2002**, *49*, 57–76. [CrossRef]
117. IAH [International Association of Hydrogeologists]. The UN-SDGs for 2030: Essential Indicators for Groundwater. 2017. Available online: https://iah.org/wp-content/uploads/2017/04/IAH-Groundwater-SDG-6-Mar-2017.pdf (accessed on 30 June 2020).

Article

Long Term Effectiveness of Wellhead Protection Areas

Joel Zeferino [1,*], Marina Paiva [2], Maria do Rosário Carvalho [1], José Martins Carvalho [2,3] and Carlos Almeida [1]

[1] IDL and Department of Geology, Faculty of Sciences, University of Lisbon, 1749-016 Lisbon, Portugal; mdrcarvalho@fc.ul.pt (M.d.R.C.); calmeida96@gmail.com (C.A.)

[2] TARH—Terra, Ambiente e Recursos Hídricos, Lda., R. Forte Monte Cintra 1B3, 2685-137 Sacavém, Portugal; marina.paiva@tarh.pt (M.P.); jmc@tarh.pt (J.M.C.)

[3] Centre GeoBioTec, University of Aveiro, Campus Universitário de Santiago, 3810-193 Aveiro, Portugal

* Correspondence: jfzeferino@fc.ul.pt; Tel.: +351-963971910

Abstract: A preventive instrument to ensure the protection of groundwater is the establishment of wellhead protect areas (WPA) for public supply wells. The shape of the WPA depends on the rate of pumping and aquifer characteristics, such as the transmissivity, porosity, hydraulic gradient, and aquifer thickness. If any parameter changes after the design of the WPA, it will no longer be effective in protecting the aquifer and its catchment. With population growth in urban areas, the pressure on groundwater abstraction increases. Changes in flow, drawdowns and hydraulic gradients often occur. The purpose of this work is to evaluate the effectiveness of the WPA after a long period of establishment, in public wells with continuous pumping, located in densely populated urban area of the municipality of Montijo (Portugal). Considering the aquifer scenario in 2019, new extended WPAs were calculated using the combined results of three analytical methods and numerical modelling. In 2009 the aquifer presented hydraulic gradients varying between 0.0005 and 0.002, giving rise to a protection area with essentially circular shape. Although there was no increase in extraction flow, in 2019 the hydraulic gradients vary from 0.0008 to 0.008, and the flow directions have changed because of the water level decline. The shape of the WPA in this case is essentially elliptical and longer upstream and it can pose difficulties in the protection of public water catchments, in an urban area with already defined and consolidated land use. The best protection of the public supply wells in disturbed aquifers is obtained through numerical modeling.

Keywords: wellhead protection area; numerical modeling; analytical methods

Citation: Zeferino, J.; Paiva, M.; Carvalho, M.d.R.; Carvalho, J.M.; Almeida, C. Long Term Effectiveness of Wellhead Protection Areas. *Water* **2022**, *14*, 1063. https://doi.org/10.3390/w14071063

Academic Editor: Yangxiao Zhou

Received: 15 February 2022
Accepted: 25 March 2022
Published: 28 March 2022

Publisher's Note: MDPI stays neutral with regard to jurisdictional claims in published maps and institutional affiliations.

1. Introduction

Groundwater is an important source of water, both potential and actual, at the regional and local level, providing almost 50% of all drinking water worldwide [1]. In Europe, more than 70% of the population's water supply comes from groundwater [2]. However, over the past decades, global groundwater demands have more than doubled. These demands will continue to increase due to population growth and climate change. It is therefore imperative to protect the quality of groundwater [3,4].

The quality of groundwater can be affected by socio-economic activities, in particular land uses and occupations in urban areas [5]. Groundwater contamination is, in most situations, persistent, and the recovery of the quality is very slow and difficult. Groundwater protection is a key strategic objective for balanced and sustainable socio-economic development [6,7]. Therefore, to protect and preserve groundwater quality, a variety of institutional, economic, and technological instruments can be used. A preventive instrument to ensure the protection of groundwater is the establishment of wellhead protect areas (WPA) for public supply wells [8,9]. The delineation of WPA is intended to reduce the risk of contamination, or, if it does occur, to prevent it from reaching the catchments in concentrations considered dangerous, or to detect it by the aquifer's monitoring system in time to prevent it from entering the distribution network.

Concerns about drinking water protection began in the early 20th century in France, but it was only in the 1950s that industrialized countries changed their legislation to better address the degradation of their water sources by establishing WPA [10]. The U.S.A. legislated the delimitation of WPA in 1986 [8]. The European Union Countries (EC) have jointly developed a common strategy for supporting the implementation of the Directive 2000/60/EC (WFD) [9]. One of the objectives of the strategy is to preserve water resources and protect them in quality and quantity. The Annex IV of the WFD defines protected areas as areas designated for the abstraction of water intended for human consumption, under Article 7 of the WFD—Drinking Water Protected Areas. The U.S. Environmental Protection Agency [8] defines a WPA as the "surface or subsurface area surrounding a water well or wellfield supplying a public water system, through which contaminants are reasonably likely to move toward and reach such well or wellfield". Delineation of the wellhead protection area is the process of determining what geographic area should be included in a wellhead protection program. This area of land is then managed to minimize the potential of groundwater contamination by human activities that occur on the land surface or in the subsurface.

Each country has regulated the design of WPA differently, even in the EU countries that follow the WFD [9]. The most frequent configuration comprises at least three different protection zones, often referred to as the wellhead protection area (zone I), inner protection area (zone II), and outer protection area (zone III). The areas have different restrictions, based on fixed distances or transit time [2,8,11]. The transit or propagation time is the time it takes a given particle of groundwater to flow to a pumping well and is directly related to the distance that the water must travel to arrive at the well once it starts pumping. However, for any given transit time, the distance will vary from well to well depending on the rate of pumping and aquifer characteristics, such as the transmissivity, porosity, hydraulic gradient, and aquifer thickness [12]. The transit time is one of the most accurate criteria on the WPA delineation because it considers important factors of the pollutant's propagation process. However, it only considers pollutant propagation velocities that are moving at the same speed as water, which is not true for many of the contaminants [13].

In Portugal the delineation of WPA is regulated by the Dec.-Law 382/99 of 22 September 1999, transposing the WFD [9]. It requires the delimitation of three WPA in all water abstractions for human consumption, with a flow rate greater than 100 m^3/day or supplying more than 500 inhabitants: (a) Immediate protection area—the area of land surface contiguous to the catchment in which, for the direct protection of the abstraction facilities and the abstracted water, all activities are prohibited at a minimum distance from the capture of 20 and 60 m from the capture, depending on the aquifer type; (b) Intermediate protection area—an area of the surface of the land contiguous to the immediate protection zone, identified by a travel time of 50 days. Depending on vulnerability and hazard conditions, the protection radius must have a minimum value from 40 to 280 m; (c) Extended protection area—the area of land surface contiguous to the intermediate protection area, intended to protect the groundwater from persistent pollutants, such as organic compounds, radioactive substances, heavy metals, hydrocarbons, and nitrates. Certain activities and facilities may be prohibited or restricted under the risk of polluting the water, considering the nature of the terrain traversed and the type and number of pollutants that may be emitted. This protection area is defined by the transit time of contaminants in the aquifer equivalent to 3500 or by a minimum distance from the catchment of 330 to 2400 m.

The methods for delineating the WPA are divided into different categories according to their complexity and available hydrogeological data that can be used singly or collectively [14,15]. The most representative categories are: (a) geometric methods (e.g., arbitrarily fixed radius) that involve the use of a pre-determined shape and aquifer geometry, without any special consideration of the aquifer system [15]; (b) analytical methods (e.g., calculated fixed radius, Wyssling, Jacob and Bear, Krijgsman and Lobo-Ferreira, uniform flow equation) that allow calculation of distances for wellhead protection using simple equations that can be easily solved, but only valid for aquifers not affected by pumping [16,17]; (c) hydrogeologic mapping, which involves the identification of the catchment zone, based on geological, hydrogeological and hydro-chemical characteristics of the exploited aquifer [18]; (d) numerical models (e.g., MODFLOW, FEFLOW, HYDRUS), which involve the use complex numerical solutions to groundwater flow, particle tracking or contaminant transport [19,20].

In all cases, the representativeness of the defined protection zone is subject to the knowledge of local hydrogeological conditions. In addition, the defined WPZs are influenced by uncontrolled flow abstraction rates, sporadic or seasonal well operation [21], proximity to other pumping wells, the occurrence of hydraulic connections with other water bodies (surface water or groundwater), etc. Therefore, neglecting these situations render WPAs ineffective as a strategy to ensure the quality of water supplies.

The purpose of this work is to evaluate the effectiveness of WPZ, after a long period of establishment, in drinking water wells with continuous pumping located in densely populated urban areas where the conflict of interests occupation and management of the territory is often put aside from the basic necessary for the preservation of underground water systems. Many of the studies promote a comparison between different methodologies to delimit the WPA (e.g., [15,16,22]), however, a critical analysis of their efficiency in the medium and long term is lacking. Uncertainty prevails in the calculations given the heterogeneity of hydrogeological parameters or changes in the hydraulic behavior of the aquifer [20,23,24], therefore a continuous reassessment of local hydrogeological conditions may be necessary to reduce the sensitivity of the models.

2. Case Study: Public Groundwater Supply of Montijo Municipality

The municipality of Montijo (Portugal) is located on the south bank of the Tagus River, bordered by the Tagus Estuary, in the north of the Setubal Peninsula (Figure 1). Montijo is one of the territorially discontinuous municipalities in Portugal, and is geographically divided into two parts, western and eastern. As a result, the eastern division is not included in the study area. Within the Lisbon Metropolitan Area, Montijo (+29.7%) is among the municipalities that grew the most in terms of population between 2001 and 2011, registering 51,222 inhabitants to date (2011) [25]. More recent estimates [26], indicating that Montijo has 55,689 residents, demonstrate a new growing trend.

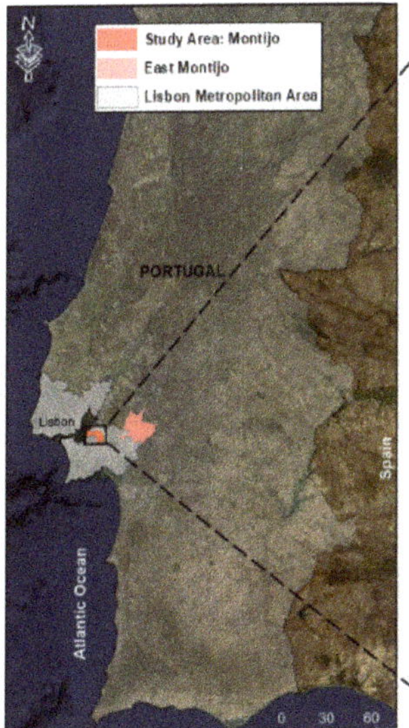

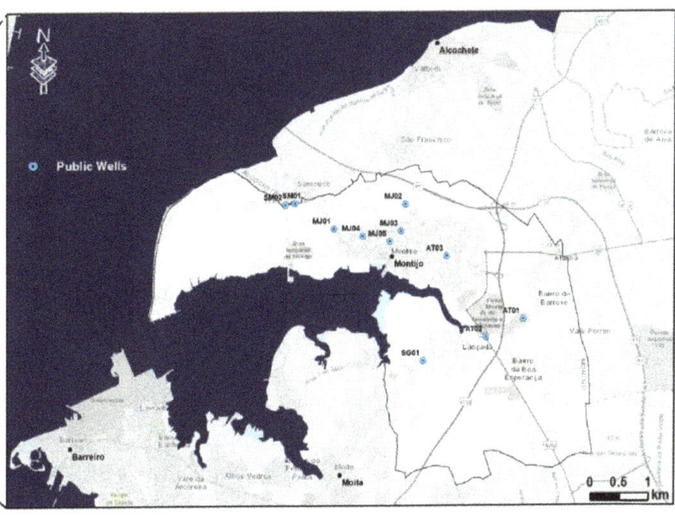

Figure 1. Geographical framework of the study area and location of the public wells supply.

2.1. Geological and Hydrogeological Framework

The study area is part of the Tagus-Sado Tertiary Basin, which forms an extensive depression elongated in the northeast-southwest direction, filled by materials from peripheral zones [27,28]. The Basin is flanked to the west and north by the Mesozoic formations of the Western Rim, to the northeast and the latter by the Hercynian substrate and communicating to the south with the Atlantic Ocean, in the Setubal Peninsula [27,28]. Over time, the Basin underwent a complex evolution controlled by the interaction of tectonic and eustatic movements, which resulted in the subsidence of lands located between faults [29–32]. The movements created by the orogenic activity gave rise to several sedimentary cycles, marked by constant alternations of depositional environments, continental, marine, and brackish facies, which reveal a succession of regressive and transgressive episodes during the Neogene [16,33–35]. The filling of the Basin is composed of Paleogenic and Neogene deposits (Miocene and Pliocene) with thicknesses that can reach 1200 m [36–39], covered discontinuously by Quaternary sediments [27,28].

The Pliocene sands of Santa Marta dominate the landscape of the study area (Figure 2). These materials extend for more than a hundred meters of depth, being composed of sands, from fine to coarse, arkhosic, with interspersed clayey levels. Occasionally, they may be covered by alluvium and river terrace deposits, consisting essentially of clayey sludge and sand of different sizes. Pliocene sediments lie in unconformity over Miocene deposits, constituted by calcareous, bioclastic, sometimes marly sandstones [28]. These occur at depths greater than 700 m, contacting in unconformity in the Montijo area with Paleogene formations or in the Pinhal Novo area with the Dagorda Formation, from the Lower Jurassic, through the Pinhal Novo—Alcochete fault zone [40]. The fault has a predominant NNW—SSE orientation, involving a deformation zone about 2 km wide, in which it presents a pattern of branched and anastomosed faults [31,40–42]. An important deformation is recognized in the Pliocene sediments that implies the occurrence of further activity in the fault zone, maintaining a predominant left-facing regime, with the basal surfaces witnessing a relative descent of the eastern fault block [28,40]. We consider the fault zone as highly unproductive, being filled with black clays that contrast with the sandstone nature of the aquifer, considering the water levels measured in the field.

The Montijo area is part of the Tagus-Sado/Left Bank aquifer system. The system is quite complex, characterized by several lateral and vertical variations of facies that significantly alter the characteristics of the aquifer. This is formed by several porous layers, confined or semi-confined, and by clayey intercalations of low permeability [43]. In the Setubal Peninsula, the system is composed of (Figure 3): (1) an unconfined aquifer, installed in the alluvial layers of the Pleisto-Holocene and the upper layers of the Pliocene, and in hydraulic connection with the Tagus River; (2) a confined multi-layered aquifer, hosted in the Pliocene basement layers and the Late to Middle Miocene limestone layers [27,44,45]. Separating the aquifer units are semi-permeable clay lenticules that form an almost continuous aquitard level, on average at 100 m in-depth, and with varying thicknesses between 1 and 80 m [46]. At the base of the confined aquifer is a marly level that assumes an aquiclude behavior and separates this aquifer from another, deeper and of lower quality, located in the Miocene base layers [27]. Recharge occurs by direct infiltration throughout the Basin, with a higher incidence in the Pliocene and Quaternary deposits of the highlands and plateaus that border the river, yielding part of this recharge through drainage to the underlying deposits [47]. At the Basin scale (Tagus-Sado Tertiary Basin), the flow occurs preferentially, in its transverse component, towards the Tagus River, the main drainage axis, yielding discharges in the alluvium, and according to a longitudinal component, towards the Atlantic Ocean. At a regional scale (Setubal Peninsula), the Tagus estuary and its tributaries are the primary flow directions of shallow and phreatic aquifers [27,47].

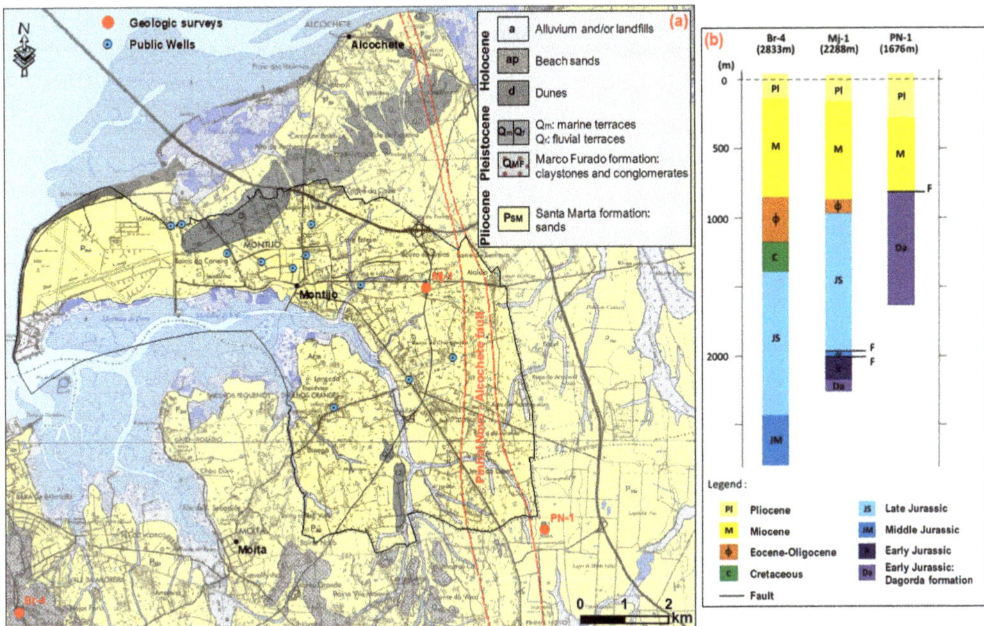

Figure 2. (**a**) Geological framework of the study area on sheet 34-D (Lisbon) of the Geological Chart of Portugal at 1:50,000 scale, (**b**) Synthetic stratigraphic columns from deep wells in the northern sector of the Setubal Peninsula [40].

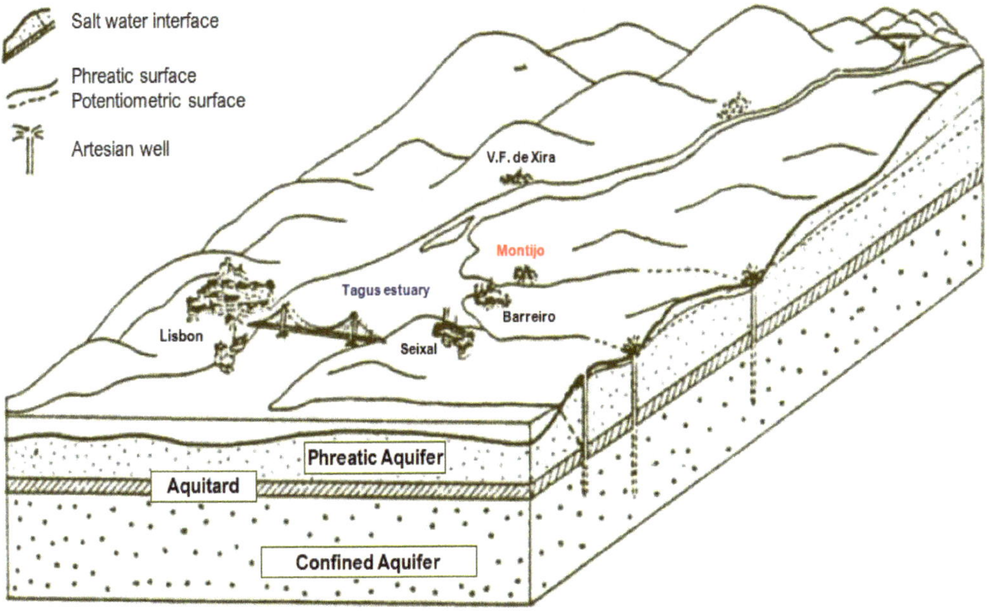

Figure 3. Schematic three-dimensional representation (conceptual model) of the Tagus-Sado Mio-Pliocene aquifer system [31].

2.2. Public Water Supply and Wellhead Protection Zones

The population of Montijo and Alcochete is supplied by groundwater. The public wells are divided into two water collection poles (Figures 1 and 4): Samouco (SM01 and SM02) and Montijo (MJ01, MJ02, MJ03, MJ04 and MJ05). All public wells explore the sandy layers of the base of the Pliocene, along with limestone layers of the top of the Miocene. This means that they only extract in the confined aquifer, a factor that significantly reduces vulnerability. According to the United States Environmental Protection Agency (USEPA) [8], in a truly confined aquifer, the WPA around the catchment does not play any protective function, and in that case, it is sufficient to define an area immediately underneath the catchment to prevent the transport of pollutants along the catchment column. However, Moinante [48] describes some factors that may put the groundwater quality at risk, even in confined aquifers. Aquifers can present local discontinuities (facies variation, fractures, or faults), or poorly constructed or abandoned wells that may hydraulically connect several aquifers.

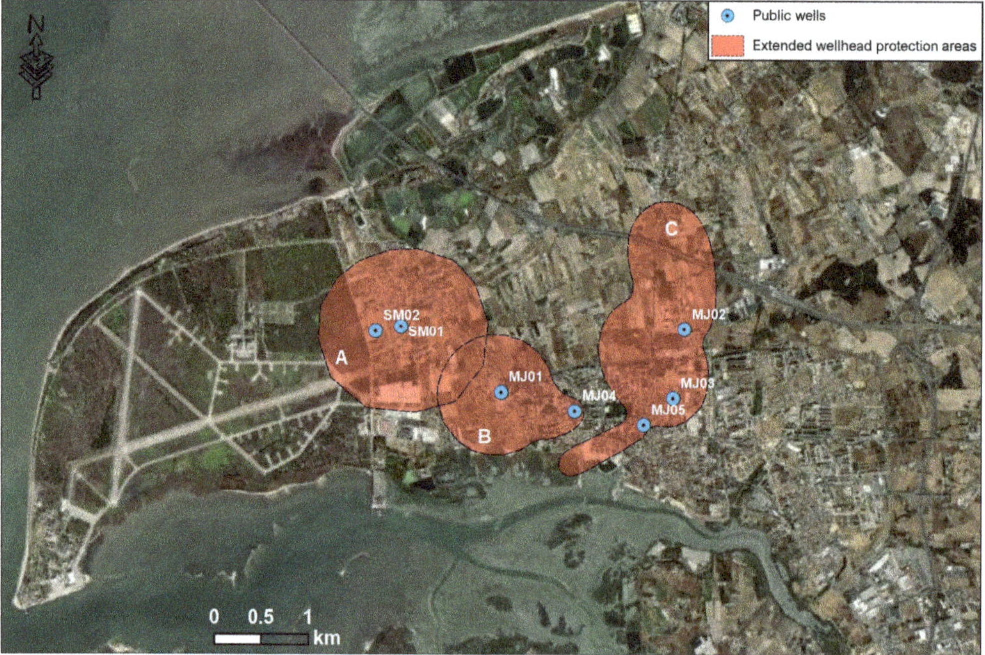

Figure 4. Current extended wellhead protection areas.

The WPAs for the public wells of Samouco and Montijo were implemented in 2009 and 2011 by the respective municipal councils (Alcochete and Montijo). These were obtained by applying three analytical methods: calculated fixed radius (CFR), Wyssling [49], and Bear and Jacobs [50]. At this time the aquifer was already exploited by several wells for public and private supply, and it was impossible to identify the regional gradient undisturbed by extractions. Therefore, the hydraulic gradient was estimated at a considerable distance from the public wells. Nevertheless, it was planned to combine the results of these three methods in the delimitation of the WPA, however, the results obtained by the Wyssling and CFR methods were poor and the Jacob and Bear method prevailed in their definition. Three extended WPAs were implemented with 2.32, 1.31, and 2.30 km^2, respectively (Figure 4). The Samouco pole is not within the administrative borders of Montijo, but within those of Alcochete. However, the extended WPA (A) implemented for the two public wells in this sector is a transboundary one, mostly delimited in the study area and intercepts one of the WPA; (B) implemented in the Montijo council. A third WPA; (C) is also delimited in a neighboring municipality (Alcochete). These perimeters, including immediate and intermediate WPAs, have been in effect since 2011 and remain unchanged to this day.

Table 1 contains the hydrogeological parameters considered in the calculations. As it was not possible to measure water levels in wells in situ, the hydraulic gradients for the Samouco pole were estimated using the piezometric levels of the closest monitoring stations of the national observation network. For the Montijo pole, a complete piezometric surface of the aquifer, already affected by public and private extractions, was developed through the interpolation of the values of the piezometric levels by ordinary kriging without anisotropy. The remaining parameters were obtained in the interpretation of pumping tests. As these protection perimeters are the result of separate studies, significant differences can be observed in the values of some parameters (porosity for example), with the authors opting for different interpretations.

Table 1. Hydrogeological parameters and wells technical features used in the application of analytical methods to calculate the WPAs in force.

Extended WPZ	Area (km^2)	Wells	Drains (m)	Flow Rate (m^3/Day)	Hydraulic Gradient	Aquifer Thickness (m)	Transmissivity (m^2/Day)	Hydraulic Conductivity (m/Day)	Porosity
A	2.32	SM01	206–259	3456	0.001–	72	370	5	0.10
		SM02	185–273	5184	0.0008	109	1140	10	
B	1.31	MJ01	110–220	3636	0.00047	43	614	14	0.25
		MJ04	121–233	774	0.00130	44	1008	23	
C	2.30	MJ02	141–248	3289	0.00203	43	1407	33	0.25
		MJ03	114–237	3589	0.00122	43	1653	38	
		MJ05		776	0.00208	43	1250	29	

3. Methodology

A simple approach to analyze the long-term effectiveness of WPAs is to recalculate them at present hydrogeological conditions. Within the calculations, most of the variables remain constant, particularly the hydraulic parameters of the aquifer. However, factors such as well flow rate, piezometric levels, and hydraulic gradients are dynamic and constantly changing over time. With population growth, water demand can increase significantly and well capacity may need to be adjusted to public needs. In turn, the piezometric surface tends to suffer alterations with the continuous exploitation of the system, being able to change the hydraulic behavior of the aquifer and the natural directions of the groundwater flow. Considering the aquifer scenario in 2019, new extended WPAs were calculated using the same three analytical methods that led to the delimitation of these in 2009–2011. It is intended to review whether the perimeters in force are still valid for the current environment, maintaining the hydraulic parameters used in the previous calculations and just changing the hydraulic gradient to current values and the azimuth that represents the flow direction. The analytical methods are valid for a regional hydraulic gradient before extraction, but for this case study, as in most urban areas of the Setubal Peninsula (Portugal), it is impossible in present times to estimate the hydraulic gradient before pumping. As an alternative to analytical methods, a numerical model was developed for the case study. Using the particle tracker tool to compute the catchment areas of each public well can lead to a better solution to the problem, anticipating some limitations of the proposed analytical methods.

3.1. Analytical Methods

An automatic program developed by C. Almeida (adapted from [51]) was used to solve the mathematical equations referring to the analytical methods. This program allows calculating the WPAs, as a function of the transit time of a contaminant. Considering the same approach used to delimit the current WPA, calculations were performed for the following methods:

(a) The Fixed Radius methods (CFR) (Dec.-Law 382/99), (Figure 5) define the WPA through a volumetric equation which calculates the volume of water that reaches the catchment in a certain time, which is considered necessary to reduce the contamination to an admissible level before reaching the catchment. It is assumed that the catchment is the only draining element of the aquifer, where all flow lines converge, and that there are no privileged flow directions. In this case, the WPAs are bounded by circles around the well, with radius calculated from Equation (1), where r is the radius of WPA (m), Q is the well flow rate (m^3/day), t is the time required for a pollutant to reach the well (days), n the effective porosity (%), and b is the saturated thickness (m).

$$r = \sqrt{\frac{Q\,t}{\pi\,n\,b}} \tag{1}$$

(b) The Wyssling method [49] (Figure 5) consists of calculating the catchment zone of a well whose size is a function of the propagation time of a contaminant in the aquifer. It is a simple method, applicable to homogeneous porous aquifers, that has the disadvantage of not considering the heterogeneities of the aquifer. The use of this method presupposes knowledge of the hydraulic gradient (i), the well capacity (Q), the hydraulic conductivity (K) or Transmissivity (T), the effective porosity (n) and the aquifer thickness (b). The variables that allow for the drawing of the WPAs are the height of the capture zone (B), the width of the capture zone front to the height of the well (B′), the capture zone radius (Xo), the Darcy velocity (V), the distance (x) corresponding to time t, in the direction of flow (upstream of capture) (So) and in the opposite direction to the flow (downstream of the catchment) (Su).

$$B = \frac{Q}{K\,b\,i} \tag{2}$$

$$B' = \frac{B}{2} \tag{3}$$

$$Xo = \frac{Q}{2\pi\,K\,b\,i} \tag{4}$$

$$V = \frac{K\,i}{n} \tag{5}$$

$$x = V\,t \tag{6}$$

$$So = \frac{+x + \sqrt{x(x + 8Xo)}}{2} \tag{7}$$

$$Su = \frac{-x + \sqrt{x(x + 8Xo)}}{2} \tag{8}$$

(c) The Bear and Jacobs method [50] is based on the definition of the capture zone induced by the capture to be protected. A capture zone is the volume of the aquifer through which groundwater flows to a pumped well during a given time of travel. To simplify the analytical model, a series of simplifying assumptions are made. Thus, the Bear and Jacobs method is applied to the case of a single catchment located in a homogeneous, isotropic aquifer of infinite extent, subjected to a uniform regional gradient. This area is delineated using the capture zone curve. The equations for the capture zone curve were derived by Bear and Jacobs [50] and is as follows:

$$\exp(X_R - t_R) = \cos(Y_R) + \frac{X_R}{Y_R}\sin(Y_R) \tag{9}$$

$$X_R = \frac{2\pi\,V\,b}{Q}x \tag{10}$$

$$Y_R = \frac{2\pi\,V\,b}{Q}y \tag{11}$$

$$t_R = \frac{2\,\pi\,V^2\,b}{n\,Q}t \tag{12}$$

where V represents the Darcy velocity (Equation (5); m/d), Q the flow rate (m^3/day), n the effective porosity, b the aquifer thickness (m), Y_R, X_R and t_R the reduced variables, x and y the real distances (m). Equation (9) cannot be solved in an explicit form in order to X_R e Y_R, and so iterative methods must be used. The solution was based on the method proposed by [51]. The relative Cartesian coordinates x and y are computed from Equations (10) and (11). These relative Cartesian coordinates then are rotated by the directional angle relative to the north of the hydraulic gradient translated by a distance equal to the well's positional coordinates and converted to longitude and latitude pairs. These longitude and latitude pairs can be plotted and connected by hand but were designed to be used by a geographic information system to generate a line that can be used to delineate the WPA.

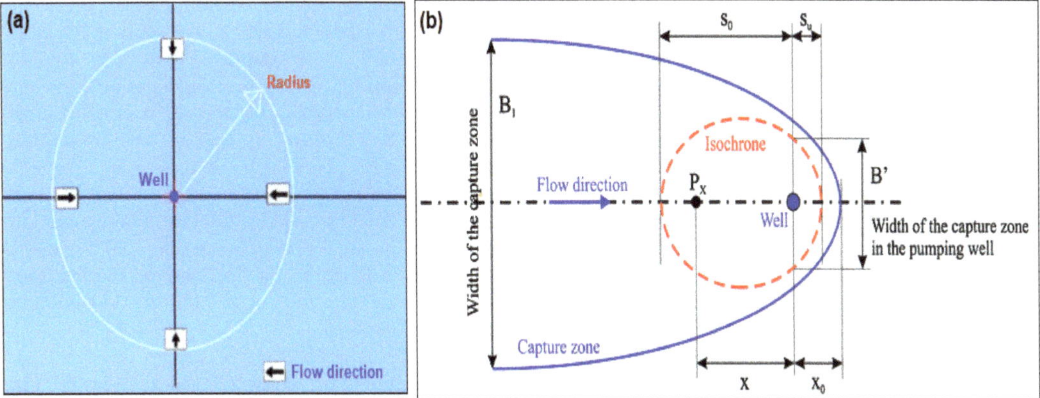

Figure 5. (**a**) The Fixed Radius methods and (**b**) WPA according to Wyssling's method [2].

3.2. Numerical Modeling

Numerical modeling was the methodology used as an alternative to WPA calculation and to simulate the current piezometric surface of the aquifer. The exercise was carried out using the FEFLOW software [52], a three-dimensional and two-dimensional simulation tool, which uses the finite element numerical method (FEM) to solve the partial differential equations that describe the groundwater flow.

The domain of the model corresponds to a small portion of the entire aquifer system located at the northern end of the Setubal Peninsula. It is delimited by the Tagus Estuary, a discharge boundary in natural conditions, and by the Pinhal Novo—Alcochete Fault, due to the potential waterproofing in the fracture zone. The model was discretized in five layers (Figure 6) with 3 aquifer subsystems: (1) a phreatic, superficial subsystem, whose thickness should not exceed two or three tens of meters that were traditionally exploited by shallow wells; (2) an intermediate semi-confined subsystem, fully inserted in the Pliocene formations, and used mostly by private abstractions; (3) a lower confined subsystem, installed in Pliocene base deposits and upper Miocene limestone formations, and exploited by public abstractions. To separate each of the three aquifer subsystems, two aquitard-type layers were considered, allowing different flow directions for each subsystem, confinement, and the need to change the boundary conditions that characterize each layer.

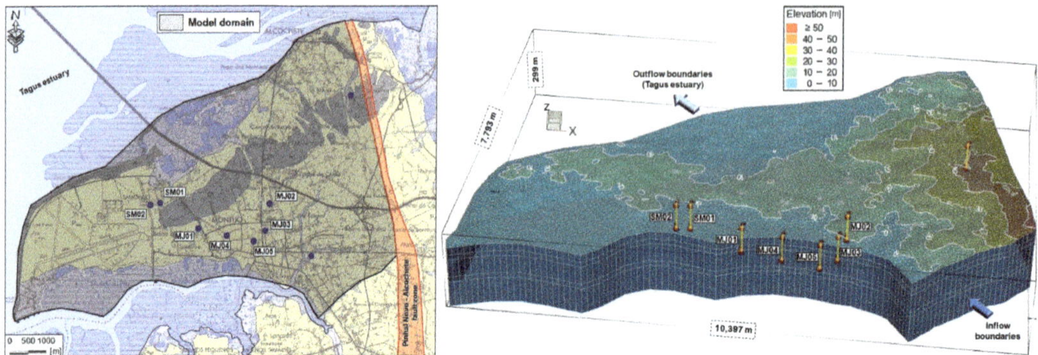

Figure 6. Schematic representation of the model domain in two and three dimensions.

Constant potential boundary conditions (Dirichlet, type 1) were imposed on the model boundaries. In the first layer (phreatic aquifer) a constant potential of 0 m was positioned in contact with the Tagus estuary, while the potential of the remaining surface water lines was defined as a function of topography (altitude). In the remaining layers, the boundary conditions are positioned to the southeast, a lateral feeding area, and to the north or northwest, in an area of potential discharge that can extend to the middle of the Tagus Estuary (Table 2). The initial conditions seek to simulate the groundwater flow in a natural regime without exploitation. The initial gradient was obtained through a set of observations of the static levels in 2019, which is quite low (0.0016). Typical of a discharge area, the hydraulic potential to the south is higher than the potential next to the estuary. The recharge in the top layer, that is, in the phreatic aquifer subsystem, was obtained through a zoning for the study area of recharge values determined by LABCARGA [53], a new methodology employed to evaluate the recharge of groundwater bodies in mainland Portugal. Due to different scales used in the geological and land use maps, data entered in the numerical model had to be slightly adjusted compared with the values obtained in that research, which spatially vary between 102 and 156 mm/year.

Table 2. Boundary conditions applied in the numerical model; BC is Boundary Condition; ✗ means not applied; ✓ is applied.

| Layer | Hydraulic-Head BC-Dirichlet | | | | Well BC |
	South	Southeast	East (Fault)	Northwest	Nodal Source/Sink Type
1 (phreatic)	✓ (estuary)	✓ (estuary)	✗	✓ (estuary)	✗
2 (aquitard)	✗	✓	✗	✓	✗
3 (aquifer)	✗	✓	✗	✓	✓
4 (aquitard)	✗	✓	✗	✓	✗
5 (aquifer)	✗	✓	✗	✓	✓

The flow rates entered in the model for each public well are the daily mean values from 2018, indicated in Table 3 as current values. It was impossible to calibrate the model with the values used to previously delimit the WPAs, also known as maximum values for a worst-case scenario. These maximum values are never reached by the abstractions, therefore they could never be used to simulate the current piezometric surface of the lower aquifer sub-system. The model calibration was based on the adjustment of boundary conditions of each aquifer, by trial and error, and calibration by the inverse method (FEPEST) of hydraulic conductivity. For this purpose, 21 observation points were selected, seven for the unconfined aquifer, eight for the intermediate subsystem, and six for the lower confined aquifer. In Figure 7 it is possible to observe the piezometric surface of the lower subsystem, captured by the public supply schemes and the main object of study of this work. The influence of water exploitation on water levels is notorious. The catchment area of the public wells is clearly drawn down in relation to the periphery, developing a radial shape flow towards these. This scenario is much more aggressive than the one observed when the WPAs were implemented. As an example, the water depth in MJ01, MJ02, and MJ03 increases from 21.5 to 33.3, 26.5 to 50.5, and 24.3 to 30.6 m, respectively, between 2011 and 2019. The continuous decrease in piezometric levels because of overexploitation of the aquifer has been a problem in several areas of the Setubal Peninsula [45,54,55].

Table 3. Minimum and maximum flowrates for the wells and comparison between the hydraulic gradients and groundwater flow direction, used for delimitation of the actual WPAs (2011), and data from 2019.

Extended WPA	Wells	Minimum Flow Rate (m³/Day)	Maximum Flow Rate (m³/Day)	Hydraulic Gradient (2011)	Flow Direction	Hydraulic Gradient (2019)	Flow Direction
A	SM01	351	3456	0.001–0.0008	–	0.00789	SE
	SM02	505	5184		–	0.00781	SSE
B	MJ01	1047	3636	0.00047	NE	0.00178	NNE
	MJ04	57	774	0.00130	ENE	0.00083	NW
C	MJ02	2089	3289	0.00203	SSE	0.00495	SW
	MJ03	2828	3589	0.00122	SE	0.00383	WNW
	MJ05	503	776	0.00208	NE	0.00228	NW

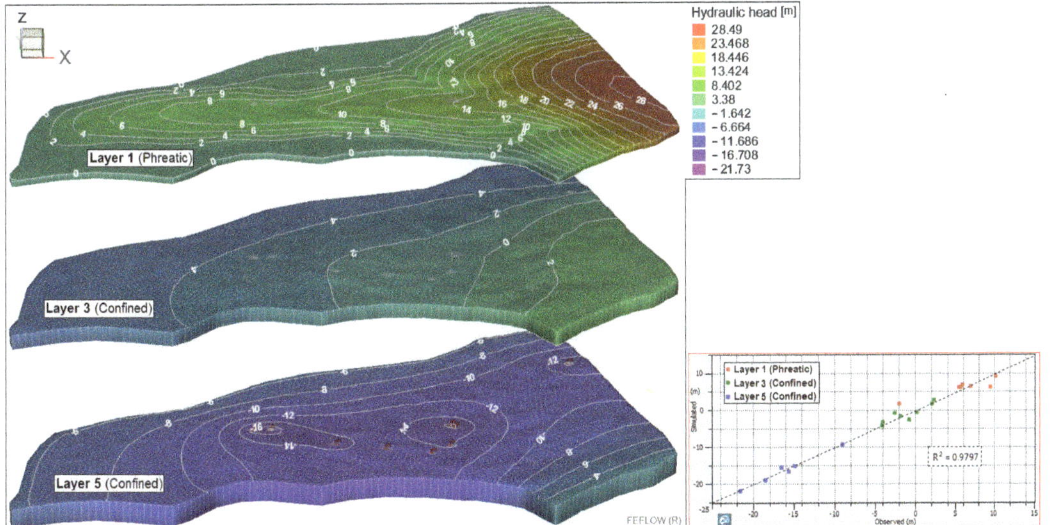

Figure 7. Groundwater flow model and calibration values.

4. Results

The results presented are divided into two chapters: analytical and numerical methods. It is important to mention that the WPAs of the Samouco pole (A) cannot be defined together with the protection perimeters of Montijo (MJ01 to MJ05), even when they intersect, since they belong to different municipal councils. On the other hand, any public well included in the extended WPAs B and C may have their perimeters delimited separately or together when they intersect.

4.1. Redefining Extended WPAs Using Analytical Methods

The hydraulic gradient (and the azimuth) is the only parameter to be changed from the previous calculations to be able to establish a comparison. In the first half of the last century, artesian wells were observed in many areas of the Setubal Peninsula [27]. Currently, due to the intensive exploitation of aquifers, drawdowns are accentuated in many urban areas.

Considering that the piezometric surface presents a radial shape around the public wells, it is difficult to identify the main flow direction for the hydraulic gradient. Hence the importance of using numerical modeling first to identify the hydraulic gradient and the catchment areas of each well. In Figure 8 it is possible to observe the flow direction towards wells, for which the hydraulic gradient was calculated. The corresponding values are shown in Table 3.

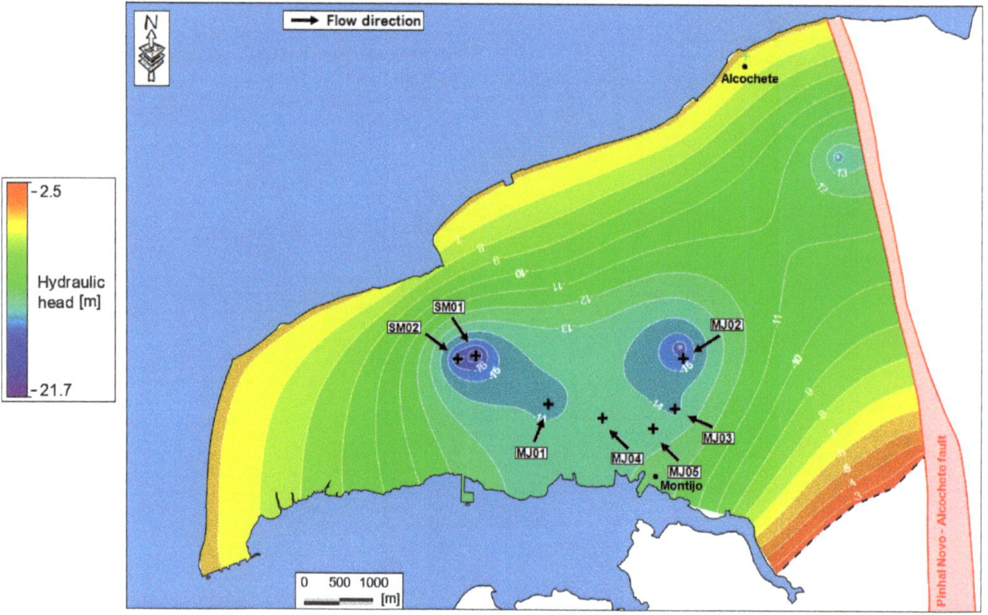

Figure 8. Piezometric surface of confined aquifer and main flow direction captured by public wells.

In general, the flow direction with the wells producing in 2019 is different from the one in 2011, with higher hydraulic gradients. These values were expected due to the strong drawdown observed in the last years, due to the aquifer exploitation. The greater the drawdown around the public well, the greater the hydraulic gradient generated. Considering the location of the wells, the Montijo pole is essentially fed by water from E, mostly from SE, while the Samouco pole is fed by water from NW. Together, the wells develop a joint drawdown zone, causing a depression in the piezometric surface of the aquifer in relation to less exploited areas. The well MJ01 is centered in this depression, which makes it difficult to identify the main flow direction in the surrounding area. The catchment area, visible in Figure 8, dictates that the flow is radial, although there is a slight increase towards SW, which is the direction for which the hydraulic gradient was calculated. Keeping the remaining variables constant used to determine the WPAs in 2011, the new extended WPAs were analytically calculated (Figures 9–11) considering the current hydraulic gradients. The CFR method does not have the hydraulic gradient as a variable, therefore, if the flow rate (Q) is the same, the results obtained are similar to the previous ones. The Wyssling method presents interesting results if we do not consider the extended WPA obtained for well MJ01. The high flow rate together with a low hydraulic gradient and permeability produce a very flat protection area for this well. Considering that Jacob and Bear's method was used in 2009–2011 to delimit the WPAs in force, the differences in the results obtained are substantial. The hydraulic gradient has a strong influence on the calculations, causing more elongated protection areas that almost never intersect.

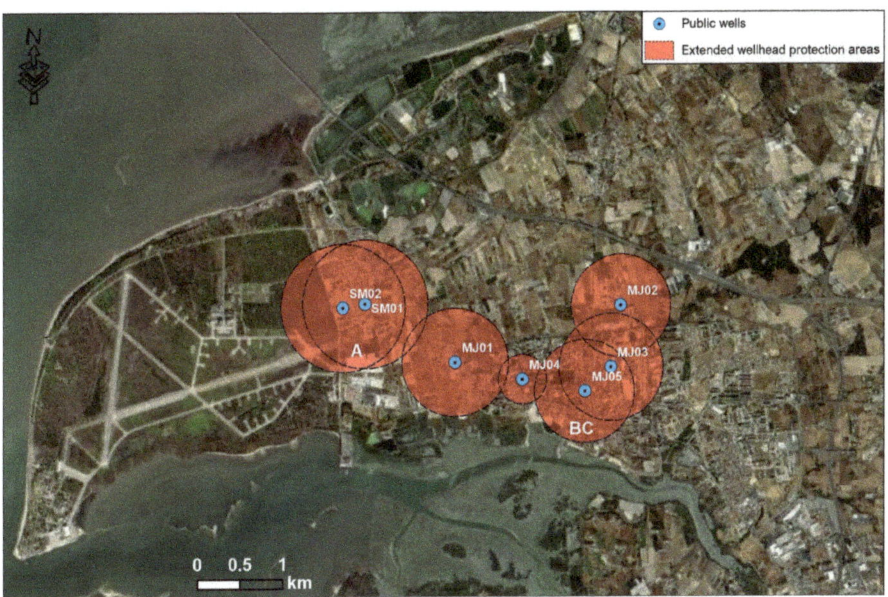

Figure 9. Extended wellhead protection zones delimited by the Fixed Radius method.

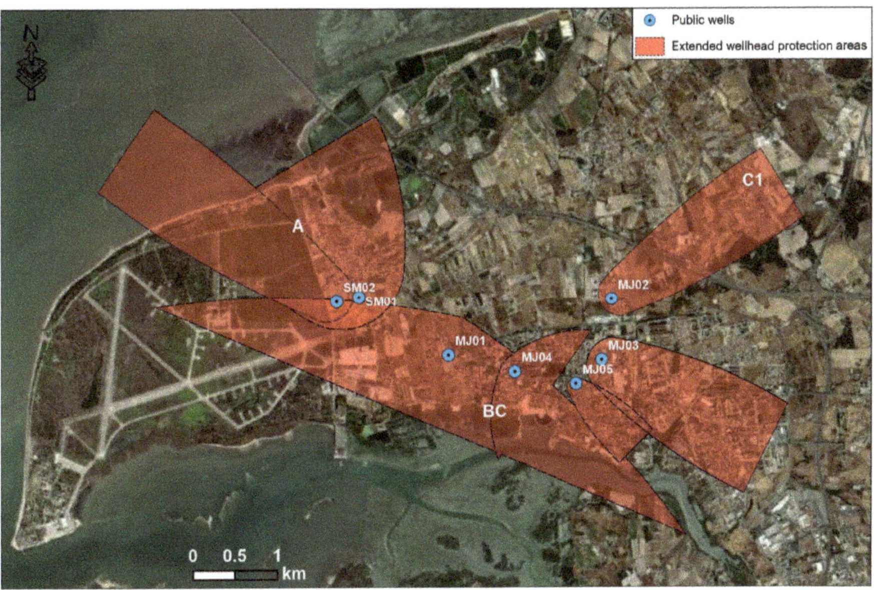

Figure 10. Extended wellhead protection zones delimited by the Wyssling method.

Figure 11. Extended wellhead protection zones delimited by the Jacob and Bear method.

4.2. Redefining Extended WPAs Using Numerical Modelling

The use of a transport model dependent on the piezometric values of a flow model allows for the defining of the trajectory of particles launched at a given point. This tool is known as a particle tracker and returns a set of regressive flow lines over a time interval. Considering the flow model developed, this exercise was carried out for a period equivalent to 3500 days, which delimits the extended WPAs. The confined aquifer, the target of exploration, corresponds in numerical terms to layer 5, which is delimited by slices 5 and 6. The particle tracker performs the numerical calculations on the slices, although the hydraulic properties are assigned to the layers. The lower aquifer subsystem is represented by layer 5 but delimited by slices 5 and 6 where numerical calculations are carried out; consequently, two solutions (options) are generated. This means that in a two-dimensional perspective we have two catchment areas, one for each slice, with different dimensions. Slice 5 is superimposed by a layer of low permeability (aquitard), which significantly reduces the flow velocity, something not observed for slice 6. In Figure 12 it is possible to observe the catchment area of the seven public wells obtained in the simulation performed. The results (Figure 13) demonstrate new formats for extended WPAs and a possible division of zone B in two (1st option), although a unification between zones B and C (2nd option) is also possible, grouping all public wells in Montijo (MJ01 to MJ05). Numerical modeling offers a solution that is more compatible with the hydraulic reality of the aquifer, with protected areas more centered around the radial drawdown zones caused by public wells.

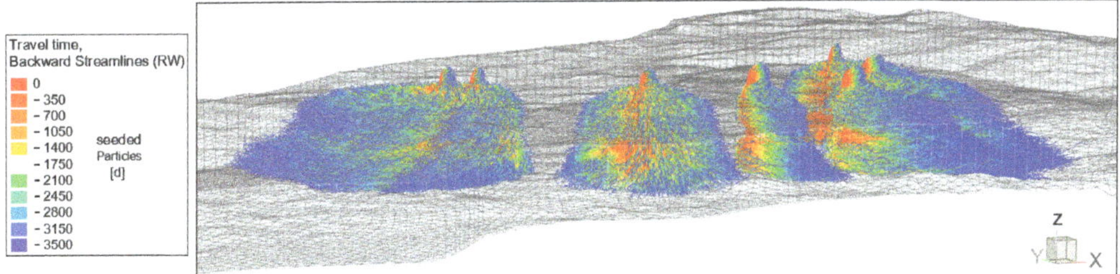

Figure 12. Using the particle tracker to calculate a catchment area for public wells in 3500 days.

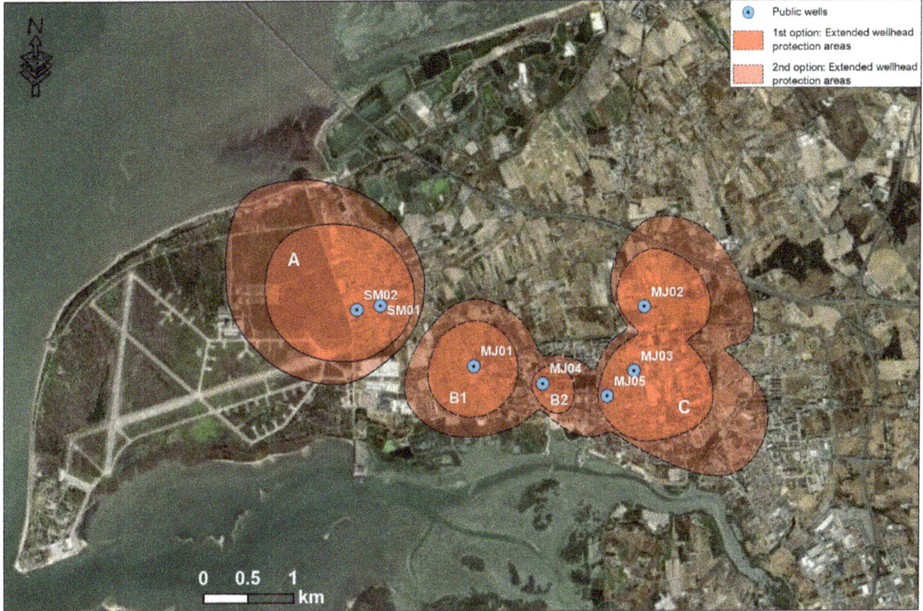

Figure 13. Extended wellhead protection zones calculated by numerical modelling.

5. Discussion

The CFR is the only method that does not depend on the hydraulic gradient for its calculation. As the properties of the aquifer do not change in this model, the results of this method could only vary if we considered another well capacity. If the flow rate increases, the protection radius and circumference also increase. Although analytical methods are only valid in non-disturbed regional gradient situations, their use here was mainly for the purpose of comparison with those calculated in 200–2011 and currently in force. The WPAs obtained for the 2019 scenario occupy a total protected area of about 5.78 km^2, similar in size to the perimeters in force (5.93 km^2) but with different formats in the wells MJ02 to MJ05. The extended WPAs outlined by the Wyssling method significantly increase the protection area, particularly for the Samouco public wells (SM01 and SM02) and MJ01. This method was also rejected in 2009–2011 due to poor results. The width of the catchment area presents an exaggerated dimension for these wells, given the combination of a flow rate that is too high for the hydraulic gradient in question. The Jacob and Bear method is the one that can provide a better comparison with the protection perimeters in force, given that it was the method that prevailed previously. The WPAs calculated better reflect the hydraulic gradient of the aquifer, however, the width of the protected areas is much narrower than before. Because of this, the WPA appears more individualized for each well, with single areas varying between 0.25 to 3.14 km^2, for an increase in the protection area to 7.04 km^2 (Table 4).

Table 4. Comparison of the areas occupied by the extended WPAs determined by the different numerical and analytical methods used.

Method	WPA	Wells	Area (km^2)	Total (km^2)
2011 (Jacob and Bear predominant)	A	SM01 and SM02	2.32	
	B	MJ01 and MJ04	1.31	5.93
	C	MJ02, MJ03 and MJ05	2.3	
Fixed Radius	A	SM01 and SM02	2.06	
	BC	MJ01, MJ02, MJ03, MJ04 and MJ05	3.72	5.78
Wyssling	A	SM01 and SM02	5.34	
	BC	MJ01, MJ03, MJ04 and MJ05	8.01	15.37
	C1	MJ02	2.02	
Jacob and Bear	A	SM01 and SM02	3.12	
	B1	MJ01	1.18	
	B2	MJ04	0.25	
	C1	MJ02	1.07	7.04
	C2	MJ03	1.17	
	C3	MJ05	0.25	
Numerical (first option)	A	SM01 and SM02	1.99	
	B1	MJ01	0.83	
	B2	MJ04	0.19	4.98
	C	MJ02, MJ03 and MJ05	1.97	
Numerical (second option)	A	SM01 and SM02	3.97	
	BC	MJ01, MJ02, MJ03, MJ04 and MJ05	6.27	10.24

Therefore, according to the results obtained, are the extended 2009–2011 WPAs aligned with the current reality of the aquifer system? None of the analytical methods used were even close to the original solution, so it is safe to say that they are not suitable for the current situation of the aquifer. It is in the best interest to question the long-term effectiveness of these methods, particularly in urban areas where operating conditions are frequently changing. The method applied to simulate the aquifer piezometry, and consequently the hydraulic gradients, is different from the one used in 2011 (kriging). Therefore, the results may be based in this direction. Even so, both methods are valid to interpolate the piezometric levels and manage to replicate the hydraulic gradients around the catchments. It is not intended to compare methodologies here, only to simulate the piezometric surface of this aquifer.

The WPAs generated by numerical modeling consider a possible division of current zone B in two (1st option), with the protected areas varying from 0.19 km^2 to 1.99 km^2, adding up to a total of 4.98 km^2 (Table 4). If a unification between zones B and C are considered (2nd option) the total protected area has around 10.98 km^2. As with analytical methods, numerical model weight variables such as permeability, porosity, hydraulic gradient, or well capacity, also add an important parameter previously unweighted, such as recharge. It has the advantage that it is not necessary to define a uniform flow direction towards the catchment since they ponder the effects from the combined pumping wells in the hydraulic gradients and the superposition of the depression cones. Therefore, the results obtained by numerical modeling can better replicate what the extended WPAs should be for the current situation of the aquifer.

Time of travel calculations for homogenous aquifers with significant secondary porosity and heterogeneous aquifers may be significantly protected wellhead areas because hydrodynamic dispersion tends to be more significant than retardation in such aquifers. Hydrodynamic dispersion is significant in these aquifers for several reasons (1) highly permeable porous zones and fracture/conduit flow result in localized velocities that are significantly higher than the average groundwater velocity, (2) retardation processes are reduced in permeable zones (gravels, sands, fractures, conduits) because permeable aquifer materials tend to be less geochemically reactive. For example, the cation exchange capacity (CEC) of a sandy permeable zone in an aquifer will be significantly lower than the CEC of less permeable fine-grained sediments. It is necessary to choose higher-than-measured hydraulic conductivity values or use values in the upper range of similar aquifer materials when the potential for hydrodynamic dispersion is higher.

6. Conclusions

The delineation of WPA is still one of the best tools to reduce the risk of contamination, or, if it does occur, to prevent it from reaching the catchments in concentrations considered dangerous. The shape and size of the WPA depend on the hydraulic parameters of the aquifer, the flow rate of the public supply wells, and the groundwater level. Any changes of these parameters undermine the effectiveness and purpose of the WPA. With population growth in urban areas, and considering climate change, the pressure on groundwater abstraction increases. Changes in the flow direction, hydraulic behavior, and hydraulic gradients occur more often. Therefore, it is imperative to review the WPA with adequate frequency and to check its effectiveness in the medium and long term. Good governance of groundwater resources assumes the active participation of all relevant stakeholders, varying from mandated government institutions to end-users of groundwater and those who value groundwater-related ecosystems.

In the present case study, the hydraulic gradients of the aquifer changed from 0.0005–0.002 in 2009–2011, to 0.0008–0.008 in 2019, and consequently the direction of the groundwater flow in some areas. New WPA must be defined, however, integrated methodologies are needed for the delimitation of the new protected areas, taking into consideration the already established land use in the city. The WPAs delineated by the CFR method do not provide adequate protection. The Wyssling method calculates very large areas, particularly in cases where the hydraulic gradients are low. The Jacob and Bear method may be a valid solution. However, there is no uniform flow direction towards the wells, as these methods require, and the local hydraulic gradients are affected by neighboring wells. The best protection of the public supply wells seems to be obtained through numerical modeling. In a balance between well protection and land use, i.e., to maintain the quality of the water captured in the wells and at the same time not creating problems in the land use planning, the best WPA delineation for the current scenario is represented in Figure 13. Considering the two proposed solutions, the protected areas can range from four smaller zones with a total of 4.98 km^2 to two larger zones with 10.24 km^2.

Many WPA for public supply were and are delineated by analytical methods; however, situations where the wells are in aquifers with piezometry not affected by other extractions should be rare. The best methodology for delineating the WPA is using numerical flow modelling together with particle tracking simulation. This method considers intrinsic aquifer parameters, but also the water balance. Considering that aquifer recharge is an important factor in the infiltration and transport of contaminants into the aquifer, any methodology for estimating WPAs should consider it and not only parameters that influence the movement of contaminants within the aquifer. In urban areas, where the hydrostatic level concept no longer applies and the catchment areas of public wells overlap each other, numerical modeling may even be the only viable solution. Once the model is built, it can be reused multiple times, varying only the exploration conditions, with the opening of new wells or the increase/decrease in the well's capacity.

Thus, the combination of WPA methods, preferably based on groundwater flow models, with methodologies for assessing the vulnerability of aquifers to contamination seems to be a good solution. The effectiveness of the WPA should also be verified through a close monitoring of the physical-chemical composition of the abstracted water, either at the abstraction itself or at sampling points (piezometers, boreholes, wells) within the WPA.

Author Contributions: All the authors contributed to the write of the paper. This paper was completed under the supervision of M.d.R.C., J.M.C. and C.A.; J.Z. conducted the calculation and numerical modeling, figures, and analyzed the results and wrote the paper; C.A. developed a software for analytical methods; M.P. contributed to the analysis discussion; M.d.R.C. helped with reviewing the manuscript. All authors have read and agreed to the published version of the manuscript.

Funding: This research received no external funding.

Acknowledgments: The authors are grateful for the financial support of the Portuguese Environment Agency (APA) and Portuguese Foundation for Science and Technology (FCT) projects (UIDB/50019/2020-IDL). This study was carried out partially under the framework of the Labcarga I ISEP re-equipment program (IPP-ISEP I PAD'2007/08) and GeoBioTec I UA (UID/GEO/04035/2020). We acknowledge the anonymous reviewers and editors for their constructive comments which helped to improve the focus of the manuscript.

Conflicts of Interest: The authors declare that they have no conflict of interest.

References

1. UNESCO World Water Assessment Programme. *The United Nations World Water Development Report 2021: Valuing Water*; UNESCO: Paris, France, 2021; p. 187. ISBN 978-92-3-100434-6.
2. Martínez Navarrete, C.; Grima Olmedo, J.; Durán Valsero, J.J.; Gómez Gómez, J.D.; Luque Espinar, J.A.; de la Orden Gómez, J.A. Groundwater protection in Mediterranean countries after the European water framework directive. *Environ. Geol.* **2008**, *54*, 537–549. [CrossRef]
3. Boretti, A.; Rosa, L. Reassessing the projections of the World Water Development Report. *NPJ Clean Water* **2019**, *2*, 15. [CrossRef]
4. Herbert, C.; Döll, P. Global Assessment of Current and Future Groundwater Stress with a Focus on Transboundary Aquifers. *Water Resour. Res.* **2019**, *55*, 4760–4784. [CrossRef]
5. Bertrand, G.; Hirata, R.; Pauwels, H.; Cary, L.; Petelet-Giraud, E.; Chatton, E.; Aquilina, L.; Labasque, T.; Martins, V.; Montenegro, S.; et al. Groundwater contamination in coastal urban areas: Anthropogenic pressure and natural attenuation processes. Example of Recife (PE State, NE Brazil). *J. Contam. Hydrol.* **2016**, *192*, 165–180. [CrossRef] [PubMed]
6. Shultz, S.D.; Lindsay, B.E. The willingness to pay for groundwater protection. *Water Resour. Res.* **1990**, *26*, 1869–1875. [CrossRef]
7. Velis, M.; Conti, K.I.; Biermann, F. Groundwater and human development: Synergies and trade-offs within the context of the sustainable development goals. *Sustain. Sci.* **2017**, *12*, 1007–1017. [CrossRef] [PubMed]
8. U.S. Environmental Protection Agency. 1987 Final Rule: Emergency and Hazardous Chemical Inventory Forms and Community Right-To-Know Reporting Requirements 52 FR 38344. Available online: https://www.epa.gov/sites/default/files/2013-09/documents/52fr38364.pdf (accessed on 7 February 2022).
9. EU—European Commission. Report from the Commission. Implementation of Council Directive 91/676/EEC Concerning the Protection of Waters against Pollution Caused by Nitrates from Agricultural Sources. Synthesis from Year 2000 Member States Reports. Available online: https://eur-lex.europa.eu/legal-content/EN/TXT/PDF/?uri=CELEX:52002DC0407&from=EN (accessed on 7 February 2022).
10. Colon, M.; Richard, S.; Roche, P.A. The evolution of water governance in France from the 1960s: Disputes as major drivers for radical changes within a consensual framework. *Water Int.* **2018**, *43*, 109–132. [CrossRef]
11. García-García, A.; Martínez-Navarrete, C. Protection of groundwater intended for human consumption in the water framework directive: Strategies and regulations applied in some European countries. *Pol. Geol. Inst. Spec. Pap.* **2005**, *18*, 28–32. Available online: https://www.pgi.gov.pl/en/docman-tree-all/publikacje-2/special-papers/79-garcia/file.html (accessed on 2 February 2022).
12. Cleary, T.; Cleary, R.W. Delineation of Wellhead Protection Areas: Theory and Practice. *Water Sci. Technol.* **1991**, *24*, 239–250. [CrossRef]
13. Taylor, R.; Cronin, A.; Pedley, S.; Barker, J.; Atkinson, T. The implications of groundwater velocity variations on microbial transport and wellhead protection—Review of field evidence. *FEMS Microbiol. Ecol.* **2004**, *49*, 17–26. [CrossRef] [PubMed]
14. Foster, S.; Hirata, R.; Gomes, D.; Paris, M.; D'Elia, M. *Groundwater Quality Protection—A Guide for Water Utilities, Municipal Authorities, and Environment Agencies*; World Bank GW-MATe Report 2007; World Bank: Washington, DC, USA, 2007; ISBN 0-8213-4951-1.
15. Fileccia, A. Some simple procedures for the calculation of the influence radius and well head protection areas (theoretical approach and a field case for a water table aquifer in an alluvial plain). *Acque Sotter. Ital. J. Groundw.* **2015**, *4*, 7–23. [CrossRef]
16. Liu, Y.; Weisbrod, N.; Yakirevich, A. Comparative Study of Methods for Delineating the Wellhead Protection Area in an Unconfined Coastal Aquifer. *Water* **2019**, *11*, 1168. [CrossRef]
17. Krijgsman, B.; Lobo-Ferreira, J.P.C. *A Methodology for Delineating Wellhead Protection Areas*; INCH 7; Laboratório Nacional de Engenharia Civil: Lisbon, Portugal, 2001; ISBN 972-49-1882-3.
18. Expósito, J.L.; Esteller, M.V.; Paredes, J.; Rico, C.; Franco, R. Groundwater Protection Using Vulnerability Maps and Wellhead Protection Area (WHPA): A Case Study in Mexico. *Water Resour. Manag.* **2010**, *24*, 4219–4236. [CrossRef]

19. Piccinini, L.; Fabbri, P.; Pola, M.; Marcolongo, E. Numerical modeling to well-head protection area delineation, an example in Veneto Region (NE Italy). *Rend. Online Soc. Geol. Ital.* **2015**, *35*, 232–235. [CrossRef]

20. Thibaut, R.; Laloy, E.; Hermans, T. A new framework for experimental design using Bayesian Evidential Learning: The case of wellhead protection area. *J. Hydrol.* **2021**, *603*, 126903. [CrossRef]

21. Ramanarayanan, T.S.; Storm, D.E.; Smolen, M.D. Seasonal Pumping Variation Effects on Wellhead Protection Area Delineation. *Water Resour. Bull.* **1995**, *31*, 421–430. [CrossRef]

22. Tosco, T.; Sethi, R. Comparison between backward probability and particle tracking methods for the delineation of well head protection areas. *Environ. Fluid Mech.* **2010**, *10*, 77–90. [CrossRef]

23. Evers, S.; Lerner, D.N. How Uncertain Is Our Estimate of a Wellhead Protection Zone? *Groundwater* **2005**, *36*, 49–57. [CrossRef]

24. Theodossiou, N.; Fotopoulou, E. Delineating well-head protection areas under conditions of hydrogeological uncertainty. A case-study application in northern Greece. *Environ. Process.* **2015**, *2*, 113–122. [CrossRef]

25. INE. Censos 2011 Resultados Definitivos—Região Lisboa. Available online: https://censos.ine.pt/ngt_server/attachfileu.jsp?look_parentBoui=156652050&att_display=n&att_download=y (accessed on 18 January 2022).

26. INE. Censos 2021 Resultados Provisórios. Available online: https://www.ine.pt/ngt_server/attachfileu.jsp?look_parentBoui=536757700&att_display=n&att_download=y (accessed on 18 January 2022).

27. Almeida, C.; Mendonça, J.J.L.; Jesus, M.R.; Gomes, A.J. *Sistemas Aquíferos de Portugal Continental*; Centro de Geologia da Universidade de Lisboa; Instituto Nacional da Água: Lisboa, Portugal, 2000. [CrossRef]

28. Pais, J.; Moniz, C.; Cabral, J.; Cardoso, J.; Legoinha, P.; Machado, S.; Morais, M.A.; Lourenço, C.; Ribeiro, M.L.; Henriques, P.; et al. *Notícia Explicativa da Carta Geológica 1:50.000, n° 34-D, Lisboa*; Instituto Nacional de Engenharia, Tecnologia e Inovação: Lisboa, Portugal, 2006. Available online: https://novaresearch.unl.pt/en/publications/not%C3%ADcia-explicativa-da-carta-geol%C3%B3gica-150000-n%C2%BA-34-d-lisboa (accessed on 15 December 2021).

29. Galopim de Carvalho, A.; Ribeiro, A.; Cabral, J. *Evolução Paleogeográfica da Bacia Cenozoica do Tejo-Sado*; Boletim da Sociedade Geológica de Portugal: Lisboa, Portugal, 1983–1985; Volume XXIV, pp. 209–212. Available online: https://www.researchgate.net/publication/292390457_Evolucao_paleogeografica_da_bacia_cenozoica_do_Tejo-Sado (accessed on 10 December 2021).

30. Kullberg, M.C.; Kullberg, J.C.; Terrinha, P. Tectónica da Cadeia da Arrábida. In *Tectónica das Regiões de Sintra e Arrábida. Memórias Geociências*; Museu Nacional de História Natural, Universidade de Lisboa: Lisboa, Portugal, 2000; pp. 35–84. Available online: https://research.unl.pt/ws/portalfiles/portal/325791/Kullberg+et+al+%28Mem+Geoc+MNHN+2000%29.pdf (accessed on 10 December 2021).

31. Kullberg, J.C.; Rocha, R.B.; Soares, A.F.; Rey, J.; Terrinha, P.; Callapez, P.; Martins, L. A Bacia Lusitaniana: Estratigrafia, Paleogeografia e Tectónica. In *Geologia de Portugal no Contexto da Ibéria*; Dias, R., Araújo, A., Terrinha, P., Kullberg, J.C., Eds.; Universidade de Évora: Évora, Portugal, 2006; pp. 317–368. Available online: http://hdl.handle.net/10362/1487 (accessed on 10 December 2021).

32. Kullberg, J.C.; Rocha, R.B.; Soares, A.F.; Rey, J.; Terrinha, P.; Azerêdo, A.C.; Callapez, P.; Duarte, L.V.; Kullberg, M.C.; Martins, L.; et al. A Bacia Lusitaniana: Estratigrafia, Paleogeografia e Tectónica. In *Geologia de Portugal*; Dias, R., Araújo, A., Terrinha, P., Kullberg, J.C., Eds.; Escolar Editora: Lisboa, Portugal, Volume II—Geologia Meso-cenozóica de Portugal; pp. 195–347. Available online: https://docentes.fct.unl.pt/omateus/files/kullberg_et_al_2013_a_bacia_lusitaniana.pdf (accessed on 10 December 2021).

33. Antunes, M.T.; Elderfield, H.; Legoinha, P.; Pais, J. A stratigraphic framework for the Miocene from the Lower Tagus Basin (Lisbon, Setúbal Peninsula, Portugal): Depositional sequences, biostratigraphy and isotopic ages. *Rev. Soc. Geológica España* **1999**, *12*, 3–15. Available online: https://run.unl.pt/handle/10362/4167 (accessed on 15 December 2021).

34. Antunes, M.T.; Legoinha, P.; Cunha, P.; Pais, J. High resolution stratigraphy and Miocene facies correlation in Lisbon and Setúbal Peninsula (Lower Tagus basin, Portugal). *Ciências Da Terra* **2000**, *14*, 183–190. Available online: http://hdl.handle.net/10362/4707 (accessed on 15 December 2021).

35. Cunha, P.; Pais, J.; Legoinha, P. Evolução geológica de Portugal continental durante o Cenozóico—Sedimentação aluvial e marinha numa margem continental passiva (Ibéria ocidental). In Proceedings of the 6th Symposium on the Atlantic Iberian Margin, Oviedo, Spain, 1–5 December 2009. Available online: https://run.unl.pt/handle/10362/2351 (accessed on 15 December 2021).

36. Antunes, M.T.; Chevalier, J.P. Notes sur la Géologie et la Paléontologie du Miocène de Lisbonne—VII—Observations compémentaires sur les madréporaires et les faciès rècifaux. *Rev. Fac. Ciências Lisb.* **1971**, *16*, 291–305.

37. Mendes-Victor, L.A.; Hirn, A.; Veinante, J.L. A seismic section across the Tagus valley, Portugal: Possible evolution of the crust. *Ann. Géophysique* **1980**, *36*, 469–476.

38. Azevedo, T.M. O Sinclinal de Albufeira. Evolução Pós-Miocénica e Reconstituição Paleogeográfica. Ph.D. Thesis, Centro de Geologia da Faculdade de Ciências de Lisboa, Lisboa, Portugal, 1982.

39. Pais, J. The Neogene of the Lower Tagus Basin (Portugal). *Rev. Esp. Paleontol.* **2004**, *19*, 229–242. [CrossRef]

40. Moniz, C. Contributo para o conhecimento da Falha de Pinhal Novo—Alcochete, no âmbito da Neotectónica do Vale Inferior do Tejo. Master's Thesis, Faculdade de Ciências da Universidade de Lisboa, Lisbon, Portugal, 2010.

41. Ribeiro, A.; Kullberg, M.C.; Kullberg, J.C.; Manuppella, G.; Phipps, S. A review of Alpine tectonics in Portugal: Foreland datachment in basement and cover rocks. *Tectonophysics* **1990**, *184*, 357–366. Available online: https://repositorio.lneg.pt/handle/10400.9/1877 (accessed on 10 December 2021). [CrossRef]

42. Cabral, J.; Moniz, C.; Ribeiro, P.; Terrinha, P.; Matias, L. Analysis of seismic reflection data as a tool for the seismotectonic assessment of a low activity intraplate basin—The Lower Tagus Valley (Portugal). *J. Seismol.* **2003**, *7*, 431–447. [CrossRef]

43. Simões, M. Contribuição Para o Conhecimento Hidrogeológico do Cenozóico na Bacia do Baixo Tejo. Ph.D. Thesis, Universidade Nova de Lisboa, Caparica, Portugal, 1998.
44. PNUD—Programme Des Nations Unies Pour Le Developpement. *Étude des Eaux Souterraines de la Péninsule de Setúbal (Système Aquifère Mio-Pliocène du Tejo et du Sado)*; Rapport Final sur les Résultats du Project, Conclusions et Recommendations; Programme des Nations Unies Pour le Développement; Direcção Geral dos Recursos e Aproveitamentos Hidráulicos: Lisbon, Portugal, 1981.
45. Zeferino, J. Modelação Numérica (FEFLOW) e Contaminação por Intrusão Salina do Sistema Aquífero Mio-Pliocénico do Tejo, na Frente Ribeirinha do Barreiro. Master's Thesis, Faculdade de Ciências e Tecnologia Universidade Nova de Lisboa, Caparica, Portugal, 2016.
46. HP—Hidrotécnica Portuguesa. *Estudo de Caracterização dos Aquíferos e dos Consumos de Água na Península de Setúbal*; Estudo Realizado Para Empresa Portuguesa de Águas Livres, S.A., (EPAL): Lisbon, Portugal, 1994.
47. Lopo Mendonça, J.P. Caracterização geológica e hidrogeológica da Bacia Terciária do Tejo-Sado. In *Tágides 7, Os Aquíferos das Bacias Hidrográficas do Rio Tejo e das Ribeiras do Oeste, Saberes e Reflexões*; ARH do Tejo, I.P.: Lisbon, Portugal, 2010; pp. 59–65. Available online: https://www.researchgate.net/publication/307205223_Caracterizacao_geologica_e_hidrogeologica_da_bacia_terciaria_do_Tejo_Sado (accessed on 12 December 2021).
48. Moinante, M.J. Delimitação de Perímetros de Protecção de Captações de Águas Subterrâneas. Estudo Comparativo Utilizando Métodos Analíticos E Numéricos. Master's Thesis, Universidade Técnica de Lisboa, Lisbon, Portugal, 2003.
49. Wyssling, L. Eine nene form el zur Berechnung der Zustromungsdaner der grunduaners zu einem Grunduaner Pumpuern. *Eclogee Geol. Helv.* **1979**, *72*, 401–406.
50. Bear, J.; Jacobs, M. On the Movement of Water Bodies Injected into Aquifer. *J. Hydrol.* **1965**, *3*, 37–57. [CrossRef]
51. McElwee, C.D. Capture zones for simple aquifers. *Ground Water* **1991**, *29*, 587–590. Available online: https://pubs.er.usgs.gov/publication/70016701 (accessed on 8 January 2022). [CrossRef]
52. Diersch, H.J. *FEFLOW: Finite Element Modeling of Flow, Mass and Heat Transport in Porous and Fractured Media*; Springer: Berlin/Heidelberg, Germany, 2014. [CrossRef]
53. LABCARGA. *Desenvolvimento de Métodos Específicos Para a Avaliação da Recarga das Massas de Água Subterrâneas, Para Melhorar a Avaliação do Estado Quantitativo*; Relatório Final; Laboratório de Cartografia e Geologia Aplicada: Porto, Portugal, 2017; 141p.
54. Simões, M. Modelos e Balanços do Aquífero Sedimentar da Bacia do Tejo—Margem Esquerda, na Península de Setúbal. In Proceedings of the 8º Seminário Sobre Águas Subterrâneas, FCUL, Lisbon, Portugal, 10–11 March 2011; pp. 119–122. Available online: https://novaresearch.unl.pt/en/publications/pmodelos-e-balan%C3%A7os-do-aqu%C3%ADfero-sedimentar-da-bacia-do-tejo-marge (accessed on 10 September 2021).
55. Gonçalves, P. Sustentabilidade do Aquífero Tejo-Sado (Exploração eficiente na Margem Sul do Tejo). In Proceedings of the 12º Seminário Sobre Águas Subterrâneas, Coimbra, Portugal, 7–8 March 2019; Livro de Atas. Universidade de Coimbra: Coimbra, Portugal, 2019.

Article

Wildfire Effects on Groundwater Quality from Springs Connected to Small Public Supply Systems in a Peri-Urban Forest Area (Braga Region, NW Portugal)

Catarina Mansilha [1,2,*], **Armindo Melo** [1,2], **Zita E. Martins** [3], **Isabel M. P. L. V. O. Ferreira** [3], **Ana Maria Pereira** [1] **and Jorge Espinha Marques** [4]

[1]   National Institute of Health Dr. Ricardo Jorge, Department of Environmental Health, Rua Alexandre Herculano 321, 4000-055 Porto, Portugal; armindo.melo@insa.min-saude.pt (A.M.); ana.maria.pereira.95@gmail.com (A.M.P.)

[2]   LAQV/REQUIMTE, University of Porto, Praça do Cel. Pacheco 42, 4050-083 Porto, Portugal

[3]   LAQV/REQUIMTE, Department of Chemical Science, Food Science Laboratory and Hydrology, Faculty of Pharmacy, University of Porto, R. Jorge de Viterbo Ferreira 228, 4050-313 Porto, Portugal; zmartins@ff.up.pt (Z.E.M.); isabel.ferreira@ff.up.pt (I.M.P.L.V.O.F.)

[4]   Institute of Earth Sciences and Department of Geosciences, Environment and Spatial Planning, Faculty of Sciences, University of Porto, Rua do Campo Alegre 1021, 4169-007 Porto, Portugal; jespinha@fc.up.pt

*   Correspondence: catarina.mansilha@insa.min-saude.pt; Tel.: +351-223401100

Received: 11 March 2020; Accepted: 15 April 2020; Published: 17 April 2020

Abstract: Peri-urban areas are territories that combine urban and rural features, being particularly vulnerable to wildfire due to the contact between human infrastructures and dense vegetation. Wildfires may cause considerable direct and indirect effects on the local water cycle, but the influence on groundwater quality is still poorly understood. The aim of this study was to characterize the chemistry of several springs connected to small public supply systems in a peri-urban area, following a large wildfire that took place in October 2017. Groundwater samples were collected in four springs that emerged within burned forests, while control samples were from one spring located in an unburned area. Sampling took place from October 2017 until September 2018, starting 15 days after the wildfire occurrence, to evaluate the influence of the time after fire and the effect of precipitation events on groundwater composition. Groundwater samples collected in burned areas presented increased content of sulfate, fluoride and nitrogen and variability in pH values. Iron, manganese and chromium contents also increased during the sampling period. Post-fire concentrations of polycyclic aromatic hydrocarbons (PAHs), mainly the carcinogenic ones, increased especially after intense winter and spring rain events, but the levels did not exceed the guideline values for drinking water.

Keywords: wildfire; peri-urban area; groundwater quality; polycyclic aromatic hydrocarbons; major ions; metals

---

1. Introduction

Wildfires are one of the main obstacles to the sustainability of forests and related ecosystems. The resulting devastation may cause severe economic and social costs, with the loss of lives and infrastructures, as well as the disturbance of the provision of goods and services, including ecosystem services, together with environmental damages such as the loss of carbon sequestration [1–3].

The urbanization of forest areas constitutes a new fire risk scenario. Over the years, throughout the world, the contact areas between human infrastructures and wild vegetation increased (the so-called peri-urban or wildland-urban interfaces) [4–6]. In Portugal, these areas result mainly from the abandonment of croplands since the seventies, with fewer people to manage the land, creating great quantities of highly flammable combustible material, mostly from non-native species. This spatial pattern of human presence in the territory leads to particularly vulnerable areas to wildfire impacts, with disastrous consequences for populations and the environment [7,8]. Climate-related droughts and extreme weather are also propitious conditions for wildfires.

In 2017, Portugal suffered the greatest devastation caused by wildfires in a single year. Although in the last decades there have been a high number of fires, when compared to other countries and similar conditions, the wildfires that took place in 2017 largely exceeded the suppression capacity of the emergency services and were disastrous. Consequently, great damages occurred to population, with the loss of 115 lives, as well as to forests, rangelands, rural, industrial and urban areas. These catastrophic wildfires were worsened by climate change (it was the driest summer in nearly 90 years) and adverse meteorological conditions, and by the vegetation cover of fire-prone species, where the eucalyptus *(Eucalyptus globulus)* and maritime pine *(Pinus pinaster)* are dominant [2,9,10]. The burnt area was 539,921 ha, representing 498% of the average of the previous decennium, which was 90,269 ha, and nearly 60% of the total area burnt in the entire European Union, in which Portugal only represents about 2.1% of total landmass. The most critical month was October, with 3234 rural wildfires (15.4% of total annual rural wildfires) and 289,124 ha burnt (53.5% of total area). Wildfire occurrence prevailed mostly in the urban districts, often in peri-urban areas, which registered 55.6% of the total number of fires [10].

Regarding surface water and groundwater resources, the burning of forest catchments may result in a long-lasting legacy of water quality deterioration, whose magnitude and persistence can be observed from a few months to several years after the wildfire. Post-fire water quality concerns are complex and vary significantly from place to place depending on the severity, intensity, and duration of the fire, the soil and vegetation cover characteristics, the geological and geomorphological nature of the terrain, and the amount and intensity of precipitation during post-fire rain events [11,12]. Water quality impacts may also result from indirect effects associated with smoke and aerial deposition of ash [13,14].

According to the literature, wildfires may cause changes in several water quality parameters of interest or concern to water systems [11,15,16]. Contamination of streams and water reservoirs by post-fire inputs of suspended sediments and various trace elements present in ash may be problematic for both health and aesthetic reasons. High concentrations of iron (Fe), manganese (Mn), zinc (Zn), sodium (Na^+) and cloride (Cl^-) cause organoleptic problems (taste, color, staining of pipes and fittings). Sulfates (SO_4^{2-}) have purgative effects for concentrations over 500 mg/L. Poisoning may occur from continued consumption of water containing high concentrations of copper (Cu), with gastrointestinal symptoms. Arsenic (As) and cromium (Cr) (specially hexavalent Cr) may be carcinogenic, while aluminium (Al), lead (Pb) and mercury (Hg) are toxic when consumed in sufficient quantities for prolonged periods. Following wildfire, increased exports of nitrogen (N) and phosphorous (P), in various forms, can also be problematic for managers of water supply catchments. High concentrations of nitrates (NO_3^-) and nitrites (NO_2^-) also present a potential risk to human health, primarily through reduction of NO_3^- to NO_2^-, which may affect oxygen transport in red blood cells, while high concentrations of ammonium NH_3^+/NH_4^+ may corrode copper pipes and fittings. N and P are limiting nutrients for growth of aquatic plants, algae and cyanobacteria in water bodies. Eutrophication increases the risk of potentially toxic blooms, with implications for human health, aesthetic problems (taste, odor and color), and aquatic ecosystem function disturbance [14,17].

Polycyclic aromatic hydrocarbons (PAHs), such as benzo[*a*]pyrene, are known for their potential teratogenicity, carcinogenic and mutagenic properties, explained by the formation of adducts between the DNA bases and epoxides derived from hydrocarbons after an oxidizing process in the liver [18]. In addition to cancer, long-term exposition can also cause chronic bronchitis, skin problems and allergies. Some compounds are also classified as potential endocrine disruptors because they have estrogenic activity, generally coupled with a high potential for bioaccumulation [15,16,19,20]. The United States Environmental Protection Agency (USEPA) designated several PAHs as priority pollutants that should be regulated due to their high toxicity and adverse effects [21]. PAHs were also designated as priority hazardous substances by the European Commission, in Directive on Environmental Quality Standards (Directive 2008/105/EC) [22].

The effects of fire on surface water are more evident that those on groundwater. Recharge rates, net infiltration and water balance can change after a wildfire, leading to difficulties in maintaining the supply of potable groundwater to populations [23]. Nowadays, despite the widespread access to tap water provided by the Portuguese municipalities, many people still prefer to drink groundwater from public fountains supplied by nearby springs, as it is associated with high purity and pleasant organoleptic characteristics. However, the impact of wildfires on the composition of groundwater from nearby springs is unknown.

The goal of this study was to identify the impact of the wildfire that occurred on October 2017, on the chemical characteristics of water from springs connected to small public supply systems used for human consumption in the peri-urban area of the city of Braga, in NW Portugal. The study included the analyses of major ions and trace elements, namely PAHs, which were carried out during one year after the wildfire.

2. Materials and Methods

2.1. Hydrogeological Framework

The city of Braga (41°32′ N; 8°25′ W) is located in NW Portugal (Figure 1), in the Minho Province. The municipality has a resident population of 181,494 inhabitants (in 2011) [24], representing the seventh largest municipality in Portugal (by population) whit an area of 183.40 km². The city is surrounded by peri-urban areas consisting of agroforestry systems, especially in the higher and steeper terrain (Figures 1 and 2). To the southeast, the urban area is bordered by a mountainous ridge (Figure 1) with altitude above 500 m at the summits of Santa Marta (562 m), Monte Frio (548 m) and Sameiro (572 m). To the north, this region falls into the Cávado river catchment, while to the south it falls into the Ave river catchment.

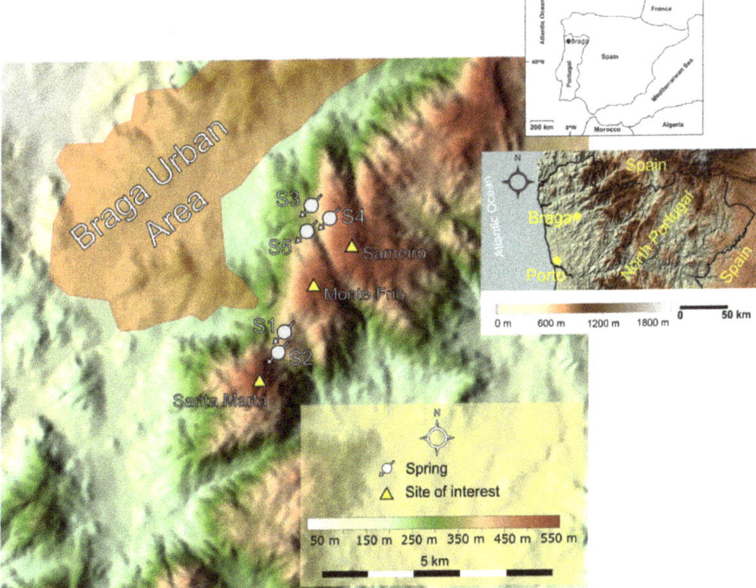

Figure 1. Location of Braga region in Iberia and in NW Portugal; hypsometric features of the study area; location of the studied springs.

Figure 2. Some aspects of the study region: (**a**) burnt area of Eucalyptus globulus (first plan) and the Braga urban area (background); (**b**) Santa Maria Madalena sampling point (S1); (**c**) Monte de Dadim spring (S5); (**d**) Quercus suber burnt area at the vicinity of Santa Marta de Leão spring (S2); (**e**) Eucalyptus globulus burnt plantation at the recharge area of the Mina Spring; ash covered burnt soil immediately after the fire (**f**) and soil showing intense erosion one year after the fire (**g**); (**h**) deciduous forest at Monte de Dadim (S5, control point).

The Braga climate has Atlantic features, with a mean value of annual precipitation around 1449 mm at the local climatic station (located at 190 m of altitude). The temporal distribution pattern of precipitation is similar to the one usually observed in NW Portugal: December is the rainiest month (about 220 mm) and July is the driest month (about 22 mm). The mean annual air temperature is around 15 °C, with maximum in July and August (21 °C) and minimum in January (9 °C). The Braga Köppen–Geiger climate classification is Csb, which is dominant in Northwestern Iberia and corresponds to warm temperate (C), with dry and warm summer (sb) [25,26].

Braga region is located in the following geomorphological units [27]: Central System (1st level unit) and Entre Douro e Minho Open Valleys and Hills and Atlantic Front of NW Iberia Mountains (2nd level units). The region is also located in the Central-Iberian Zone of the Iberian Massif [28]. The dominant regional geological units are granitic rocks and metasedimentary rocks; sedimentary cover areas are residual. Therefore, the major processes in the local groundwater cycle take place mostly in fractured circulation media and, to a minor extent, in porous media.

The unsaturated zone depth often reaches more than 20 m at hilltops and less than 10 m at valley bottoms. In addition, the structure of the unsaturated zone encompasses a soil cover (mainly Leptosols, Regosols and Cambisols, with an umbric A horizon) overlying granite or metasedimentary rock.

2.2. Wildfire Description

Between 13 and 18 October of 2017, more than 7900 wildfires affected Northwestern Iberia. During this month, 15,440 wildfires were active, 33 of which were of important size, which spread quickly due to the drought and to particular meteorological conditions: strong winds caused by the Ophelia hurricane that swept the western coast of the Iberian Peninsula and unusual air temperatures, above 30 °C. September 2017 was the driest month in 87 years [29]. Around 81% of the territory was under severe drought and 7.4% in extreme drought. In the district of Braga, 23 wildfires were registered on 14 October and 43 wildfires on 15 October. The wildfire of greatest impact on the study region started in Leitões, Guimarães, and quickly reached Braga (particularly the mountain area between Santa Marta and Sameiro, see Figure 2), consuming over 1200 ha of forest [29].

The wildfire started at 13:00h on 15 October and was controlled within in a few hours. However, a strong reactivation projected the fire in several directions, with a very significant speed of propagation, and by 17:17h the authorities requested additional means to protect the houses, since the fire had two very long and intense fronts. Only by 10:09h, on 16 October, the wildfire was finally controlled [30].

2.3. Water Sampling

Groundwater samples were collected in springs that supply non-treated water used for human consumption, located in near Braga city (Figure 1): (i) Santa Maria Madalena Fountain (S1); (ii) Santa Marta de Leão Fountain (S2); (iii) Tanque de Dadim Fountain (S3); (iv) Depósitos Spring (S4); (v) Monte de Dadim Spring (S5). The recharge areas of S1 to S4 springs were affected by the wildfire, while S5 corresponds to a spring located in an unburned area, which was used as control.

All springs are characterized by relatively shallow water circulation. In fact, water is abstracted by means of hand dug galleries which, in this region, are typically associated with springs connected to the upper saturated zone and, sometimes, also to interflow. Moreover, springs S1 and S2 are located at hilltops and, therefore, have very short circulation paths. Finally, S3, S4, and S5 springs are all located at slopes and abstract water circulating in the granite weathering mantle.

The site selection criteria were the existence of permanent flow throughout the year, the location of the recharge areas regarding the burnt region, and the sampling feasibility. These sampling points were also chosen to be preserved as much as possible from other anthropogenic impacts, reducing the risk of water contamination by sources of pollution other than wildfires (Figures 1 and 2 and Table 1). The control spring analytical results were used to provide a framework of reference for contaminant concentrations on burned areas. The discharge flow in all springs was under 0.5 L/s.

Table 1. Features of the sampling points and the corresponding spring recharge areas.

Sampling Point	Altitude (m a.s.l.)	Lithology	Soil Types	Land Cover
S1—Sta. Maria Madalena Fountain	425	Metasedimentary rocks	Leptosol and Regosol	Public garden with *Quercus robur* and *Quercus suber*
S2—Sta. Marta de Leão Fountain	415	Granite	Leptosol and Regosol	Public garden and forest with *Quercus robur, Quercus suber, Pinus pinaster* and *Eucalyptus globulus*
S3—Tanque de Dadim Fountain	350	Granite	Regosol, Cambisol and Anthrosol	Deciduous forest
S4—Depósitos Spring	400	Granite (dominant) and metasedimentary rocks (residual)	Leptosol, Regosol and Cambisol	*Eucalyptus globulus* forest
S5—Monte de Dadim (control point)	390	Granite	Regosol and Cambisol	Deciduous forest

Samples collected at S1, S2, S3 and S4 were designated as Burnt samples (BR), and the ones collected at S5 as unburnt samples (NB). The sampling plan included five campaigns, carried out during one year, from October 2017 until September 2018, starting 15 days after the wildfire occurrence (before the first post-fire rain event), and included 25 samples.

Samples were collected according to ISO 5667-3:2003(E) Water quality—Sampling—Part 3: guidance on the preservation and handling of water samples, during non-storm conditions, before and after the first overland flow event and aquifer recharge following the fire.

The influence of wildfire on water quality constituents was investigated. The physicochemical analyses of key water quality constituents included the major ions, organic matter and heavy metals, which may be mobilized according to fire intensity, post-fire precipitation, geological and geomorphological conditions and vegetation cover of each site. Compounds that can serve as specific markers, indicative of wildfire, such as the high molecular weight PAHs, were also monitored. Water pH, electrical conductivity (EC) and temperature were measured *in situ* during sampling.

The temporal distribution of precipitation and climate data (measured at Braga meteorological station) during the study period, provided the knowledge of the influence of rainfall on the input of pollutants into groundwater [31].

2.4. Laboratory Analyses

Analyses were performed according to procedures outlined in *Standard Methods for the Examination of Water and Wastewater 23rd edition* and in *Le Rodier—L'analyse de l'eau 10e édition*. The laboratory has been accredited under ISO/IEC 17025 since 2007. Precision and accuracy were calculated for all analytical methods with values <10%. Uncertainties were also calculated with results varying from 2% to 10%.

Water turbidity was measured in a Hach 2100N Laboratory Turbidity Meter. Electrical conductivity (EC) and pH were determined in a Crison MultiMeter MM 41. Total alkalinity and total hardness were analyzed by titration and reported as milligrams per liter of calcium carbonate (mgL^{-1} $CaCO_3$). Color, PO_4^{2-} and total phosphorus, expressed as P (TP) were analyzed in a Shimadzu UV-1601 Spectrophotometer (Shimadzu Corporation, Kyoto, Japan). Total nitrogen, expressed as N (TN) and chemical oxygen demand (COD), were evaluated in a Hach DR 2800 Spectrophotometer (Hach Company, Loveland, CO, USA). Major inorganic ions (Na^+, K^+, Mg^{2+}, Ca^{2+}, Li^+, Cl^-, NO_3^-, F^- and SO_4^{2-}) were analyzed by ion chromatography (DionexTM system DX-120/ICS-1000, Dionex Corporation, Sunnyvale, CA, USA). Total organic carbon (TOC) was analyzed in a Shimadzu TOC-V (TOC-ASI-V, Shimadzu Corporation, Kyoto, Japan), heavy metals (Cr, Mn, Ni, Cu, Zn, As, Cd and Pb) and other components, such as Al, Fe, NO_2^-, NH_4^+ and SiO_2, were, analyzed in a Varian AA240 Atomic Absorption Spectrometer (Varian Inc., Palo Alto, CA, USA) and in a Continuous Segmented Flow Instrument (San-Plus Skalar, Skalar Analytical, Breda, The Netherlands), respectively. PAHs were analyzed by dispersive liquid-liquid microextraction coupled to gas chromatography/mass spectrometry (DLLME–GC/MS) methodology in a Shimadzu GCMS-QP2010 gas chromatograph mass spectrometer equipped with an auto injector AOC5000 (Shimadzu Corporation, Kyoto, Japan), according the procedure described in Borges et al. [32].

Analytical standards were supplied by Sigma–Aldrich (Steinheim, Germany) and Merck (Darmstadt, Germany). The reference standard mixture containing the 15 EPA PAHs (acenaphthylene, Acy; acenaphthene, Ace; fluorene, Flu; phenanthrene, Phe; anthracene, Ant; fluoranthene, Flt; pyrene, Pyr; benz[*a*]anthracene, BaA; chrysene, Chr; benzo[*b*]fluoranthene, BbF; benzo[*k*]fluoranthene BkF; benzo[*a*]pyrene, BaP; dibenz[*a,h*]anthracene, DahA; benzo[*ghi*] perylene, BghiP; and indeno[*1,2,3-cd*]pyrene, Ind) was purchased from Sigma-Aldrich (Steinheim, Germany).

Methanol, dichloromethane and acetonitrile were organic trace analysis grade SupraSolv and were supplied by Merck (Darmstadt, Germany). Ultrapure water was highly purified by a Milli-Q gradient system (18.2 mΩ/cm) from Millipore (Milford, MA, USA).

2.5. Statistical Studies

Data on chemical concentration was analyzed throughout time, location, and throughout time in burned areas. All dependent variables from every analyzed parameter were tested for distribution of the residuals with the Shapiro–Wilk's test. Chemical concentrations were studied using a one-way analysis of variance (ANOVA), if normal distribution of the residuals was confirmed. Welch correction was applied when the homogeneity of variances was not verified. Whenever statistical significances were found, Tukey's test or the Tamhane's test post-hoc tests were applied for mean comparison, depending on variances assumption or not.

If normal distribution of the residuals was not found, parameters analyzed were studied using a Kruskal–Wallis test. Whenever statistical significances were found, Dunn's post-hoc test was applied for median comparison.

All analyses were performed at 5% significance level, using XLSTAT for Windows version 2014.5 (Addinsoft, Paris, France).

3. Results and Discussion

The effects of wildfires on the catchment's hydrologic responses were noted worldwide, although the specific impacts are unpredictable in terms of both the magnitude of potential effects and the persistence of the influence. So, there is no clear pattern for wildfire effects on water bodies, which may not be affected or, on the other hand, experience fire-related changes that can range from aesthetic concerns (taste or appearance) to potential toxicity or carcinogenicity with prolonged exposure, as well as environmental damages [14,33].

The concentration of major and trace elements monitored in groundwater collected at the peri-urban area of Braga, after the wildfire of October 2017, shown significant differences between samples for several compounds. However, for others, a clear tendency that could be attributed to the wildfire was not observed, or the variation over time was similar to that observed in control samples.

Post-fire analytical results are shown in Table 2 and corresponding statistical analysis on Supplementary Table S1.

The analyses of the results show a slight decrease followed by an increase in pH values in S1 and S2, and an increase in pH values in S3 and S4 during the sampling period, wherein 18 January and April 2018 were grouped, whereas the remaining sampling times were independent of each other (p = 0.006). This is in accordance with studies that described wood ash as being alkaline and rich in carbonates and metal oxides [14]. However, ash chemical composition is highly variable as it reflects the type of vegetation and the part of the plant burned, as well as the soil type and combustion conditions, with different implications on water quality parameters [14,34]. No statistical differences (p > 0.050) were found for turbidity, color and silica. The electrical conductivity, which is a water quality indicator for estimating the amount of mineralization and total dissolved solids, remained also constant during the study period, with no significant variations in BR samples (p > 0.050).

Wildfires often induce quantitative and qualitative changes in soil organic matter, sometimes with significant losses due to the partial or total removal of the litter layer and, possibly, some organics from the upper few centimeters of mineral soil [35]. However, in this study, the total organic carbon concentrations appeared to be unaffected by fire, with mean values of 0.37 ± 0.15 mgL^{-1} and 0.33 ± 0.17 mgL^{-1} in BR and NB samples, respectively, and the chemical oxygen demand increased immediately after the fire, reaching a peak during spring, but then slowly declined as vegetation re-established. Nevertheless, these variations could not be attributed to the wildfire, as they occurred in the NB samples too (p > 0.050).

Table 2. Physicochemical data of water samples from the studied region.

Sampling Point	Sampling Date	pH	COD	EC	Si	TP	HCO$_3^-$	F$^-$	Cl$^-$	SO$_4^{2-}$	PO$_4^{2-}$	NO$_3^-$	NO$_2^-$	NH$_4^+$	Na$^+$	K$^+$	Ca^{2+}	Mg^{2+}
			(mg/L)	(µS/cm)	(mg/L)	(mg/L)	(mg/L)	(mg/L)	(mg/L)	(mg/L)	(mg/L)	(mg/L)	(mg/L)	(mg/L)	(mg/L)	(mg/L)	(mg/L)	(mg/L)
S1—Sta. Maria Madalena Fountain	17 October	6.0	2.5	37.1	8.5	0.05	10.1	0.02	6.6	0.4	0.067	<LD	0.007	0.006	5.7	0.6	1.7	0.5
	18 January	5.7	8.3	37.6	7.9	0.04	7.9	0.03	7.7	0.7	0.066	1.0	0.001	<LD	5.6	0.5	1.8	0.6
	18 April	5.5	10.1	41.7	6.9	0.04	5.0	0.09	9.0	1.5	<LD	2.0	0.005	0.002	6.8	0.4	0.5	0.4
	18 May	5.6	7.1	35.2	7.1	0.03	5.7	0.27	8.3	1.0	0.035	1.1	0.001	<LD	6.2	0.5	1.4	0.9
	18 September	6.3	5.6	34.9	9.3	0.12	8.8	<LD	6.9	1.3	0.041	0.8	0.002	<LD	5.2	0.8	2.0	0.5
S2—Sta. Marta de Leão Fountain	17 October	6.1	2.8	37.8	8.5	0.03	10.1	0.02	6.8	0.5	0.064	0.2	0.003	<LD	6.0	1.0	1.6	0.4
	18 January	5.8	8.6	37.2	7.8	0.06	7.3	0.04	7.7	0.7	0.070	1.1	<LD	<LD	5.6	0.5	1.8	0.6
	18 April	5.4	7.8	40.2	6.8	0.02	4.4	0.10	9.0	1.6	<LD	1.8	0.004	0.003	6.2	0.4	0.5	0.4
	18 May	5.6	5.4	34.2	7.1	0.24	5.7	0.28	8.5	1.1	0.014	1.6	<LD	<LD	5.8	0.5	1.3	0.7
	18September	6.4	3.9	37.5	9.1	0.11	8.8	<LD	6.9	1.3	0.021	0.7	0.004	<LD	5.0	0.8	2.0	0.5
S3—Tanque de Dadim Fountain	17 October	5.6	2.9	48.4	15.4	0.05	15.7	0.02	7.7	<LD	0.260	0.5	0.003	<LD	7.9	0.8	2.0	0.6
	18 January	5.8	7.1	48.2	16.4	0.25	17.0	0.04	7.9	<LD	0.266	1.2	<LD	<LD	7.7	0.7	2.3	0.8
	18 April	6.1	9.1	47.0	16.3	0.07	12.6	0.05	7.7	0.5	0.118	0.8	0.004	0.011	8.5	0.7	1.8	0.5
	18 May	6.0	6.9	42.8	16.0	0.29	14.0	0.26	7.6	0.1	0.124	0.8	0.001	<LD	7.5	0.9	2.4	0.8
	18 September	6.3	5.9	46.7	17.0	0.19	15.3	<LD	7.4	0.8	0.146	0.7	0.002	<LD	6.7	0.8	2.3	0.7
S4—Depósitos Spring	17 October	5.6	2.1	61.6	10.6	0.05	26.5	0.02	8.5	<LD	0.121	4.7	<LD	<LD	7.5	0.6	4.0	1.1
	18 January	5.8	9,.1	59.7	11.9	0.08	14.1	0.03	8.9	<LD	0.083	6.8	<LD	<LD	7.4	0.5	3.5	1.3
	18 April	5.9	8.5	61.8	12.1	0.04	12.6	0.05	9.3	0.8	0.282	7.5	0.006	0.005	9.0	0.6	2.9	1.2
	18 May	6.0	2.8	58.4	11.7	0.07	12.2	0.31	9.6	0.4	0.021	6.9	0.001	<LD	8.2	0.8	4.0	1.2
	18 September	6.7	2.9	58.7	11.9	0.06	13.5	<LD	8.8	1.0	0.060	5.5	0.004	0.001	6.5	0.8	3.2	1.2
S5—Monte de Dadim (control point, NB)	17 October	6.4	2.0	65.6	23.4	0.04	26.5	0.03	8.4	<LD	0.332	0.2	0.002	<LD	9.2	0.9	4.0	1.1
	18 January	6.7	7.1	65.0	23.4	0.36	28.4	0.06	8.3	<LD	0.337	0.8	0.001	<LD	8.7	1.0	4.5	1.2
	18 April	6.6	10.0	64.0	23.7	0.11	26.5	0.07	8.7	0.2	0.187	0.6	0.004	<LD	10.1	1.0	3.8	1.0
	18 May	6.4	5.8	59.8	22.6	0.11	23.6	0.11	8.4	<LD	0.200	0.6	<LD	<LD	7.5	0.6	3.0	1.5
	18 September	6.5	8.2	61.1	23.5	0.15	25.7	<LD	8.0	0.6	0.196	0.3	0.001	<LD	7.3	0.9	3.9	1.1

Analytical methods limits of detection (LD): 1.0 mg/L for COD; 0.8 µS/cm for EC; 1.78 mg/L for Si; 0.02 mg/L for TP and F; 0.3 mg/L for HCO$_3$; 0.03 mg/L for Cl, SO$_4$, NO$_3$, Na, K, Ca, Mg; 0.01 mg/L for PO$_4$; 0.001 mg/L for NO$_2$ and NH$_4$.

Few compositional changes regarding major ions were observed. The Piper diagram presented in Figure 3 highlights the main changes: all sampling points reveal a minor shift from the first campaign (October 2017, at the end of the dry season) to the third campaign (April 2018, after the intense March precipitation events). The cations content shift results mainly from a decrease in calcium and an increase in sodium, while the anions content shift results from an increase in chloride and sulfate. However, changes in calcium and chloride ions observed in the burnt area are similar to those observed in the unburnt area and, thus, may not be connected to the wildfire. A decrease in bicarbonate is also observed in S1 to S4 points, but not in S5 point. Bicarbonate is originated by water–rock interaction and depends on the nature of the groundwater flowpaths that, in this case, are poorly known.

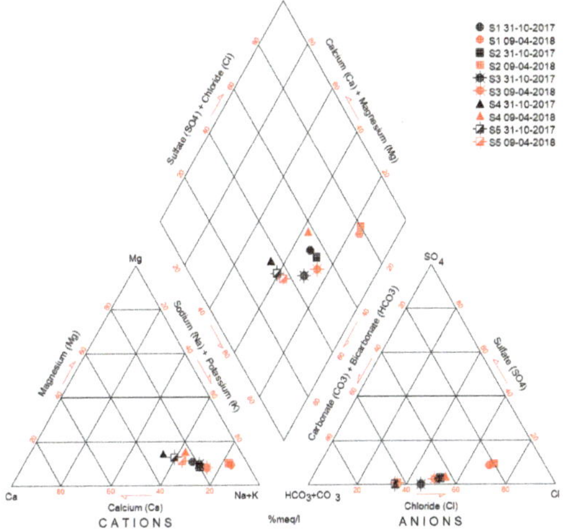

Figure 3. Piper diagram comparing the October 2017 campaign to the April 2018 campaign.

Regarding the sampling date, ANOVA analyses of major ions also reveal that chloride, sodium, potassium, calcium and magnesium concentrations did not vary significantly ($p > 0.050$). In contrast, the increase of sulfate in groundwater from the burnt areas was statistically significant ($p = 0.034$) and is probably due to the oxidation of sulfur in soil organic matter after the wildfire. Similar results were already observed in an earlier study carried out in Caramulo region, in Central Portugal [36] and reported in the literature by other authors [14].

The levels of fluoride have also risen ($p = 0.003$) in May by approximately two orders of magnitude regarding the control point (S5), decreasing to levels similar to the initial ones in the last campaign.

Nutrient export from burnt soils usually increases after wildfires, and this process may affect groundwater composition. However, wildfire effects on stream exports of total phosphorus (TP) and total nitrogen (TN) vary significantly [14,37,38]. In the present study, it was observed a small decline (despite this, with no statistical significance) of phosphate, but an increase of TP with multiple change of 1.2 to 4.2 times the initial values. The wildfire effect on nitrogen should be examined with caution, because small differences regarding pre-fire values or control samples could be very important [39,40]. During the first six months after fire, the concentration of nitrogen species increased significantly ($p < 0.001$), especially after precipitation, compared with concentrations in the reference samples. Combined concentrations of post-fire nitrite, nitrate and ammonium, which are dissolved forms, varied from 0.52 mgL^{-1} in control to 6.28 mgL^{-1} (mean values) in S4 sample. The land cover at the S4 recharge area consists only of forest, without other pollution sources besides the wildfire. The higher nitrate content observed in S4 should be a result of the wildfire and is in agreement with the thicker layer of ash observed in the recharge area of this spring. Several factors could explain the increase of nitrogen exports in post-fire situations. On one hand, there is a lower plant demand and the nitrogen mineralization is stimulated (due to changes in pH and electrolytes). On the other hand, the nitrate form is mobile in soil-water systems and leaches through soil into catchments and drainages after heavy overland flow events that follow the wildfire [39,41]. In addition, because the wildfire removes forest cover and litter, rain interception decreases and nutrient transport via infiltration may increase. As well as nitrate, ammonium loading may increase as it is volatilized during fire and can dissolve into water. This compound may be retained in soil in its exchangeable form and subsequently be leached.

Hazardous chemicals, such as metals, were also monitored in Braga, as wildfires may influence the concentration of trace metals differently, with potentially harmful effects to human health and the environment. In this study no significant differences between samples collected in NB and BR locations were observed for cadmium (Cd), arsenic (As), lead (Pb), nickel (Ni), copper (Cu) and zinc (Zn). In contrast, the mean levels of iron (Fe), manganese (Mn) and chromium (Cr) were about 1.3, 7.0 and 2.8 times above the NB sample levels, respectively. Substantial post-fire increases in total iron and total manganese have been reported in the literature, indicating an added influx of these metals as part of an increase in particulates [42]. Even at values below the criteria established for aquatic systems, results must be considered as iron and manganese are related to aesthetic issues of the water (taste and color), and chromium, namely the hexavalent form, is carcinogenic. Concentrations of metals in groundwater samples are displayed in Table 3.

Table 3. Descriptive statistics of selected trace elements in groundwater samples after the wildfire.

		S1	S2	S3	S4	S5
Cd	Avg	0.07	0.02	0.02	0.06	<LD
	Min-Max	<LD–0.15	<LD–0.1	<LD–0.1	<LD–0.2	<LD
As	Avg	0.3	0.4	1.9	1.2	0.7
	Min-Max	<LD–0.8	<LD–0.6	1.6–2.4	0.7–2.3	0.2–1.3
Pb	Avg	0.04	<LD	0.5	<LD	<LD
	Min-Max	<LD–0.2	<LD	<LD–2.5	<LD	<LD
Ni	Avg	0.6	0.1	1.5	0.1	0.7
	Min-Max	<LD–3.0	<LD–0.4	<LD–6.7	<LD–0.4	<LD–3.0
Cu	Avg	3.1	0.8	2.0	2.3	2.4
	Min-Max	<LD–7.7	<LD–2.2	<LD–5.6	<LD–6.5	<LD–8.1
Zn	Avg	0.02	0.05	0.02	0.1	0.03
	Min-Max	<LD–0.1	<LD–0.2	<LD–0.1	<LD–0.1	<LD–0.2
Cr	Avg	0.8	0.26	0.2	0.2	0.1
	Min-Max	<LD–2.7	<LD–1.1	<LD–0.7	<LD–0.7	<LD–0.6
Fe	Avg	19.0	5.8	13.0	21.8	11.6
	Min-Max	6.1–55.5	<LD–13.5	<LD–13.8	<LD–56.0	<LD–25.5
Mn	Avg	2.6	2.9	0.5	2.4	0.3
	Min-Max	1.7–3.7	1.8–4.2	<LD–1.0	0.9–4.8	<LD–0.8

Min, Max, Avg, for minimum, maximum and average. Values of trace elements are in µg/L. Limits of detection (LD): 0.1 µg/L for Cd, As, Pb, Ni, Cu, Zn and Cr; 5 µg/L for Fe.

PAH Analyses

This study also included the analyses of the 15 PAHs designated as priority hazardous pollutants. Thirteen hydrocarbons were found during the sampling period. The most abundant compounds in samples collected in burnt areas were Ant (26%), Acy (17%), BaA (15%) and BaP (14%), while in control samples Acy was the dominant compound (33%), followed by BaA (21%) and Ant (18%) (Figure 4). The sum of total concentrations, as well as the variety of PAHs, has increased throughout the year in all sampling points, with maximum values in May, after the wet season precipitation events. Mean values ranged from negligible amounts in S5 sample in January to 0.029 µgL^{-1} in S3 in May. Comparing the initial to the last campaign, total values increased by factors of 1.3 to 2.2 times the initial concentrations. Furthermore, in control samples, only light PAHs, with three to four rings (Acy, Ace, Flu, Phe, Ant, Flt, Pyr, BaA, and Chr), were detected, in contrast to the samples of burnt areas, where a different profile was observed throughout the study period. Heavy PAHs, with five (BbF, BkF, BaP, and DahA) and six rings (InP, BghiP), have increased until May followed by a decrease in September, which probably indicates a natural remediation concerning these compounds. The compositional pattern of PAHs by ring size for the water samples is shown in Figure 5.

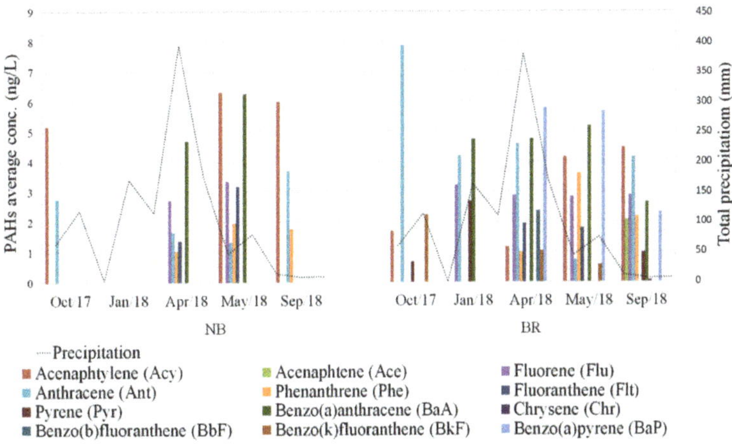

Figure 4. Average concentrations of polycyclic aromatic hydrocarbons (PAHs) (individual fractions) in S5 unburnt samples (NB) and in burnt areas (BR) (n = 4) and monthly precipitation.

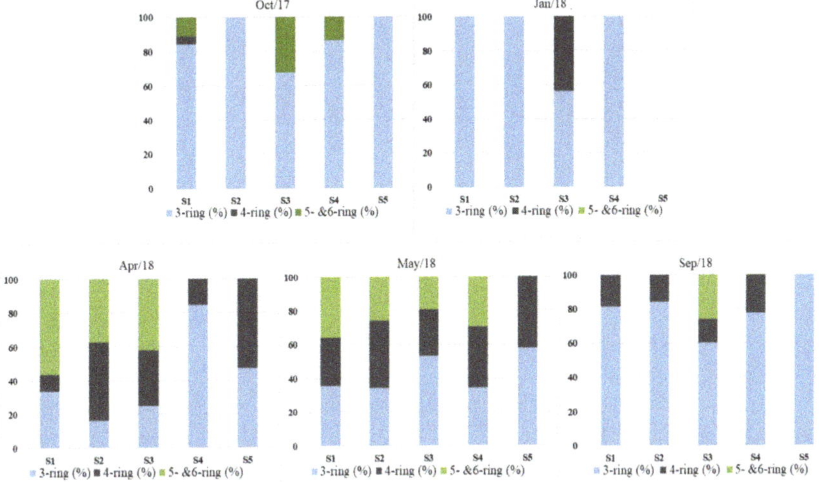

Figure 5. PAHs profiles of the samples collected in burned and control areas in the five sampling campaigns, according to their structural composition (number of benzene rings).

Significant differences for some PAHs were observed if the date of sampling was considered. Results from Shapiro–Wilk's test shown that PAHs do not have a normal distribution of the residuals.

Kruskal–Wallis analysis for the sum of PAHs revealed that October 2017 and May 2018 sample results were significantly different. (Table S1).

Regarding the carcinogenic PAHs (BaA, Chr, BbF, BkF, BaP, Ind and DahA), the concentrations were also significantly higher in May 2018 (p = 0.022), at levels around 0.010 μgL^{-1} (0.0098 to 0.013 μgL^{-1}) in BR samples, representing 60.4%, 52.9%, 36.2% and 62.6% of total PAHs in S1, S2, S3 and S4, respectively. Approximately half of these values correspond to BaP, with concentrations ranging from 0.0046 μgL^{-1} in S4 to 0.0077 μgL^{-1} in S1. In S3, concentrations of BaP increased in April 2018 and remained approximately constant up to September 2018, with a mean value of 0.0052 ± 0.0006 μgL^{-1}. In S5 samples (NB), BaP was not detected during the study period (Figure 6).

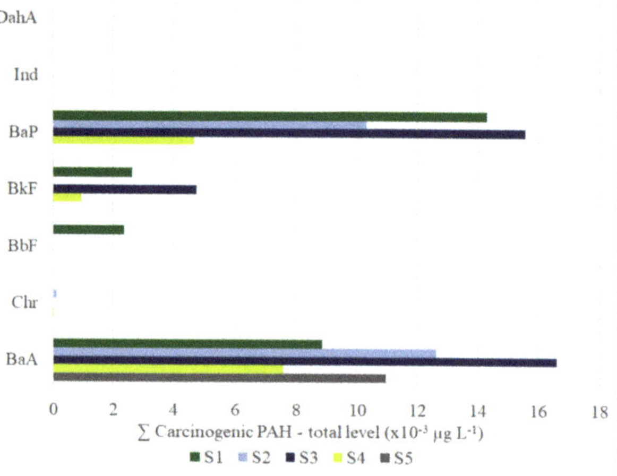

Figure 6. Total levels of carcinogenic PAHs (benz[*a*]anthracene (BaA), chrysene (Chr), benzo[*b*]fluoranthene (BbF), benzo[*k*]fluoranthene (BkF), benzo[*a*]pyrene (BaP), indeno[*1,2,3-cd*]pyrene (Ind) and dibenz[*a,h*]anthracene (DahA)) in NB (S5-control) and BR (S1 to S4) samples in the five campaigns.

Groundwater quality standards for PAHs (referred to as "threshold values") were established by several European Member States taking into account identified risks [43]. Ranges of threshold values throughout Europe varied from 0.005 to 0.03 μgL^{-1} for BaP, 0.01 to 0.1 for BghiP and BbF, and 0.05 to 0.1 μgL^{-1} for BkF. Regarding these values, our results reveal several nonconformities for BaP, namely in samples collected in April and May in S1, S2 and S3. BaP, is the most extensively studied carcinogenic PAH, classified by IARC as a *Group 1* or a known human carcinogen [44], and is in the top ten priority pollutants designated by the Agency for Toxic Substances and Disease Registry (ATSDR) in 2015 [45]. BaP is also considered a powerful endocrine disruptor compound [46]. Despite the parametric value for drinking water established by the European Union Council Directive 98/83/EC for BaP has not been exceeded (0.010 μgL^{-1}) [47], attention is needed especially regarding mixture cancer potency, as individual PAHs occur as part of environmental mixtures and cumulative risk assessment should be considered [48]. The established limit of 0.100 μgL^{-1} for the sum of concentrations of BbF, BkF, BghiP and Ind was also not exceeded.

Regarding PAHs sources, they may have a pyrogenic origin, linked to processes of incomplete combustion of organic matter (in particular, vegetation or fossil fuels) or a petrogenic origin, resulting from the transformation of organic matter that occurs in geological materials [49]. Another source of PAH is the activity of plants, algae/phytoplankton, and microorganisms (biogenic PAHs) [50,51]. Diagnostic ratios based on PAHs physical and chemical properties and stability against photolysis can be used to distinguish between petrogenic and pyrogenic origins. Congener ratios of (Ant/Ant + Phe)

> 0.1 and (Phe/Ant) < 10 indicate pyrogenic sources. The (BaA/BaA + Chr) and (Flt/Flt + Pyr) ratios are used to further distinguish combustion sources: high ratios (>0.35 and >0.50, respectively) indicate grass, wood or coal combustion, intermediate ratios (0.20–0.35 and 0.40–0.50, respectively) indicate liquid fossil fuel combustion or mixed petrogenic and pyrogenic origin and low ratios (<0.20 and <0.40) usually imply petrogenic sources [52,53].

The ratios of (Ant/Ant + Phe) and (Phe/Ant) were calculated with results between 0.68–0.82 and between 0.23–0.50, respectively, for samples in which PAHs were detected. The application of (BaA/BaA + Chr) and (Flt/Flt+Pyr) ratios was also performed, with values > 0.35 and >0.50, also suggesting a pyrogenic source of PAHs (Figure 7).

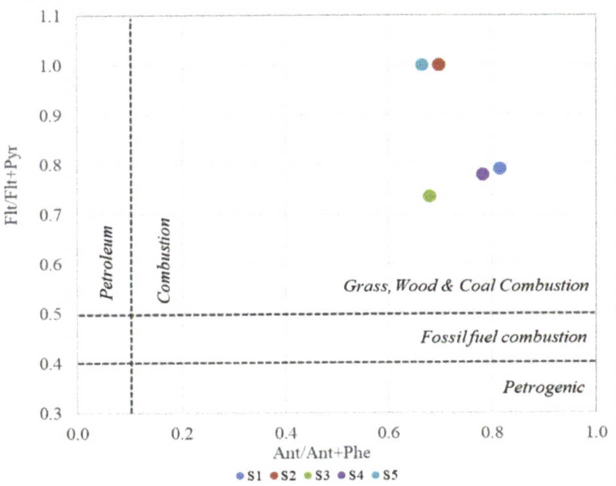

Figure 7. Ratios of (anthracene (Ant)/Ant + phenanthrene (Phe)) and (fluoranthene (Flt)/Flt + pyrene (Pyr)) for samples in which PAHs were detected.

4. Conclusions

This study identified the main impacts of a large forest wildfire on groundwater quality of springs connected to small public supply systems in a peri-urban area. Results pointed out that the best practice for assessing wildfire hydrochemical effects is to start monitoring programs immediately after the wildfire event, and proceed with sampling campaigns for at least 12 months, since several parameters show considerable variation through time. It was also found that extreme events, like intense precipitation, were much more important for groundwater contamination than long-term average changes.

An increase in several parameters, such as sulfate, fluoride, phosphorous, nitrogen compounds, iron, manganese and chromium was observed in Braga peri-urban aquifers. Concerning PAHs, the results reflected the fire impact mainly through the profile of the compounds that appear in the BR springs, which differ significantly from the control spring. Six months after the wildfire, and after the first intense rain event, carcinogenic PAHs, including BaP, began to be detected in considerable concentrations, corroborating the idea of it being difficult to predict the long-term impacts of wildfires on groundwater quality.

Although the connection between groundwater depletion and destructive wildfires might seem tenuous at first glance, and the parametric values for drinking water established by international guidelines have not been exceeded, the results clearly demonstrate the vulnerability of aquifers to wildfires, especially for PAHs, which constitutes an issue yet poorly understood in terms of both the magnitude and persistence.

Water **2020**, *12*, 1146

More information is needed on appropriate monitoring strategies in order to identify a standard set of trace and major compounds to be analyzed and establish protocols to effectively assess water quality, which is essential for developing sustainable water resources management practices.

Supplementary Materials: The following are available online at http://www.mdpi.com/2073-4441/12/4/1146/s1, Table S1: Values for statically analyses trends in the different sampling time for burned areas.

Author Contributions: Conceptualization, C.M. and J.E.M; Data curation, C.M., A.M., Z.E.M., J.E.M.; Formal analysis, A.M., A.M.P.; Writing—original draft, C.M., A.M., J.E.M.; Writing—review and editing, C.M., I.M.P.L.V.O.F. All authors have read and agreed to the published version of the manuscript.

Funding: This work received financial support from the European Union (FEDER funds POCI/01/0145/FEDER/007265) and National Funds (FCT/MEC, Fundação para a Ciência e Tecnologia and Ministério da Educação e Ciência) under the Partnership Agreement PT2020 UID/QUI/50006/2013. The author J. Espinha Marques acknowledges the funding provided by the Institute of Earth Sciences (ICT), under contracts UIDB/04683/2020 with FCT (the Portuguese Science and Technology Foundation), and COMPETE POCI-01-0145-FEDER-007690.

Conflicts of Interest: The authors declare no conflict of interest.

References

1. Chou, Y.H. Management of Wildfires with a Geographical Information-System. *Int. J. Geogr. Inf. Syst.* **1992**, *6*, 123–140. [CrossRef]

2. Turco, M.; Rosa-Cánovas, J.J.; Bedia, J.; Jerez, S.; Montávez, J.P.; Llasat, M.C.; Provenzale, A. Exacerbated fires in Mediterranean Europe due to anthropogenic warming projected with non-stationary climate-fire models. *Nat. Commun.* **2018**, *9*, 3821. [CrossRef] [PubMed]

3. Salis, M.; Ager, A.A.; Alcasena, F.J.; Arca, B.; Finney, M.A.; Pellizzaro, G.; Spano, D. Analyzing seasonal patterns of wildfire exposure factors in Sardinia, Italy. *Environ. Monit. Assess.* **2015**, *187*, 4175. [CrossRef] [PubMed]

4. Tartaglia, E.S.; Aronson, M.F.J.; Raphael, J. Does Suburban Horticulture Influence Plant Invasions in a Remnant Natural Area? *Nat. Area. J.* **2018**, *38*, 259–267. [CrossRef]

5. Bardsley, D.K.; Weber, D.; Robinson, G.M.; Moskwa, E.; Bardsley, A.M. Wildfire risk, biodiversity and peri-urban planning in the Mt Lofty Ranges, South Australia. *Appl. Geogr.* **2015**, *63*, 155–165. [CrossRef]

6. Esposito, G.; Parodi, A.; Lagasio, M.; Masi, R.; Nanni, G.; Russo, F.; Alfano, S.; Giannatiempo, G. Characterizing Consecutive Flooding Events after the 2017 Mt. Salto Wildfires (Southern Italy): Hazard and Emergency Management Implications. *Water* **2019**, *11*, 2663. [CrossRef]

7. Chas-Amil, M.L.; Touza, J.; Garcia-Martinez, E. Forest fires in the wildland-urban interface: A spatial analysis of forest fragmentation and human impacts. *Appl. Geogr.* **2013**, *43*, 127–137. [CrossRef]

8. Font, M.; Chauvin, S.; Plana, E.; Garcia, J.; Gladiné, J.; Serra, M. *Forest Fire Risk in the Wildland-urban Interface, Elements for the Analysis of the Vulnerability of Municipalities and Homes at Risk*; FIRECOM project (DG ECHO 2014/PREV/13); CTFC Editions: Solsona, Spain, 2016.

9. Vyklyuk, Y.; Radovanović, M.M.; Pasichnyk, V.; Kunanets, N.; Petro, S. Forecasting of Forest Fires in Portugal Using Parallel Calculations and Machine Learning. In *Recent Developments in Data Science and Intelligent Analysis of Information*; Springer: Cham, Switzerland, 2018; pp. 39–49.

10. JRC. *Forest Fires in Europe, Middle East and North Africa 2017*; Joint Research Centre (JRC): Ispra (VA), Italy, 2018.

11. Tecle, A.; Neary, D. Water Quality Impacts of Forest Fires. *J. Pollut. Eff. Cont.* **2015**, *3*, 1–7. [CrossRef]

12. Carignan, R.; D'Arcy, P.; Lamontagne, S. Comparative impacts of fire and forest harvesting on water quality in Boreal Shield lakes. *Can. J. Fish Aquat. Sci.* **2000**, *57*, 105–117. [CrossRef]

13. Miller, M.E.; Billmire, M.; Elliot, W.J.; Endsley, K.A.; Robichaud, P.R. Rapid Response Tools and Datasets for Post-Fire Modeling: Linking Earth Observations and Process-Based Hydrological Models to Support Post-Fire Remediation. *Int. Arch. Photogramm.* **2015**, *47*, 469–476. [CrossRef]

14. Smith, H.G.; Sheridan, G.J.; Lane, P.N.J.; Nyman, P.; Haydon, S. Wildfire effects on water quality in forest catchments: A review with implications for water supply. *J. Hydrol.* **2011**, *396*, 170–192. [CrossRef]

15. Yuan, H.; Tao, S.; Li, B.; Lang, C.; Cao, J.; Coveney, R.M. Emission and outflow of polycyclic aromatic hydrocarbons from wildfires in China. *Atmos. Environ.* **2008**, *42*, 6828–6835. [CrossRef]

16. Nunes, B.; Silva, V.; Campos, I.; Pereira, J.L.; Pereira, P.; Keizer, J.J.; Gonçalves, F.; Abrantes, N. Off-site impacts of wildfires on aquatic systems—Biomarker responses of the mosquitofish Gambusia holbrooki. *Sci. Total Environ.* **2017**, *581–582*, 305–313. [CrossRef] [PubMed]

17. WHO. *Guidelines for Drinking-Water Quality, Fourth Edition, Incorporating the 1st Addendum*; World Health Organization: Geneva, Switzerland, 2017.

18. Zuo, J.; Brewer, D.S.; Arlt, V.M.; Cooper, C.S.; Phillips, D.H. Benzo pyrene-induced DNA adducts and gene expression profiles in target and non-target organs for carcinogenesis in mice. *BMC Genom.* **2014**, *15*, 880. [CrossRef]

19. Albinet, A.; Leoz-Garziandia, E.; Budzinski, H.; Villenave, E. Polycyclic aromatic hydrocarbons (PAHs), nitrated PAHs and oxygenated PAHs in ambient air of the Marseilles area (South of France): Concentrations and sources. *Sci. Total. Environ.* **2007**, *384*, 280–292. [CrossRef]

20. Abdel-Shafy, H.I.; Mansour, M.S.M. A review on polycyclic aromatic hydrocarbons: Source, environmental impact, effect on human health and remediation. *Egypt. J. Petrol.* **2016**, *25*, 107–123. [CrossRef]

21. Keith, L.H. The Source of U.S. EPA's Sixteen PAH Priority Pollutants. *Polycycl. Aromat. Comp.* **2015**, *35*, 147–160. [CrossRef]

22. EC. *Directive 2008/105/EC of the European Parliament and of the Council on Environmental Quality Standards in the Field of Water Policy, Amending and Subsequently Repealing Council Directives 82/176/EEC, 83/513/EEC, 84/156/EEC, 84/491/EEC, 86/280/EEC and Amending Directive 2000/60/EC of the European Parliament and of the Council*; Official Journal of the European Union: Aberdeen, UK, 2008.

23. Moody, J.A.; Ebel, B.A.; Nyman, P.; Martin, D.A.; Stoof, C.R.; McKinley, R. Relations between soil hydraulic properties and burn severity. *Int. J. Wildland Fire* **2016**, *25*, 279–293. [CrossRef]

24. INE. *Censos 2011 Resultados Definitivos-Região Norte*; Instituto Nacional de Estatística, I.P.: Lisboa, Portugal, 2012.

25. AEMET-IM. *Atlas Climático Ibérico. Temperatura do Ar e Precipitação (1971–2000)*; Departamento de Producción da Agência Estatal de Meteorologia de Espanha (Área de Climatología y Aplicaciones Operativas) e Departamento de Meteorologia e Clima (Divisão de Observação Meteorológica e Clima), do Instituto de Meteorologia: Madrid, Spain, 2011.

26. Peel, M.C.; Finlayson, B.L.; McMahon, T.A. Updated world map of the Köppen-Geiger climate classification. *Hydrol. Earth Syst. Sci.* **2007**, *11*, 1633–1644. [CrossRef]

27. Pereira, D.M.I.; Pereira, J.S.P.; Santos, L.J.C.; França da Silva, J.M. Unidades geomorfológicas de Portugal Continental. *Rev. Bras. Geomorf.* **2014**, *15*, 567–584. [CrossRef]

28. Ribeiro, A.; Munhá, J.; Dias, R.; Mateus, A.; Pereira, E.; Ribeiro, L.; Fonseca, P.; Araújo, A.; Oliveira, T.; Romão, J.; et al. Geodynamic evolution of the SW Europe Variscides. *Tectonics* **2007**, *26*, TC6009. [CrossRef]

29. ICNF. *9.º Relatório Provisório de Incêndios Florestais -2017*; Departamento de Gestão de Áreas Públicas e de Proteção Floresta: Lisboa, Portugal, 2017.

30. Guerreiro, J.; Fonseca, C.; Salgueiro, A.; Fernandes, P.; Lopez Iglésias, E.; de Neufville, R.; Mateus, F.; Castellnou Ribau, M.; Sande Silva, J.; Moura, J.M.; et al. *Avaliação dos incêndios ocorridos entre 14 e 16 de outubro de 2017 em Portugal Continental. Relatório Final*; Comissão Técnica Independente, Assembleia da República.: Lisboa, Portugal, 2018.

31. IPMA. *Boletim Climatológico*; Instituto Português do Mar e da Atmosfera, I.P.: Lisboa, Portugal, 2018.

32. Borges, B.; Armindo, M.; Ferreira, I.M.P.L.V.O.; Mansilha, C. Dispersive liquid–liquid microextraction for the simultaneous determination of parent and nitrated polycyclic aromatic hydrocarbons in water samples. *Acta Chromatog.* **2018**, *30*, 119–126. [CrossRef]

33. Martin, D.A. At the nexus of fire, water and society. *Philos. Trans. R. Soc. B* **2016**, *371*, 20150172. [CrossRef]

34. Andreu, V.; Rubio, J.L.; Cerni, R. Effect of Mediterranean shrub on water erosion control. *Environ. Monit. Assess.* **1995**, *37*, 5–15. [CrossRef] [PubMed]

35. Certini, G.; Nocentini, C.; Knicker, H.; Arfaioli, P.; Rumpel, C. Wildfire effects on soil organic matter quantity and quality in two fire-prone Mediterranean pine forests. *Geoderma* **2011**, *167*, 148–155. [CrossRef]

36. Mansilha, C.; Duarte, C.G.; Melo, A.; Ribeiro, J.; Flores, D.; Marques, J.E. Impact of wildfire on water quality in Caramulo Mountain ridge (Central Portugal). *Sustain. Water Resour. Manag.* **2017**, *5*, 319–331. [CrossRef]

37. Townsend, S.A.; Douglas, M.M. The effect of a wildfire on stream water quality and catchment water yield in a tropical savanna excluded from fire for 10 years (Kakadu National Park, North Australia). *Water Res.* **2004**, *38*, 3051–3058. [CrossRef]

38. Mast, M.A.; Clow, D.W. Effects of 2003 wildfires on stream chemistry in Glacier National Park, Montana. *Hydrol. Process.* **2008**, *22*, 5013–5023. [CrossRef]

39. Delwiche, J. After the Fire, Follow the Nitrogen. *JFSP Briefs.* **2010**, *80*. Available online: http://digitalcommons.unl.edu/jfspbriefs/80 (accessed on 15 January 2020).

40. Ranalli, A.J. *A Summary of the Scientific Literature on the Effects of Fire on the Concentration of Nutrients in Surface Waters*; U.S. Geological Survey: Reston, VA, USA, 2004.

41. Rust, A.J.; Hogue, T.S.; Saxe, S.; McCray, J. Post-fire water-quality response in the western United States. *Inter. J. Wildland Fire* **2018**, *27*, 203–216. [CrossRef]

42. Sham, C.H.; Tuccillo, M.E.; Rooke, J. *Report on the Effects of Wildfire on Drinking Water Utilities and Effective Practices for Wildfire Risk Reduction and Mitigation*; Water Research Foundation and U.S. Environmental Protection Agency: Waltham, MA, USA, 2013.

43. EC. *Commission Staff Working Document Accompanying the Report from the Commission in Accordance with Article 3.7 of the Groundwater Directive 2006/118/EC on the Establishment of Groundwater Threshold Values*; European Commission: Brussels, Belgium, 2010.

44. IARC. *Agents Classified by the IARC Monographs, Volumes 1 to 113*; International Agency for Research on Cancer: Lyon, France, 2015; Volume 1–113.

45. ATSDR. *Priority List of Hazardous Substances*; Agency for Toxic Substances and Disease Registry: Atlanta, Georgia, 2015.

46. Hyzd'alova, M.; Pivnicka, J.; Zapletal, O.; Vazquez-Gomez, G.; Matthews, J.; Neca, J.; Pencikova, K.; Machala, M.; Vondracek, J. Aryl Hydrocarbon Receptor-Dependent Metabolism Plays a Significant Role in Estrogen-Like Effects of Polycyclic Aromatic Hydrocarbons on Cell Proliferation. *Toxicol. Sci.* **2018**, *165*, 447–461. [CrossRef]

47. EC. *Directive 98/83/EC on the Quality of Water Intended for Human Consumption*; Official Journal of the European Union: Aberdeen, UK, 1998.

48. MDH. *Guidance for Evaluating the Cancer Potency of Polycyclic Aromatic Hydrocarbon (PAH) Mixtures in Environmental Samples*; Minnesota Department of Health: St. Paul, MN, USA, 2016.

49. Stogiannidis, E.; Laane, R. Source Characterization of Polycyclic Aromatic Hydrocarbons by Using Their Molecular Indices: An Overview of Possibilities. *Rev. Environ. Contam. Toxicol.* **2015**, *234*, 49–133.

50. Denis, E.H.; Toney, J.L.; Tarozo, R.; Scott Anderson, R.; Roach, L.D.; Huang, Y. Polycyclic aromatic hydrocarbons (PAHs) in lake sediments record historic fire events: Validation using HPLC-fluorescence detection. *Org. Geochem.* **2012**, *45*, 7–17. [CrossRef]

51. Nasher, E.; Heng, L.Y.; Zakaria, Z.; Surif, S. Concentrations and Sources of Polycyclic Aromatic Hydrocarbons in the Seawater around Langkawi Island, Malaysia. *J. Chem.* **2013**, *2013*, 975781. [CrossRef]

52. Edokpayi, J.; Odiyo, J.; Popoola, O.; Msagati, T. Determination and Distribution of Polycyclic Aromatic Hydrocarbons in Rivers, Sediments and Wastewater Effluents in Vhembe District, South Africa. *Int. J. Environ. Res. Public Health* **2016**, *13*, 387. [CrossRef] [PubMed]

53. Yunker, M.B.; Macdonald, R.W.; Vingarzan, R.; Mitchell, R.H.; Goyette, D.; Sylvestre, S. PAHs in the Fraser River basin: A critical appraisal of PAH ratios as indicators of PAH source and composition. *Org. Geochem.* **2002**, *33*, 489–515. [CrossRef]

Catarina Mansilha [1,2,*], Armindo Melo [1,2], Deolinda Flores [3,4], Joana Ribeiro [3,5], João Ramalheira Rocha [4], Vítor Martins [4], Patrícia Santos [3,4] and Jorge Espinha Marques [3,4]

[1] National Institute of Health Doutor Ricardo Jorge, Department of Environmental Health, University of Porto, 4050-083 Porto, Portugal; armindo.melo@insa.min-saude.pt

[2] LAQV/REQUIMTE, Associated Laboratory for Green Chemistry of the Network of Chemistry and Technology, University of Porto, 4050-083 Porto, Portugal

[3] Institute of Earth Sciences, Porto Pole, University of Porto, 4169-007 Porto, Portugal; dflores@fc.up.pt (D.F.); joana.ribeiro@uc.pt (J.R.); patricia.santos@fc.up.pt (P.S.); jespinha@fc.up.pt (J.E.M.)

[4] Department of Geosciences, Environment and Spatial Planning, University of Porto, 4169-007 Porto, Portugal; up201206901@fc.up.pt (J.R.R.); up199702778@fc.up.pt (V.M.)

[5] Department of Earth Sciences, University of Coimbra, 3030-790 Coimbra, Portugal

* Correspondence: catarina.mansilha@insa.min-saude.pt

Citation: Mansilha, C.; Melo, A.; Flores, D.; Ribeiro, J.; Rocha, J.R.; Martins, V.; Santos, P.; Espinha Marques, J. Irrigation with Coal Mining Effluents: Sustainability and Water Quality Considerations (São Pedro da Cova, North Portugal). *Water* **2021**, *13*, 2157. https://doi.org/10.3390/w13162157

Academic Editor: William Frederick Ritter

Received: 21 June 2021
Accepted: 31 July 2021
Published: 5 August 2021

Publisher's Note: MDPI stays neutral with regard to jurisdictional claims in published maps and institutional affiliations.

Abstract: Two water effluents that drain from the abandoned coal mine of São Pedro da Cova (NW Portugal) were characterized in terms of their physic-chemical properties and suitability for irrigation purposes. Samples were also collected in a local surface stream, upstream and downstream from the mine drainage points, also used for irrigation by local farmers. Water samples were analyzed for major and minor ions and for trace element concentrations. Sampling campaigns started in 2017 and ended in 2019 and there were 46 water quality parameters tested. There were also proposed all-inclusive indices (the Water Quality Index and the Contamination Index, and also the Trace Element Toxicity Index) based on specific groups of 18 and 17 physic-chemical parameters, respectively, to achieve adequate monitoring requirements for mine effluents and surface water from coalfield. From the physical and chemical aspects of mine water it is inferred that the mine is not producing acid mine drainage. The coal mine water is of medium to high salinity, having almost neutral pH and a high thermal stability during the year, which is a distinguishing feature of the effluents. When compared to international irrigation water quality standards, as Food and Agriculture Organization of the United Nations admissible concentrations, the impacted waters are unsuitable for irrigation. The major outliers to the guidelines were iron, manganese, potassium, magnesium and bicarbonates, being also detected carcinogenic polycyclic aromatic hydrocarbons. Cost-effective ways of monitoring water quality parameters are needed to help control and manage the impact of coal mine effluents that should be treated before releasing into a ditch system that could be then used by local farmers to irrigate their crops.

Keywords: coal mine wastewater quality; irrigation; heavy metals; water quality index; environmental impact

1. Introduction

Irrigation is fundamental for agriculture but policies that push towards a restrained use of water are not popular among farmers, who are also not prepared to respond to drastic increases in water costs, which could decrease the economic profitability of their activities. Therefore, the use of unconventional free water sources, such as mine wastewaters, is regarded as a possible choice for irrigation. Effluents from coal mines are often considered severe and persistent forms of pollution, with environmental impact not only throughout the mine's life cycle, but also long after the end of mining activities. Some deleterious impacts on the environment include the disruption of hydrological pathways, contamination of surface and groundwater, depression of the water table, soil contamination and loss of biodiversity [1,2]. The concentration of contaminants in coal mine drainage waters vary greatly and depends on a series of geological, hydrological and mining conditions, which are different from mine to mine. Therefore, the effluents can be alkaline, acidic, ferruginous, highly saline, or even clean [3]. Frequently, coal mine drainage is acid metal-rich waters with high concentrations of iron (Fe), copper (Cu), manganese (Mn) and nickel (Ni), formed during water–rock interaction involving sulfur-bearing minerals, such as pyrite (FeS_2). These processes may cause red, orange, or yellow sediments with negative impacts on ecosystems and water resources. Additionally, mine effluents often contain high levels of total dissolved and suspended solids. The dissolved cations include mainly calcium (Ca), magnesium (Mg), sodium (Na) and potassium (K); the major anions are sulfate (SO_4), chloride (Cl), fluoride (F), nitrate (NO_4), bicarbonate (HCO_3) or carbonate (CO_3) [4].

The Douro Coalfield (NW Portugal) represents the most important coal-bearing deposit in Portugal [5–8] with 53 km length and width between 30 and 250 m (Figure 1).

São Pedro da Cova mining area was one of the principal centers of mining activity in Portugal (Figure 2a,b), with great economic and technological impacts, and a cultural significance from the end of the 17th century (1795), until the 20th century (1970) [9].

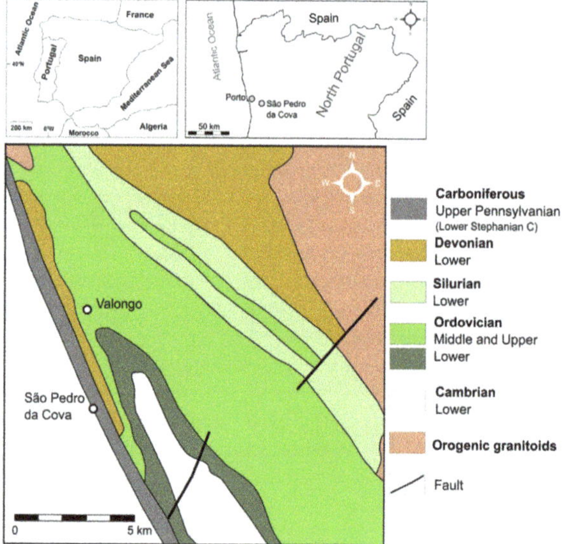

Figure 1. Some aspects of the study area: (**a**) São Pedro da Cova mine in the mid 20th century; (**b**) São Pedro da Cova mine in 2021; (**c**) Silveirinhos stream upstream from the mining effluents discharge; (**d**) mining effluents discharge; (**e**) water monitoring in Silveirinhos stream downstream from the mining effluents discharge; (**f**) irrigation with polluted water from Silveirinhos stream; (**g**) agricultural area irrigated with polluted water from Silveirinhos stream.

The São Pedro da Cova coal mine is an abandoned mine located in a peri-urban area, with a landscape consisting of a mosaic of urban, industrial, agricultural and forest areas (Figure 2a,b). The mine is located very close to a densely populated zone with several basic facilities including schools, a healthcare center and a leisure center.

The coal mining effluents from two mine drainage galleries are discharged around 1 km to SE of the mine, producing an ocher-colored sediment that is continuously accumulated in local watercourses (Silveirinhos stream and Ferreira river). Water from Silveirinhos stream is collected downstream from the mine drainage discharges and is locally used for agriculture irrigation. Local farmers have been using this polluted water for decades to produce a wide variety of products, including green vegetables, corn, barley and fruits (Figure 2c–g).

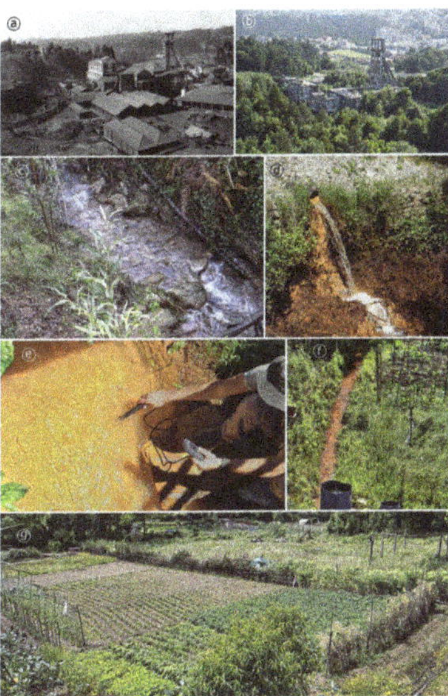

Figure 2. São Pedro da Cova location and geological setting (modified from [10]).

This study aimed at assessing the impact of coal mining effluents on the quality of irrigation water, by means of water quality individual parameters and indices.

This approach intended to contribute to the development of a specific evaluation methodology for coal mining regions in terms of i) the suitability of this type of water for irrigation purposes and ii) the environmental impacts resulting from the uncontrolled applications of these unconventional water resources in irrigated areas.

2. Assessment of Irrigation Water Quality: A Short State of the Art

Mine effluents are used for irrigation purposes worldwide, as they constitute an easily accessible and inexpensive source of water.

The quality of irrigation water is considered a key factor for safe food production [11]. However, when good quality water is scarce, water of marginal value is often considered for use in agriculture [12]. The prolonged and uncontrolled use of polluted water could result in reduced crop yields, deterioration of soil properties and severe environmental and health damages, requiring more complex management practices and more stringent monitoring procedures [13]. Therefore, the agricultural use of water resources in areas affected by mining activities requires not only a baseline water quality data, but also continuous monitoring.

There are several quality standards and guidelines for irrigation water proposed by various countries and organizations, combining conservative and liberal approaches. However, despite being useful, they are not always satisfactory due to the wide variability of hydrological and hydrogeological settings [14–17].

In 1985, the Food and Agriculture Organization of the United Nations (FAO) published a document concerning water quality for agriculture, which was reprinted in 1994, presenting a set of guidelines modified to give more practical procedures for evaluating and managing water quality-related problems, emphasizing long-term effects [18]. These general water quality classification guidelines help identify potential crop production problems associated with the use of conventional water sources, and are equally applicable to evaluate wastewaters for irrigation purposes in terms of their chemical constituents.

As wastewater effluents may contain a number of toxic compounds, FAO also presented threshold levels for some selected trace elements.

FAO's guidelines have been widely incorporated into national regulations all over the world and the following criteria have been considered the most relevant in defining quality [13,15,18]:

(i) Salinity hazard: the concentration of soluble salts in irrigation water, estimated in terms of electrical conductivity. Salinity has been deemed as the most important factor of irrigation water quality because high salinity in soil can create a hostile environment for the crop to absorb nutrients and cause specific ion toxicity;

(ii) Infiltration and permeability problems: the two most common water quality factors which influence the normal infiltration rate are water salinity (total quantity of salts in the water) and the sodium content relative to the calcium and magnesium content;

(iii) Specific toxicity hazard: certain ions and metals can accumulate in sensitive crops in concentrations high enough to cause damage and reduce yields. The ions of primary concern are boron, sodium and chloride;

(iv) Miscellaneous effects: these include high nitrogen concentrations in water, which supplies nitrogen to the crop and may cause excessive vegetative growth, lodging and delay crop maturity; unsightly deposits on fruit or leaves due to overhead sprinkler irrigation using water with high bicarbonate, gypsum, or iron contents; various abnormalities often associated with an unusual water pH.

In order to better classify water based on its specific characteristics and possible uses, a number of models of water quality, named Water Quality Indices (WQI), have also been developed since the 60s [19]. WQI are simplified representations of complex realities, used to assess the suitability of water for certain purposes based on specific characteristics. The use of WQI allows the representation of a large number of parameters in a single numerical value, facilitating the operational management of water resources and their allocation for different uses [20]. However, for this number to accurately represent the reality of a water body, the correct selection of environmental quality parameters is essential [21].

The first modern WQI was proposed by Horton (1965) [22] and was followed by numerous studies in this field. In recent years, many modifications have been considered in the WQI concept and several indices have been proposed and used by governmental agencies and researchers [19,20,23].

Misaghi et al. (2017) [24] introduced the first systematic WQI for irrigation purposes. However, this index considers a limited set of parameters for estimating water quality and does not take into account all potential impacting properties that could be critical, especially regarding wastewaters or other "marginal" quality waters.

The selection of variables is of major importance in calculating WQI as they should be independent and the most relevant ones, in order to define water quality and detect water quality deterioration. Several indices use parameters selected according to the opinion of specific expert panels, and the consequence is that the final evaluation can be highly subjective and variable. Other authors propose that the selection of parameters should be made according to the water management objectives, the location of the studied waters and the sampling periodicity [20].

To assess the impact of mining activities on irrigation water, some indices seemed to be more appropriate than others to be used as additional tools, as the Weight Arithmetic Water Quality Index method (WQI$_A$), proposed by Cude (2001) [25], which is based on Horton's principles but has been modified by introducing the normative values of the major factors of the water [20,25], the Contamination Index (CI), developed by Backman et al. (1998) [26–28] and the Trace Element Toxicity Index (TETI), by Ali et al. (2017) which is based on contaminant hazard intensity [28].

Furthermore, despite the usefulness and importance of normative guidelines, the effect of unusual or special water constituents is not always considered. An example is the contamination by polycyclic aromatic hydrocarbons (PAH), which are persistent, semi-volatile organic pollutants that can result from the oxidation and self-combustion of mine wastes and should be analyzed in mining surroundings due to their genotoxic, mutagenic and carcinogenic properties [29]. Although there are many PAHs, scientists and regulators have focused on 16 compounds that have been identified as priority control pollutants by the U.S. Environmental Protection Agency (USEPA). PAHs in underground mining environments may dissolve in mine water and eventually pollute the groundwater system. In addition, mine effluents can bring PAHs to the surface environment, polluting surface water and soils as well [30].

3. Materials and Methods

3.1. Study Area

The region of São Pedro da Cova (N 41°09′; W 8°30′) is situated in NW Portugal (Figure 1), in the city of Porto peri-urban region and has a resident population of around 16,500 inhabitants [31] and an area of 16 km^2, where the territory comprises residential, industrial, and agroforestry areas (Figure 2a,b,g).

The region is located in the Central-Iberian Zone of the Iberian Massif [32]. The regional geological units consist of metasedimentary rocks (Figure 1) with minor sedimentary cover areas. For that reason, the prevailing groundwater circulation media are fractured.

The study region is located along the western limb of the Valongo Anticline, a regional megastructure [33], which created mountainous landforms reaching 367 and 385 m of altitude at the Santa Justa and Pias summits, respectively, on its western and eastern flanks. The regional morphology is dominated by two hill alignments that originated from differential erosion and are crosscut by Douro River. This structure controlled significantly the regional drainage network, that is part of the Douro river. In São Pedro da Cova it is possible to identify different metasedimentary formations with ages between the Precambrian and/or Cambrian, Ordovician, Silurian, Devonian and Carboniferous.

The old mine of São Pedro da Cova, where exploitation of anthracite A occurred for nearly 200 years, is located in one of the multiple coal deposits hosted in Douro Coalfield (from Upper Pennsylvanian), that represents the most significant Portuguese coal-bearing deposit. This deposit is elongated along NW–SE, presenting an approximate length of 53 km and variable width, ranging from 30 to 250 m. The sedimentary sequence comprises a basal breccia, followed by fossiliferous shales, siltstones and sandstones, along with interlayered conglomerates and coal seams [10].

The regional climate is Atlantic, the normal annual precipitation is around 1254 mm (with 195 mm in December and 18 mm in July) and the normal annual air temperature is around 15°C (with 20°C in July and August and 9°C in January). The Köppen-Geiger climate classification is Csb: warm temperate, with dry and warm summers [34,35].

3.2. Water Sampling

Groundwater samples were collected from two mine drainage galleries (G1 and G2). Additionally, surface water from Silveirinhos stream was sampled in two points, one located upstream (SS-U) and the other downstream (SS-D) from the mine galleries discharge (Figure 2d). Nine sampling campaigns were carried out in two periods, from April 2017 until December 2017 (Apr, Jun, Sep, Dec 2017), and from November 2018 until December 2019 (Nov 2018; Feb, May, Sep, Dec 2019), in a total of 34 samples. In the campaigns of October 2017 and September 2019 it was not possible to collect water from Silveirinhos stream upstream from the drainage galleries because, during the dry season, this part of the stream does not flow.

The sampling points were chosen in order to be preserved from other anthropogenic impacts besides coal mining, reducing the risk of water contamination by other pollution sources. Sampling was carried out according to standard methods: ISO 5667-3:2018 (E) Water quality—Sampling—Part 3 [36]. Samples were collected in glass or polyethylene bottles, were stored at low temperature (<5 °C) in the dark, and delivered to the laboratory within 5 h. The samples were collected with as little agitation or disturbance as possible. Special preservatives were required for some parameters. In this case, care was taken not to flush any preservative out of the bottle during sampling. Conditions such as strong winds or heavy rain were avoided during sampling.

Water samples were analyzed for a range of physical and chemical constituents in the laboratory, while temperature, pH and EC were measured in situ at the moment of sampling (Figure 2e), using a multiparametric meter from Hanna Instruments, model HI-991300, Woonsocket, RI, USA.

This set of sampling points provided a monitoring network for investigating the impact of mine drainage on environment, namely on irrigation water chemistry, including its seasonal variation.

3.3. Laboratory Analysis

Analyses were performed according to procedures outlined in Standard Methods for the Examination of Water and Wastewater 23rd edition [37] and in Le Rodier—L'analyse de l'eau 10e édition [38]. The laboratory has been accredited under ISO/IEC 17025 since 2007. Precision and accuracy were calculated for all analytical methods with values <10%. Uncertainties were also calculated with results varying from 2% to 10%.

Water turbidity was measured in a Hach 2100N Laboratory Turbidity Meter (Hach Lange, Düsseldorf, Germany). Electrical conductivity (EC) and pH were determined in a Crison MultiMeter MM 41 (Hach Lange Spain, S.L.U., Barcelona, Spain). Total alkalinity, carbonates (CO_3^{2-}) and bicarbonates (HCO_3^{-}), were analyzed by titration. Phosphate (PO_4^{2-}) was analyzed in a Shimadzu UV-1601 Spectrophotometer (Shimadzu Corporation, Kyoto, Japan). Chemical oxygen demand (COD) was evaluated in a Hach DR 2800 Spectrophotometer (Hach Company, Loveland, CO, USA). Major inorganic ions (Na^+, K^+, Mg^{2+}, Ca^{2+}, Li^+, Cl^-, NO_3^-, F^- and SO_4^{2-}) were analyzed by ion chromatography (DionexTM system DX-120/ICS-1000, Dionex Corporation, Sunnyvale, CA, USA). Total organic carbon (TOC) was analyzed in a Shimadzu TOC-V (TOC-ASI-V, Shimadzu Corporation, Kyoto, Japan), heavy metals (Cr, Mn, Ni, Cu, Zn, As, Cd and Pb) and other components, such as Al, Fe, NO_2^-, NH_4^+ and SiO_2 were analyzed in a Varian AA240 Atomic Absorption Spectrometer (Varian Inc., Palo Alto, CA, USA) and in a Continuous Segmented Flow Instrument (San-Plus Skalar, Skalar Analytical, Breda, The Netherlands), respectively. PAHs were analyzed by dispersive liquid–liquid microextraction coupled to gas chromatography/mass spectrometry (DLLME–GC/MS) methodology in a Shimadzu GCMS-QP2010 gas chromatograph mass spectrometer equipped with an auto injector AOC5000 (Shimadzu Corporation, Kyoto, Japan), according the procedure described in Borges et al. 2018 [39].

Analytical standards were supplied by Sigma–Aldrich (Steinheim, Germany) and Merck (Darmstadt, Germany). The reference standard mixture containing the 16 EPA PAHs (naphthalene, Nap; acenaphthylene, Acy; acenaphthene, Ace; fluorene, Flu; phenanthrene, Phe; anthracene, Ant; fluoranthene, Flt; pyrene, Pyr; benz[*a*]anthracene, BaA; chrysene, Chr; benzo[*b*]fluoranthene, BbF; benzo[*k*]fluoranthene BkF; benzo[*a*]pyrene, BaP; dibenz[*a,h*]anthracene, DahA; benzo[*ghi*]perylene, BghiP; and indeno[1,2,3-*cd*]pyrene, Ind) was purchased from Sigma-Aldrich (Steinheim, Germany).

Methanol, dichloromethane and acetonitrile were organic trace analysis grade Supra-Solv and were supplied by Merck (Darmstadt, Germany). Ultrapure water was highly purified by a Milli-Q gradient system (18.2 mΩ/cm) from Millipore (Milford, MA, USA).

3.4. Irrigation Water Quality Parameters

Water quality evaluation is necessary to assess the suitability of water to serve a specific purpose, and to determine appropriate treatments or precautions, if necessary. However, monitoring all parameters involved in a water source could be time-consuming and expensive. Therefore, reducing the subjectivity and the effective cost for assessing water quality is a great challenge.

This study focuses on the parameters adopted by FAO guidelines 29 [18] and on a set of quantitative assessment ratios which included the widely applied Sodium Adsorption

Ratio (SAR), the Total Hardness (TH), the Residual Sodium Carbonate (RSC) and the Permeability Index (PI) (Table 1).

Table 1. Water quality classification as per different water quality ratios/parameters.

Parameter		Categories	Ranges
Sodium Adsorption Ratio (SAR) [40]	$SAR = \dfrac{Na^+}{\sqrt{(Ca^{2+}+Mg^{2+})/2}}$	• Excellent • Good • Doubtful • Unsuitable	<10 10–18 18–26 >26
Total Hardness (TH) [41]	$TH = \left(Ca^{2+}+Mg^{2+}\right)\times 50$	• Soft • Moderately hard • Hard • Very hard	<75 75–150 150–300 >300
Residual Sodium Carbonate (RSC) [40]	$RSC = \left(CO_3^{2-}+HCO_3^-\right)-\left(Ca^{2+}+Mg^{2+}\right)$	• Good • Medium • Not suitable	<1.25 1.25–2.5 > 2.5
Permeability index (PI) [42]	$PI = \dfrac{Na^+ +\sqrt{HCO_3^-}}{Ca^{2+}+Na^+ +Mg^{2+}}\times 100$	• Class I • Class II • Class III	>75 25–75 <25

Moreover, in order to meet the hydrogeological and hydrogeochemical specificity of coal mining effluents, the water quality indices WQI_A, CI and TETI were considered of significant importance and, therefore, were also calculated.

The water quality index based on the weighted arithmetic method (WQI_A) was amended to be specific for irrigation, being adjusted taking into consideration the FAO recommendations, that is, the weights were defined as functions of the standards proposed in this guideline. For computation of the WQI_A index, 18 water quality parameters were used, namely, the EC to estimate the salinity hazard; 3 elements with specific ion toxicity (Na^+, Cl^- and B); 3 elements with miscellaneous effects (NO_3^-, HCO_3^- and pH) and 11 trace elements with Recommended Maximum Concentration Values (Al^{3+}, As, Cd^{2+}, Pb^{2+}, Cu^{2+}, Cr^{3+}, Fe^{2+}, Mn^{2+}, Ni^{2+}, Zn^{2+} and F^-).

WQIA was calculated by using the following equation [20,43,44]:

$$WQI_A = \sum_{i=1}^{n} Q_i W_i \Big/ \sum_{i=1}^{n} W_i \tag{1}$$

The quality rating scale (Q_i) for each parameter for a total of n water quality parameters is calculated by using this expression:

$$Qi = 100[(V_i - V_0/S_i - V_0)] \tag{2}$$

where, V_i is the actual value of the ith water quality parameter obtained from laboratory analysis, V_0 is the ideal value of that water quality parameter obtained from standard Tables ($V_0 = 0$, except for pH = 7.0) and S_i is the recommended standard value of ith parameter.

The relative unit weight (W_i) for each water quality parameter is calculated by adopting the following formula:

$$W_i = K/S_i \tag{3}$$

where, K is the proportionality constant and can also be calculated by using the following equation:

$$K = \frac{1}{\sum_{i=1}^{n}(1/S_i)} \tag{4}$$

The proposed index ranges from 0 to 100 and plain descriptions for index data were developed in order to provide a qualitative description of the index outcome [44,45]. The calculation of WQI_A following the 'weighted arithmetic index method' involves the estimation of 'unit weight', assigned to each physic-chemical parameter considered for the calculation. By assigning unit-weights, all the concerned parameters of different units and dimensions are transformed to a common scale.

Weightage of each parameter means its relative importance in the overall water quality, and it depends on the permissible limits. Those parameters which have low permissible limits and can influence water quality to a large extent allocate high weighting, while parameters having high permissible limits are less harmful to the water quality and allocate low weighting.

Table 2 shows the irrigation water quality standards and the unit weights assigned to each parameter used for calculating the WQI_A index. Maximum weights were assigned to cadmium (0.7135), arsenic and chromium (0.07135) and to copper, manganese and nickel (0.03568), thus suggesting the key significance of these trace elements in water quality assessment and their considerable impact on the index.

Table 2. Standards for irrigation water and relative weight of parameters.

No.	Parameter [1]	Standards [2] Si	$1/Si$	K	Relative Weight Wi
1	pH	6.5–8.5	0.1176		0.00084
2	Elect.Conduct.(EC)	0.7	1.429		0.01019
3	Sodium	69	0.014		0.00010
4	Chloride	141.6	0.007		0.00005
5	Boron	0.7	1.429		0.01019
6	Nitrate	22.14	0.045		0.00032
7	Bicarbonate	91.46	0.011		0.00008
8	Aluminum	5.0	0.2		0.00143
9	Arsenic	0.10	10.0		0.07135
10	Cadmium	0.01	100	0.007	0.71350
11	Chromium	0.10	10.0		0.07135
12	Copper	0.20	5.0		0.03568
13	Fluoride	1.0	1.0		0.00714
14	Iron	5.0	0.2		0.00143
15	Manganese	0.20	5.0		0.03568
16	Nickel	0.20	5.0		0.03568
17	Lead	5.0	0.2		0.00143
18	Zinc	2.0	0.5		0.00357

[1] All values are in mg/L, except pH and EC (mS·cm^{-1}); [2] FAO [18]; source: own elaboration.

The CI was also calculated as a sum of the contamination factors of individual components (the 18 water quality parameters chosen for WQI_A calculation, analyzed in the nine sampling campaigns), some of these exceeding the trigger values recommended by FAO [18]. The CI is determined by the following formula [26]:

$$CI = \sum_{i=1}^{n} \left[\left(\frac{C_{Ai}}{C_{Ni}} \right) - 1 \right] \tag{5}$$

where C_{Ai} and C_{Ni} represent the analytical value and upper permissible concentration of the ith component, respectively. Note that C_{Ni} is taken as maximum allowable concentration.

Based on the CI index, values less than 1, 1–3, and more than 3 indicate low, medium and high levels of contamination, respectively [26].

TETI [28] was calculated based on the contaminants hazard intensity. The hazard intensity, or total score, of each parameter was determined according to the Toxicological Profiles of the Priority List of Hazardous Substances prepared by the Agency for Toxic Substances and Disease Registry (ATSDR), the Division of Toxicology and Environmental

Medicine, Atlanta, GA, USA [46]. The ATSDR prioritization of substances is based on a combination of their frequency, toxicity, and potential for human exposure.

The concentration of each trace element detected was multiplied by its total score, and products were added to calculate TETI. The proposed TETI only considers toxic elements in water and is calculated by using the expression:

$$TETI = \sum_{i=1}^{n} C_i \times TS_i \tag{6}$$

where C_i is the concentration of each individual trace element and TS_i is its Total Score (ATSDR). This index clearly represents the impact of mine activities on the aquatic environment, where the lower TETI value represents better water quality.

In addition to the conventional parameters of water quality, studies on organic pollutants as PAHs are also very important, as several mining activities such as coal mining, processing or storage of coal provide the basic conditions for the generation and release of these compounds.

As PAHs can also become a source of pollution after the abandonment of coal mines, the 16 priority hydrocarbons were analyzed in São Pedro da Cova samples in order to investigate contamination levels and distribution [47].

3.5. Statistical Analysis

Data obtained for different parameters were tested for distribution of the residuals with the Shapiro–Wilk's test. Chemical concentrations were studied using a one-way analysis of variance (ANOVA), if normal distribution of the residuals was confirmed. Welch correction was applied when the homogeneity of variances was not verified. Whenever statistical significances were found, Tukey's test or the Tamhane's test post-hoc tests were applied for mean comparison, depending on variances assumption or not. If normal distribution of the residuals was not found, the analyzed parameters were studied using a Kruskal–Wallis test. Whenever statistical significances were found, Dunn's post-hoc test was applied for median comparison. All statistical analyses were performed at 5% significance level using R version 4.0.2 (R Project for Statistical Computing).

4. Results and Discussion

4.1. Water chemistry

The values of the physical and chemical parameters used to evaluate water quality from the two mine drainage galleries (G1 and G2, groundwater) and from Silveirinhos stream, collected upstream (SS-U) and downstream (SS-D) from the mine effluents discharge points are reported in Table 3. Results were compared to FAO guidelines in order to assess the water suitability for irrigation. The mining effluents G1 and G2 correspond to groundwater which circulates in the exploited rock massif as well as in the mine galleries. The SS-U water corresponds to surface water without mining influence and the SS-D water originates from the mixture of SS-U, G1 and G2 waters.

Table 3. Summary statistics of physical and chemical parameters analyzed in mining effluents and in Silveirinhos stream. Comparison with FAO Guidelines for irrigation.

Parameters	Units	FAO Guidelines for Irrigation [18]			Silveirinhos Stream (SS-U)	Mine Wastewater (G1)	Mine Wastewater (G2)	Silveirinhos Stream (SS-D)	*p*
Potential Irrigation Problem		Degree of Restriction on Use							
		None	Slight to Moderate	Severe					
Salinity (affects crop water availability)									
EC	dS/m	<0.7	0.7–3.0	>3.0	0.08 ± 0.02 c	0.84 ± 0.14 a	0.92 ± 0.18 a	0.56 ± 0.21 b	<0.0001 ##
Specific Ion toxicity (affects sensitive crops)									
Sodium (Na)	meq/L	<3	3–9	>9	0.32 ± 0.04 b	1.03 ± 0.08 a	1.03 ± 0.06 a	0.78 ± 0.22 a	<0.0001 ##
Chloride (Cl)	meq/L	<4	4–10	>10	0.39 ± 0.03	0.87 ± 0.18	0.90 ± 0.12	0.72 ± 0.17	0.0003 &
Boron (B)	mg/L	<0.7	0.7–3.0	>3.0	0.07 ± 0.02 a	0.24 ± 0.07 bc	0.26 ± 0.08 c	0.14 ± 0.05 ab	0.00127 #
Trace Elements (* Recommended Maximum Concentration)									
Aluminum (Al)	mg/L		5.0 *		0.19 ± 0.05 a	0.37 ± 0.06 b	0.44 ± 0.09 c	0.21 ± 0.19 ab	0.0359 #
Arsenic (As)	mg/L		0.10 *		0.001 ± 0.001 d	0.015 ± 0.005 a	0.034 ± 0.009 b	0.008 ± 0.003 c	<0.0001 ##
Cadmium (Cd)	mg/L		0.01 *		<LD	<LD	<LD	<LD	-
Chromium (Cr)	mg/L		0.10 *		<LD	<LD	<LD	<LD	-
Copper Cu)	mg/L		0.20 *		<LD	<LD	<LD	<LD	-
Fluoride (F)	mg/L		1.0 *		0.04(0.00–0.07) b	0.29(0.00–0.38) a	0.33(0.16–0.35) a	0.15(0.09–0.27) ab	0.0001 &
Iron (Fe)	mg/L		5.0 *		0.12(0.03–0.47) c	52.28(45.66–62.39) ab	84.66(57.65–115.08) a	18.24(15.72–62.79) b	<0.0001 &
Manganese (Mn)	mg/L		0.20 *		0.06 ± 0.02 b	4.96 ± 1.08 a	4.82 ± 1.16 a	2.93 ± 1.72 a	<0.0001 ##
Nickel (Ni)	mg/L		0.20 *		<LD b	0.04(0.00–0.05) a	0.02(0.00–0.03) ab	0.02(0.00–0.03) ab	0.0008 &
Lead (Pb)	mg/L		5.0 *		<LD	<LD	<LD	<LD	-
Zinc (Zn)	mg/L		2.0 *		0.01(0.00–0.90)	0.00(0.00–0.08)	0.00(0.00–0.03)	0.00(0.00–0.03)	n.s.
Miscellaneous Effects (affects susceptible crops)									
Nitrate (NO₃-N)	mg/L	<5	5–30	>30	0.42(0.23–2.20)	0.28(0.00–1.62)	0.28(0.00–0.67)	0.61(0.05–0.91)	n.s.
Bicarbonates (HCO₃)	meq/L	<1.5	1.5–8.5	>8.5	0.12 ± 0.03 c	2.99 ± 0.70 ab	3.37 ± 1.04 a	1.71 ± 1.10 b	<0.0001 ##
Carbonates (CO₃)	meq/L				<LD	<LD	<LD	<LD	-
pH			Normal range 6.5–8.4		6.26(6.17–6.56) ab	6.16(6.09–6.44) a	6.12(6.00–6.57) a	6.51(6.23–6.70) b	0.0005 &
Turbidity	NTU				1.1(0.3–4.9) b	74.2(9.0–328.5) a	111.5(1.9–342.0) a	32.5(5.3–109.6) ab	0.0008 &
Cyanide	mg/L				<LD	<LD	<LD	<LD	-

Table 3. *Cont.*

Parameters		FAO Guidelines for Irrigation [18]			Silveirinhos Stream (SS-U)	Mine Wastewater (G1)	Mine Wastewater (G2)	Silveirinhos Stream (SS-D)	p
Potential Irrigation Problem	Units	Degree of Restriction on Use							
		None	Slight to Moderate	Severe					
Salinity: Other cations and anions		*Usual range in irrigation water*							
Calcium (Ca)	meq/L		0–20		0.19 ± 0.07 b	4.08 ± 0.62 a	4.12 ± 1.12 a	2.71 ± 1.30 a	<0.0001 ##
Magnesium (Mg)	meq/L		0–5		0.25 ± 0.05 c	5.43 ± 0.82 a	5.10 ± 1.39 ab	3.46 ± 1.54 b	<0.0001 ##
Sulfate (SO_4)	meq/L		0–20		0.13(0.09–0.20) b	3.56(2.96–3.98) a	4.24(3.03–5.03) a	1.79(1.16–4.08) ab	<0.0001 &
Nutrients									
Nitrite (NO_2-N) [1]	mg/L				<LD	<LD	<LD	<LD	-
Ammonium (NH_4-N) [1]	mg/L		0–5		0.01(0.00–0.02) b	0.60(0.32–1.80) a	0.59(0.26–2.60) a	0.52(0.13–2.06) a	0.0006 &
Phosphate (PO_4-P) [1]	mg/L		0–2		0.01(0.00–0.40)	0.05(0.00–0.30)	0.01(0.00–0.33)	0.00(0.00–0.10)	n.s.
Potassium (K)	mg/L		0–2		0.51(0.00–1.17) b	4.10(3.06–16.40) a	5.14(3.32–19.40) a	2.08(0.00–13.11) ab	0.0009 &
TOC	mg/L				0.80(0.23–1.55)	0.56(0.00–33.16)	0.40(0.00–14.47)	0.63(0.60–28.56)	n.s.
COD	mg/L				5.65(0.10–7.42)	10.20(0.00–14.70)	11.60(0.00–19.10)	7.96(1.20–11.00)	n.s.
Suitability for irrigation water and soil water infiltration. Color Classification according to guidelines in Table 1									
SAR	meq/L				0.68	0.47	0.48	0.45	-
CROSS	mmol/L				1.12	0.82	0.84	0.76	-
TH	mg/L				22.0	475	461	308	-
RSC	meq/L				−0.3	−6.5	−5.8	−4.5	-
PI	%				87	26	28	30	-

[1] Results reported in terms of elemental nitrogen/phosphorus. EC—electrical conductivity in deciSiemens per metre at 25 °C; TOC—Total Organic Carbon; COD—Chemical Oxygen Demand; SAR—sodium adsorption ratio; CROSS—Cation Ratio of Structural Stability; TH—total hardness; RSC—residual sodium carbonate; PI—permeability index. Different letters (a, b, c) for each parameter in a row show statistically significant differences ($p < 0.05$) between means in normal distribution and median in non-normal distribution. Data expressed as mean $\pm$ standard deviation or as median (minimum−maximum). n.s, not significant. #—p-values from one-way ANOVA analysis. Means were compared by Tukey's test, since homogeneity of variances was confirmed by Levene's test ($p > 0.05$). ##—p-values from one-way Welch ANOVA analysis. Means were compared by Tamhane's test, since homogeneity of variances was not confirmed by Levene's test ($p < 0.05$). &—p-values from Kruskal–Wallis analysis. Medians were compared by Dunn's post-hoc test. Analytical methods limits of detection (LD): 1.0 mg/L for COD; 0.3 for TOC; 0.8µS/cm for EC; 0.2 NTU for Turbidity; 0.02 mg/L for F; 0.005 meq/L for HCO_3 and CO_3; 0.03 mg/L for Cl, SO_4, NO_3, Na, K, Ca, Mg; 0.01 mg/L for PO_4 and Cyanide; 0.001 mg/L for NO_2 and NH_4; 0.001 mg/L for Cd, As, Pb, Ni, Cu, Zn, Al and Cr; 0.025 mg/L for Fe; 0.05 for Al. For an easier analysis of the table the FAO's Guidelines are highlighted in gray.

The water samples from Silveirinhos stream collected upstream from the galleries (SS-U) showed results within FAO's permissible limits for irrigation purposes, with the exception of pH values, which were slightly below 6.5 due to the geological features of the catchment. Water samples from the mine drainage galleries G1 and G2 reflect the geochemical system of the coal seams and overlying strata. These mining effluents are not as acidic as one might expect, being neutral to slightly acidic, with pH values close to FAO's inferior permissible limit.

Acidity in coal mine waters results mainly from the dissolution of oxidized pyritic materials associated with coal, during mining operations, which explains the existence of iron and sulphate in the water. The pH values of samples G1 and G2, concomitantly with their high levels of iron and sulphate, suggest the existence of an underground neutralization process. The scarcity of pyrites in particular layers and the predominance of carbonate minerals constitute the most common explanation for neutral or alkaline mine drainages [48]. However, the calcareous materials in this region are very scarce and do not constitute a reasonable justification. In São Pedro da Cova, a plausible origin of the neutralization process could be the mixture of acid groundwater, circulating in the shallow rock massif along the mine galleries and wells, with alkaline thermomineral water following deeper circulation paths, possibly through major faults. This hypothesis is corroborated by the relatively high fluoride content in G1 and G2 samples (a hydrogeochemical signature of thermomineral circulation), when compared to SS-U samples, and by the water temperature measured *in situ* during the nine sampling campaigns. Data analysis shows that the temperatures measured in mine effluents were higher than the average annual air temperature for the study area (15 °C), ranging from 9 °C in January to 20 °C in July. G1 and G2 temperatures remained constant during the study, with mean values of 18.9 ± 0.5 and 19.1 ± 0.4 °C, respectively, which are significantly higher than the mean water temperature in SS-U (15.4 ± 4.7 °C). SS-U water temperature is seasonal, and it closely follows changes in air temperature as can be seen in Figure 3. In SS-D the influence of the mining drainages is notorious, with a mean value of temperature of 17.3 ± 1.8 °C.

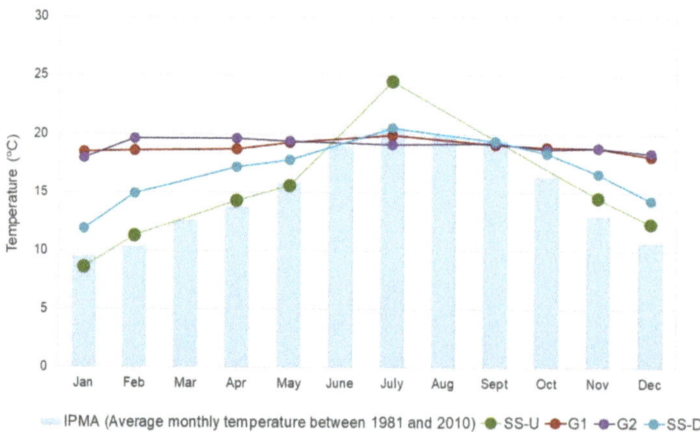

Figure 3. Average monthly air temperature measured between 1981 and 2010 in Serra do Pilar weather station (IPMA). Water temperature recorded at the sampling points SS-U, G1, G2 and SS-D on April, June, October and December 2017, January and November 2018, and February, May, September and December 2019.

The number of particles in water can be expressed by turbidity. With the exception of the SS-U samples that recorded a turbidity median value of 1.1 NTU (Table 3), all the impacted waters have values that were far above the limits proposed by the US EPA of

2 NTU for directly consumed crops and unrestricted irrigation [49], and by Spain that recommends levels lower than 10 NTU for vegetable irrigation water [15].

High levels of turbidity can affect the performance of irrigation facilities, causing the clogging of the equipment, can lower the hydraulic conductivity of the soil, pollute the soil surface, and lead to aesthetic impairment of the water and of the vegetables produced. In addition, irrigating vegetables with turbid water could affect the quality of the products since microorganisms, such as parasites, bacteria and viruses can be attached to the solid particles and contaminate the crops [18].

The EC is also a significant parameter in determining the suitability of water for irrigation use, as it affects water salinity, which subsequently affects the productivity and yield of crops. The EC levels obtained in this study were all below the 3.0 dS/m permissible limit set by FAO for irrigation water [18]. However, the levels recorded in G1 and G2 were significantly higher than those recorded in Silveirinhos stream upstream from mine drainage galleries, these waters having a slight to moderate restriction on use.

Regarding major ions, the levels in SS-U were within FAO permissible limits, and significantly lower than in the impacted waters. Coal mining pollution significantly increases mineralization as a result of a greater water–rock interaction. Mg^{2+} and K^+ were above the usual range for irrigation water, being the abundance of the ions as follows: $SO_4^{2-} > Mg^{2+} > Ca^{2+} > HCO_3^- > Na^+ > Cl^- > K^+ > CO_3^{2-}$.

The hydrogeochemical effect of coal mining, in terms of hydrogeochemical facies and major ion content, is illustrated by means of a Piper diagram (Figure 4) and a Stiff diagram (Figure 5). Surface water without mining influence (SS-U) has an intermediate SO_4/Cl-Na/Mg classification while mine drainage waters (G1 and G2) as well as the water collected downstream from the mining effluents discharge (SS-D) have a hydrogeochemical SO_4-Mg facies.

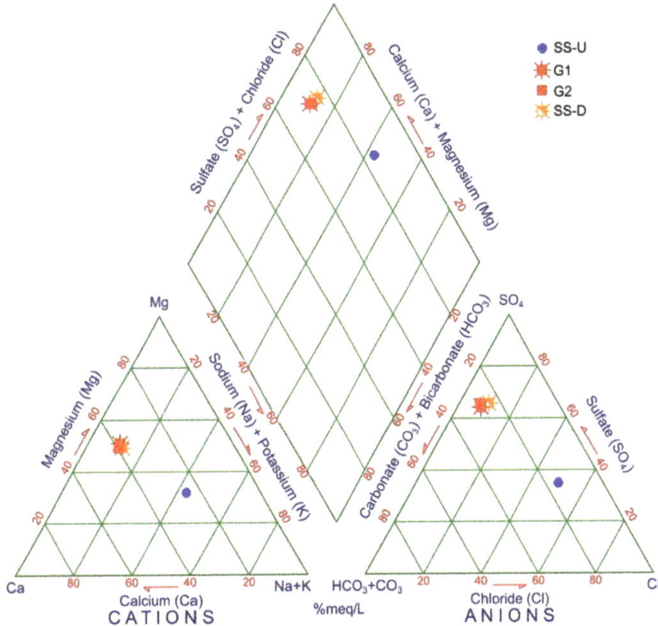

Figure 4. Piper diagram of the studied waters (average values from April 2017 to December 2019, *n* = 9).

mining activities have higher metal content, especially iron, manganese, aluminum, nickel and arsenic, which may cause various health hazards such as cancer and environmental pollution. Only iron and manganese exceeded the FAO standards, with values far above the recommended concentrations, but the levels of aluminum almost duplicated in the impacted waters, and nickel and arsenic increased significantly in G1, G2 and SS-D, in a proportion of 7, 3, 3 times and 21, 48, 11 times, respectively, in comparison with the values in water samples collected upstream from the discharges.

Iron was the most abundant metal detected in the mine wastewater samples and in the samples collected downstream from the mine galleries. The median concentration of iron ranged from 0.12 mg/L in SS-U to 52.28 and 84.66 mg/L in G1 and G2, respectively, and 18.24 mg/L in SS-D. Iron can be a complex water quality problem, which not only affects plant growth, as it can compete with other needed micro-nutrients, but also can clog irrigation equipment. For micro-irrigation systems, iron levels need to be below 0.3 mg/L to prevent clogging. Above 1.0 mg/L, iron may lead to rust stains and discoloration on foliage plants in overhead irrigation systems, and above 5 mg/L iron is toxic to plant tissues.

Manganese presents many of the same issues as iron in irrigation water. It can clog irrigation equipment and cause foliar staining. The recommended drinking water standard for manganese is 0.05 mg/L, which is also the level where black staining and irrigation clogging may occur. Concentrations above 2.0 mg/L can be directly toxic to some plant species. In this study the mean concentration of manganese ranged from 0.06 mg/L in SS-U to 4.96 mg/L in G1.

4.2. Irrigation Water Quality Indices

As the results suggest that no unique parameter can sufficiently describe water quality, thus, the chemical status of the water samples was also assessed by using Water Pollution Indices. Indices were calculated covering a wide range of variables that were gathered in a single numerical value, allowing a simplified representation of a complex reality and also the evaluation of historical trends. The most significant parameters for the water quality evaluation were selected according to FAO's guidelines, and for TETI according to the Toxicological Profiles of the Priority List of Hazardous Substances of ATSDR, in order to proceed with the calculation of the indices in a robust but simple way, using the smallest number of analytes. Indices calculations were also performed including other parameters, but no differences were found in the results and in the outcomes of the evaluation.

The results of the WQIA and the CI indices are shown in Table 4. From WQIA values waters were classified into five categories: excellent, good, poor, very poor and unsuitable, and CI values indicate a low, medium or high level of contamination.

Table 4. Calculated values of WQI$_A$ (Water Quality Status) and CI (Level of Contamination).

Sampling Date	WQI$_A$				CI Degree of Contamination			
	SS-U	G1	G2	SS-D	SS-U	G1	G2	SS-D
apr17	2.8	99.3	85.3	47.3	−13.3	32.5	32.4	6.4
jun17	3.2	110.8	107.3	94.0	−12.5	34.5	40.0	24.0
sep17	-	129.8	140.1	105.4	-	41.2	55.3	30.1
dec17	2.0	100.3	101.1	47.5	−13.2	28.9	40.8	3.9
nov18	1.7	96.1	101.0	40.9	−15.5	30.3	35.5	10.3
feb19	1.4	75.4	73.2	32.5	−15.3	18.5	27.0	−1.2
may19	1.7	81.0	81.8	35.6	−15.8	26.0	30.4	−1.7
sep19	-	87.9	86.3	86.1	-	31.6	38.8	26.5
dec19	1.5	66.2	69.4	16.2	−16.2	20.7	22.0	−7.0
Avg	2.0	94.1	93.9	56.2	−14.5	29.4	35.8	10.1

Water Quality Status (WQI$_A$): Excellent (0–25 ●); Good (26–50 ●); Poor (51–75 ●); Very Poor (76–100 ●); Unsuitable (>100 ●). Level of contamination (CI): Low (<1 ●); medium (1–3 ●); high (>3 ●).

The relation of precipitation events with the WQI_A values is represented in Figure 6, with data highlighted in different colors according to the classification for irrigation purposes (Legend Table 4).

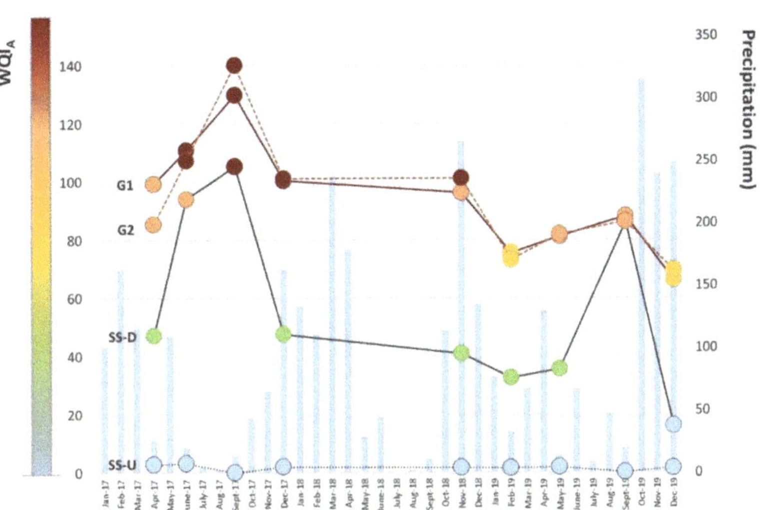

Figure 6. Relation between precipitation and water quality assessed by means of WQI_A. Precipitation measured at Porto meteorological station. Data from the Portuguese Institute for Sea and Atmosphere, I. P. (IPMA, IP) [52].

Results comparing upstream and downstream sampling sites point out the impact of mining effluents on surface water, being SS-U classified as excellent (WQI_A values ranging from 1.4 to 3.2, with a mean value of 2.0) and SS-D as poor (WQIA values ranging from 16.2 to 105.4, with a mean value of 56.2). Analysis also revealed that G1 and G2 were, as expected, the two most polluted waters, reported as very poor, with values ranging from 66.2–129.8 and 69.4–140.1, respectively. Out of the 18 parameters considered for this study, iron and manganese were the two deciding parameters, followed by arsenic and EC, which exhibit the maximum influence (Qi x Wi) in the WQI_A calculations.

Figure 6 shows fluctuations in WQI_A values in both study periods: from April 2017 to December 2017 and from November 2018 to December 2019. In the first period, the effect of draught in water quality is clear: the highest WQI_A values correspond to the driest months due to the lack of mixture of mine drainage with recently infiltrated precipitation. In the second period, a similar trend is observed in September 2019, with G1, G2 and SS-D presenting similar values. It was not possible to collect samples on SS-U because there was no streamflow due the drought conditions. The lower WQI_A values from this period were observed in February and in December 2019 as a result of a dilution effect due the infiltration of precipitation in the previous months.

The results of the computed CI index are, in general, comparable with the WQI_A values. The CI results for impacted waters exceeded the value of 3, ranging from a mean value of 10.1 in SS-D to 29.4 in G1 and 35.8 in G2, which indicates a high degree of pollution due mainly to iron, manganese and bicarbonate content, which exceed the limits of FAO guidelines. The SS-U samples have their computed CI values below 1, reflecting the absence of coal mining influence.

TETI results, based on the elemental toxicological impact, are shown in Table 5.

Table 5. Trace element toxicity index values.

	Tsi	SS-U		G1		G2		SS-D	
		Ci	Ci × Tsi	Ci	Ci × Tsi	Ci	Ci × Tsi	Ci	Ci × Tsi
K (Potassium) mg/L	607	0.61	371.14	5.16	3134.48	6.74	4089.83	3.08	1867.54
NO$_3$ (Nitrate) mg/L	605	0.71	428.69	0.46	276.96	0.26	154.61	0.56	336.11
NO$_2$ (Nitrite) mg/L	610	0.003	2.09	0.01	5.69	0.01	7.59	0.02	9.83
NH$_4$ (Ammonia) mg/L	742	0.01	3.92	0.76	560.37	0.84	625.88	0.73	543.56
PO$_4$ (Phosphate) mg/L	264	0.15	40.43	0.15	38.34	0.27	72.19	0.12	32.65
CN (Cyanide) mg/L	1070	0.001	0.79	0.00	1.12	0.00	1.08	0.00	1.11
B (Boron) mg/L	440	0.07	30.25	0.24	103.49	0.26	112.55	0.14	61.95
Al (Aluminum) mg/L	685	0.19	129.07	0.37	250.39	0.44	300.78	0.21	142.85
As (Arsenic) mg/L	1676	0.001	1.20	0.02	25.26	0.03	57.51	0.01	13.65
Cd (Cadmium) mg/L	1318	0.000	0.14	0.00	0.03	0.00	0.01	0.00	0.07
Cr (Chromium) mg/L	893	0.001	1.05	0.00	0.72	0.00	0.57	0.00	0.45
Cu (Copper) mg/L	805	0.003	2.19	0.00	0.67	0.00	1.44	0.00	0.55
F (Fluoride) mg/L	550	0.03	18.54	0.26	141.11	0.28	156.20	0.15	84.52
Mn (Manganese) mg/L	797	0.06	45.08	4.96	3956.93	4.82	3845.08	2.93	2338.42
Ni (Nickel) mg/L	993	0.01	5.38	0.04	35.00	0.02	15.46	0.02	16.85
Pb (Lead) mg/L	1531	0.003	5.31	0.00	1.39	0.00	1.16	0.00	0.83
Zn (Zinc) mg/L	913	0.14	126.65	0.02	14.25	0.01	7.46	0.01	8.01
	TETI		841		5412		5360		3591

Ci—mean concentration values.

TETI indicates that manganese and potassium had the highest impact on the toxicological profiles of the polluted waters (G1, G2 and SS-D) with a total score above 2000, according to the ATSDR assessment, followed by ammonium, nitrate and aluminum [43]. Regarding SS-U water samples, nitrate was the most important constituent in terms of water quality concerns, followed by potassium, aluminum and zinc. G1 had the highest index value followed by G2 and SS-D, which clearly demonstrates the impact of mine activities on the water environment, where higher TETI values represent lower quality.

The evaluation of these three selected indices (WQI$_A$, CI and TETI) highlights the coal mine inputs of metals and other pollutants in the study area. Although the three indices specify similar levels of contamination, their outcomes regarding the most important constituents are not uniform.

For example, for TETI calculation, potassium had a high impact with a total score 5–11 times higher than in SS-U. However, it is not considered in FAO or any international guideline for restriction on use purposes, hence it is not accounted for either the WQI$_A$ or CI indices. The same occurs regarding ammonia.

These findings clearly highlight the limitations of each index and of the international water quality guidelines that are, firstly, non-standardized between different countries and, secondly, do not provide guidelines for a number of pollutants of importance for specific matrices as is the case of coal mine effluents.

4.3. PAH Analyses

Regarding PAHs, analyses were performed in six campaigns (April 2017, September 2017, December 2017, November 2018, February 2019 and May 2019). According to the ring numbers, PAHs can be classified into three classes: 2–3 rings, 4 rings, and 5–6 rings composition, which represent low, medium, and high molecular weight hydrocarbons, respectively.

PAHs were detected in water samples at very low concentrations, with prevalence of low molecular weight compounds, with average concentration percentages that varied from 62% to 80%, which indicate a petrogenic origin consistent with the water circulation through the coal bearing rocks in the mine [53] (Figure 7).

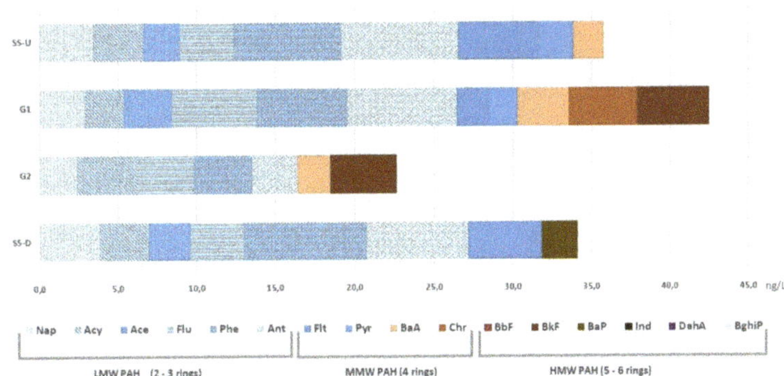

Figure 7. Composite model and average concentrations of PAHs in the studied samples.

The highest average concentration was detected in G1 (42.5 ng/L). SS-U and SS-D showed similar values of 35.8 and 34.1 ng/L, respectively.

The carcinogenic PAHs (BaA, Chr, BbF, BkF, BaP, InP, and DahA) were detected mostly in G1 and G2, accounting for 28.8% and 27.6% of the total average concentration of 16-PAHs, in contrast to 5.6% and 6.7% found in SS-U and SS-D, being the main components in mine effluents the BaA and the BkF. The carcinogenic PAHs are high-ring PAHs, i.e., 4–6 ring, which is related to the degree of coal maturation [47].

PAHs are of high environmental and human health concern, as they are toxic and persistent in the environment, susceptible to long-range atmospheric transport, and able to bioaccumulate. The uptake of PAHs by plants is important when considering their transfer from soils and water into the food chain. Recent studies also demonstrate that PAHs-metal co-contamination also alters PAHs uptake, attributing to the metal–soil or metal–root interactions [54].

As food chain is the most important pathway for pollutants' entry into the human body, the uptake of carcinogenic PAHs and heavy metals through soil-to-root system and their translocation/accumulation in plant tissues is very important, particularly for food crops cultivated on non-treated wastewater-irrigated soils, as is the case.

5. Conclusions

This study points out the suitability of the coal mining effluents and the polluted surface water for irrigation purposes in São Pedro da Cova abandoned coalfield. The results allow proposing a cost-effective assessment methodology adjusted to specific problems of the water, minimize pollution of natural watercourses and soils and increase the potential of use of these effluents.

The evaluation of the use of mining effluents for irrigation can be an issue as specific water quality guidelines or legislation does not exist, and several dangerous chemicals are not included in routine water quality assays. In an attempt to standardize decision-making regarding irrigation with mining effluents, the criteria and data have been combined in user-friendly indices, which could assist in the practical implementation of mining effluents irrigation control plans as part of optimal mine-water management and reuse strategy. Eighteen parameters including selected anions, cations and trace elements were chosen after an exhaustive analysis and were used for water quality indices calculation. Source specific water quality indices are a very helpful tool to represent water quality in a simple and understandable manner, minimizing the data volume to a great extent and simplifying the expression of water quality status, giving efficiently the overall water quality of a specific area and for a specific use.

Results revealed that samples associated with mining activities have unacceptable index values due mainly to the high concentration of iron, manganese, bicarbonates, magnesium and potassium, and are not suitable for irrigation. WQI_A indicated an overall Poor/Very Poor quality status, which is in accordance with the level of contamination (CI) of these waters. The TETI index, which only reflects the elemental toxicological impact of the waters, not considering other fundamental water quality parameters, indicated that the impacted water samples had higher index values, and manganese had the highest impact on the toxicological profiles, also showing the influence of coal activities on surface water quality. Risks to human health also arise from water pollution by organic substances such as PAHs with several carcinogenic compounds detected in G1 and G2. Levels in the water collected upstream from mine drainage points are within acceptable range for irrigation.

Scientific community and local authorities of mine areas are committed to mitigating the effects of past actions through the development of better management strategies for reducing environmental and health impacts in the mining area. The understanding of the mine water chemistry is fundamental for the design of an effective treatment system. Simple passive treatments could be an option for the contaminated drainage at this abandoned mine site according to Robert Hedin el al. 2013 (Effective Passive Treatment of Coal Mine Drainage. Robert Hedin, Ted Weaver, Neil Wolf, George Watzalaf. Paper presented at the 35th Annual National Association of Abandoned Mine Land Programs Conference, 2013). If properly designed, constructed and maintained, passive systems provide highly reliable treatment at a fraction of the cost of active alternatives.

As the overall quality of groundwater from mine drainage galleries revealed contamination, it is very important to raise awareness for rapid intervention in the area, since the mine is located near a population center and social infrastructures, as well as to mitigate the pollution on adjacent agricultural lands. Despite the evident deleterious impact for local communities, the health effects of mine drainage remain neglected in research and policy arenas, mainly because of the lack of documented evidence. Such communities suffer the effects of mine drainage principally through a perpetual risk posed by water pollution. This project intended to study the impact of mine drainage contamination towards an investment in temporary to long-term solutions, in order to reduce the risks caused by mining externalities.

Author Contributions: Conceptualization, C.M.; methodology, C.M., A.M., J.E.M.; validation, C.M., J.E.M.; formal analysis, C.M., A.M., J.E.M.; investigation, C.M., J.E.M., V.M., P.S., J.R., J.R.R., A.M. and D.F.; resources, J.E.M., C.M., D.F.; writing—original draft preparation, C.M., J.E.M., A.M., P.S., J.R.; writing—review and editing, C.M., J.E.M., A.M.; supervision, C.M., J.E.M., D.F.; project administration, D.F.; funding acquisition, D.F. All authors have read and agreed to the published version of the manuscript.

Funding: This work was supported by the project CoalMine—Coal mining wastes: assessment, monitoring and reclamation of environmental impacts through remote sensing and geostatistical analysis—financed by the Portuguese Science and Technology Foundation, FCT, call AAC n° 02/SAICT/2017 (POCI-01-0145-FEDER-030138) and framed within the activities of the ICT (Ref. UIDB/04683/2020).

Institutional Review Board Statement: Not applicable.

Informed Consent Statement: Not applicable.

Data Availability Statement: The data supporting the findings of this study are available within the article.

Acknowledgments: The authors acknowledge the Mining Museum of São Pedro da Cova for the permission to use the photograph from Figure 1a. This work received support and help from the UID/QUI/50006/2020 with funding from FCT/MCTES through national funds.

Conflicts of Interest: The authors declare no conflict of interest.

References

1. Mahato, M.K.; Singh, A.K.; Singh, G.; Mishra, L.P. Impacts of coal mine water and Damodar River water irrigation on soil and maize (*Zea mays* L.) in a coalfield area of Damodar Valley, India. *Arch. Agric. Environ. Sci.* **2017**, *2*, 293–297. [CrossRef]
2. Bott, T.L.; Jackson, J.K.; McTammany, M.E.; Newbold, J.D.; Rier, S.T.; Sweeney, B.W.; Battle, J.M. Abandoned coal mine drainage and its remediation: Impacts on stream ecosystem structure and function. *Ecol. Appl.* **2012**, *22*, 2144–2163. [CrossRef] [PubMed]
3. Johnston, D.; Potter, H.; Jones, C.; Rolley, S.; Watson, I.; Pritchard, J. *Abandoned Mines and the Water Environment*; Environment Agency: Bristol, UK, 2008; p. 31.
4. Tiwary, R.K. Environmental Impact of Coal Mining on Water Regime and Its Management. *Water Air Soil Pollut.* **2001**, *132*, 185–199. [CrossRef]
5. Lemos de Sousa, M.J.; Wagner, R.H. General Description of the Terrestrial Carboniferous Basins in Portugal and History of Investigations. In *The Carboniferous of Portugal. Memórias dos Serviços Geológicos de Portugal*; Lemos de Sousa, M.J., Oliveira, J.T., Eds.; Direção Geral de Geologia e Minas: Lisboa, Portugal, 1983; Volume 29, pp. 117–126.
6. Wagner, R.H.; Lemos de Sousa, M.J. The Carboniferous Megafloras of Portugal—A revision of identifications and discussion of stratigraphic ages. In *The Carboniferous of Portugal. Memórias dos Serviços Geológicos de Portugal*; Lemos de Sousa, M.J., Oliveira, J.T., Eds.; Direção Geral de Geologia e Minas: Lisboa, Portugal, 1983; Volume 29, pp. 127–152.
7. Eagar, R.M.C. The non marine bivalve fauna of the Stephanian C of North Portugal. In *The Carboniferous of Portugal. Memórias dos Serviços Geológicos de Portugal*; Sousa, M.J.L., Oliveira, J.T., Eds.; Direção Geral de Geologia e Minas: Lisboa, Portugal, 1983; Volume 29, pp. 179–185.
8. Fernandes, J.P.; Pinto de Jesus, A.; Teixeira, F.; Sousa, M.J.L. First palynological results in the Douro Carboniferous Basin (NW of Portugal). In *Proceedings of XIII Jornadas de Paleontologia "Fósiles de Galicia" y V Reunión Internacional Proyecto 351 PICG "Paleozoico Inferior del Noroeste de Gondowana"*; Sociedade Espanola de Paleontologia: Coruna, Spain, 1997; pp. 176–179.
9. Custódio, J. *Museum of Coal & Mines of Pejão*; Museum Programme: Castelo de Paiva, Portugal, 2004.
10. Pinto de Jesus, A. Sedimentary and tectonic evolution of the Douro Coalfield Basin (lower Stephanian C, NW Portugal). *Cad. Lab. Xeolóxico Laxe* **2003**, *28*, 107–125.
11. HLPE. Water for food security and nutrition. In *A Report by the High Level Panel of Experts on Food Security and Nutrition of the Committee on World Food Security*; Food and Agriculture Organization of the United Nations: Rome, Italy, 2015.
12. Pescod, M.B. *Wastewater Treatment and Use in Agriculture—FAO Irrigation and Drainage Paper 47*; FAO: Rome, Italy, 1992.
13. Al-Hwaiti, M.S.; Brumsack, H.J.; Schnetger, B. Suitability assessment of phosphate mine waste water for agricultural irrigation: An example from Eshidiya Mines, South Jordan. *Environ. Earth Sci.* **2016**, *75*, 276. [CrossRef]
14. Müller, K.; Cornel, P. Setting water quality criteria for agricultural water reuse purposes. *J. Water Reuse Desalin.* **2016**, *7*, 121–135. [CrossRef]
15. Jeong, H.; Kim, H.; Jang, T. Irrigation Water Quality Standards for Indirect Wastewater Reuse in Agriculture: A Contribution toward Sustainable Wastewater Reuse in South Korea. *Water* **2016**, *8*, 169. [CrossRef]
16. Alobaidy, A.; Al-Sameraiy, M.; Kadhem, A.; Majeed, A. Evaluation of treated municipal wastewater quality for irrigation. *J. Environ. Prot.* **2010**, *1*, 216–225. [CrossRef]
17. Abdul-Rahman, S.; Saoud, I.P.; Owaied, M.K.; Holail, H.; Farajalla, N.; Haidar, M.; Ghanawi, J. Improving Water Use Efficiency in Semi-Arid Regions through Integrated Aquaculture/Agriculture. *J. Appl. Aquac.* **2011**, *23*, 212–230. [CrossRef]
18. Ayers, R.S.; Westcot, D.W. Water quality for agriculture. In *FAO Irrigation and Drainage Paper Rev. 1*; FAO: Rome, Italy, 1994.
19. Alves, M.T.R.; Teresa, F.B.; Nabout, J.C. A global scientific literature of research on water quality indices: Trends, biases and future directions. *Acta Limnol. Bras.* **2014**, *26*, 245–253. [CrossRef]
20. Kachroud, M.; Trolard, F.; Kefi, M.; Jebari, S.; Bourrié, G. Water Quality Indices: Challenges and Application Limits in the Literature. *Water* **2019**, *11*, 361. [CrossRef]
21. Poonam, T.; Tanushree, B.; Sukalyan, C. Water Quality Indices- Important Tools for Water Quality Assessment: A Review. *Int. J. Adv. Chem.* **2015**, *1*, 15–29.
22. Horton, R.K. An index number system for rating water quality. *J. Water Pollut. Control Fed.* **1965**, *37*, 300–306.
23. Abbasi, T.; Abbasi, S. *Water Quality Indices*; Elsevier: Amsterdam, The Netherlands, 2012.
24. Misaghi, F.; Delgosha, F.; Razzaghmanesh, M.; Myers, B. Introducing a water quality index for assessing water for irrigation purposes: A case study of the Ghezel Ozan River. *Sci. Total Environ.* **2017**, *589*, 107–116. [CrossRef]
25. Cude, C.G. Oregon water quality index a tool for evaluating water quality management effectiveness. *J. Am. Water Resour. Assoc.* **2001**, *37*, 125–137. [CrossRef]
26. Backman, B.; Bodiš, D.; Lahermo, P.; Rapant, S.; Tarvainen, T. Application of a groundwater contamination index in Finland and Slovakia. *Environ. Geol.* **1998**, *36*, 55–64. [CrossRef]
27. Biswas, P.K.; Uddin, N.; Alam, S.; -Us-Sakib, T.; Sultana, S.; Ahmed, T. Evaluation of Heavy Metal Pollution Indices in Irrigation and Drinking Water Systems of Barapukuria Coal Mine Area, Bangladesh. *Am. J. Water Res.* **2017**, *5*, 146–151.
28. Ali, A.; Strezov, V.; Davies, P.; Wright, I. Environmental impact of coal mining and coal seam gas production on surface water quality in the Sydney basin, Australia. *Environ. Monit. Assess.* **2017**, *189*, 408. [CrossRef]
29. EFSA. Polycyclic Aromatic Hydrocarbons in Food—Scientific Opinion of the Panel on Contaminants in the Food Chain. *EFSA J.* **2008**, *6*, 724. [CrossRef]
30. Rodríguez-Eugenio, N.; McLaughlin, M.; Pennock, D. *Soil Pollution, A Hidden Reality*; FAO: Rome, Italy, 2018.

31. INE. *Censos 2011 Resultados Definitivos—Portugal*; Instituto Nacional de Estatística, I.P.: Lisbon, Portugal, 2012.
32. Julivert, M.; Fontboté, J.M.; Ribeiro, A.; Conde, L. *Mapa Tectónico de la Península Ibérica y Baleares*; IGME—Instituto Geológico y Minero de España: Madrid, Spain, 1974.
33. Medeiros, A.; Pereira, E.; Moreira, A. *Notícia Explicativa da Folha 9-D (Penafiel) da Carta Geológica de Portugal à Escala 1:50 000*; Serviços Geológicos de Portugal: Lisbon, Portugal, 1980.
34. Peel, M.C.; Finlayson, B.L.; McMahon, T.A. Updated world map of the Köppen-Geiger climate classification. *Hydrol. Earth Syst. Sci.* **2007**, *11*, 1633–1644. [CrossRef]
35. AEMET-IM. *Atlas Climático Ibérico. Temperatura do Ar e Precipitação (1971–2000)*; Departamento de Producción da Agência Estatal de Meteorologia de Espanha (Área de Climatología y Aplicaciones Operativas) e Departamento de Meteorologia e Clima (Divisão de Observação Meteorológica e Clima), do Instituto de Meteorologia: Lisboa, Portugal, 2011.
36. ISO. *ISO 5667-3:2018—Water Quality—Sampling—Part 3: Preservation and Handling of Water Samples*, ISO: Geneva, Switzerland, 2018.
37. Rice, E.W.; Baird, R.B.; Eaton, A.D. *Standard Methods for the Examination of Water and Wastewater*, 23rd ed.; American Public Health Association: Washington, DC, USA, 2017.
38. Rodier, J.; Legube, B.; Merlet, N. *L'analyse de l'eau*, 10th ed.; Dunod: Paris, France, 2016.
39. Borges, B.; Armindo, M.; Ferreira, I.M.P.L.V.O.; Mansilha, C. Dispersive liquid–liquid microextraction for the simultaneous determination of parent and nitrated polycyclic aromatic hydrocarbons in water samples. *Acta Chromatogr.* **2018**, *30*, 119–126. [CrossRef]
40. Richards, L.A. *Diagnosis and Improvement of Saline and Alkali Soils. Agricultural Handbook 60*; United States Department of Agriculture: Washington, DC, USA, 1954.
41. Sawyer, C.N. *Chemistry of Sanitary Engineers*, 2nd ed.; McGraw-Hill: New York, NY, USA, 1967.
42. Doneen, L.D. *Notes on Water Quality in Agriculture*; Department of Water Science and Engineering, University of California, Davis: Oakland, CA, USA, 1964.
43. Oni, O.; Fasakin, O. The use of water quality index method to determine the potability of surface water and groundwater in the vicinity of a municipal solid waste dumpsite in Nigeria. *Am. J. Eng. Res.* **2016**, *5*, 96–101.
44. Tyagi, S.; Sharma, B.; Singh, P.; Dobhal, R. Water Quality Assessment in Terms of Water Quality Index. *Am. J. Water Res.* **2013**, *1*, 34–38. [CrossRef]
45. Chatterjee, C.; Raziuddin, M. Determination of water quality index (WQI) of a degraded river in Asansol Industrial area, P.O. Raniganj, District Burdwan, West Bengal. *Nat. Environ. Pollut. Technol.* **2002**, *1*, 181–189.
46. ATSDR. *Priority List of Hazardous Substances*; Agency for Toxic Substances and Disease Registry: Division of Toxicology and Human Health Sciences: Atlanta, GA, USA, 2019.
47. Chen, D.; Feng, Q.; Liang, H.; Gao, B.; Alam, E. Distribution characteristics and ecological risk assessment of polycyclic aromatic hydrocarbons (PAHs) in underground coal mining environment of Xuzhou. *Hum. Ecol. Risk Assess.* **2019**, *25*, 1564–1578. [CrossRef]
48. Singh, G. Impact of coal mining on mine water quality. *Int. J. Mine Water* **1988**, *7*, 49–59. [CrossRef]
49. EPA. *Guidelines for Water Reuse*; U.S. Environmental Protection Agency: Washington, DC, USA, 2012.
50. Rengasamy, P.; Marchuk, A. Cation ratio of soil structural stability (CROSS). *Soil Res.* **2011**, *49*, 280–285. [CrossRef]
51. Oster, J.D.; Sposito, G.; Smith, C.J. Accounting for potassium and magnesium in irrigation water quality assessment. *Calif. Agric.* **2016**, *70*, 71–76. [CrossRef]
52. Climatological Bulletin. Available online: https://www.ipma.pt/pt/publicacoes/boletins.jsp (accessed on 6 April 2021).
53. Stout, S.A.; Emsbo-Mattingly, S.D. Concentration and character of PAHs and other hydrocarbons in coals of varying rank—Implications for environmental studies of soils and sediments containing particulate coal. *Org. Geochem.* **2008**, *39*, 801–819. [CrossRef]
54. Zhang, S.; Yao, H.; Lu, Y.; Yu, X.; Wang, J.; Sun, S.; Liu, M.; Li, D.; Li, Y.-F.; Zhang, D. Uptake and translocation of polycyclic aromatic hydrocarbons (PAHs) and heavy metals by maize from soil irrigated with wastewater. *Sci. Rep.* **2017**, *7*, 12165. [CrossRef]

Article

A Three-Stage Hybrid Model for Space-Time Analysis of Water Resources Carrying Capacity: A Case Study of Jilin Province, China

Tong Liu [1], Xiaohua Yang [1,*], Leihua Geng [2] and Boyang Sun [1]

[1] State Key Laboratory of Water Environment Simulation, School of Environment, Beijing Normal University, Beijing 100875, China; cdliutong@mail.bnu.edu.cn (T.L.); 201431180013@mail.bnu.edu.cn (B.S.)
[2] Nanjing Hydraulic Research Institute, Key Laboratory of Water Resource and Hydraulic Engineering, Nanjing 210029, China; lhgeng@nhri.cn
* Correspondence: xiaohuayang@bnu.edu.cn

Received: 5 January 2020; Accepted: 30 January 2020; Published: 5 February 2020

Abstract: Water shortage, water pollution, shrinking water area and water mobility are the main contents of the water resources crisis, which are widespread in the social and economic development of Jilin Province. In this paper, a three-stage hybrid model integrating evaluation, prediction and regulation is constructed by combining the load-balance method and the system dynamics method. Using this model, the current states of water resources carrying capacity (WRCC) in 2017 and the trend of water demand/available from 2018 to 2030 were obtained. Using the orthogonal test method, the optimal combination program of agricultural and industrial water efficiency regulation and water resources allocation was selected. The results show that the pressure of the human–water resources system in Changchun, Liaoyuan and Baicheng is greater than the support, and the other six cities are not overloaded. The water demand in Jilin Province and its nine cities will increase from 2018 to 2030, if the current socio-economic development pattern is maintained. Therefore, we change the water quantity carrying capacity index by controlling agriculture, industrial water efficiency and trans-regional water transfer. Compared with 2015, among the optimal program obtained, the change range of the water use per 10,000 RMB of agricultural output is (−5%, 25%), and the water use per 10,000 RMB of industrial added value is (−45%, −35%), and the maximum water transfer is 1.5 billion m^3 per year in 2030. This study analyzes the development pattern of WRCC in the process of water conservancy modernization in Jilin Province and provides reference for other provinces to make the similar plan.

Keywords: water resources carrying capacity; load-balance; system dynamics model; water resources allocation; Jilin province

1. Introduction

As a precious natural resource, water plays an important role for the survival and development of human society. As the population grows, the demand for water resources is increasing, and water resources problems such as water shortage and water pollution are becoming more and more prominent [1–3]. There are two human causes of water resources problems. First, there is too much water for living and production, but less water for ecological environment, resulting in the failure of basic ecological flow and water area of rivers and lakes. Secondly, the amount of pollutants discharged from the socio-economic system exceeds environmental capacity. Therefore, the total volume of natural water resources consisting of non-renewable water resources and renewable water resources is calculated, which is used to solve the problem of water shortage [4–6]. In addition, water quality models are often used to simulate the diffusion process of pollutants in the water environment to calculate environmental capacity, thereby solving the problem of water pollution [7–9]. For the sustainable development of human-water resources system (HWRS), the rational allocation of water resources is essential [10,11].

Jilin Province benefited from abundant natural resources and a good industrial base. Before the 1990s, it had been China's most important industrial base for a long time, and its economic development level was at the forefront of the country. However, due to the contradiction of system and structure, the economy of Jilin has shown a severe decline since the end of the last century. In recent years, with the implementation of the strategy of revitalizing Northeast China, the demand for water resources in Jilin has been increasing, and the regional water resource load balance has gradually shifted. Considering the water resources crisis and water conservancy modernization plan of Jilin Province, this study analyzes the water resources carrying capacity (WRCC) of Jilin Province. Through orthogonal experiments, a regulation plan in Jilin Province was obtained, and comprehensive suggestions for improving WRCC in Jilin Province were put forward.

The remainder of this article is arranged as follows: the second section presents the literature review; the third section includes the study area and data sources; the fourth section presents the three-stage hybrid model; the fifth section presents the results; the sixth section presents the discussion, and the last section presents the conclusions.

2. Literature Review

The term "carrying capacity" is derived from ecology, which is used to measure the largest organisms that a specific region can support for some period of time without obvious adverse effect on the local environment [12]. After that, the application field of carrying capacity continued to expand, and the ecological carrying capacity [13], environmental carrying capacity [14], land carrying capacity [15], and mineral resources carrying capacity [16] appeared one after the other. WRCC is an extension of the concept of carrying capacity in the field of water resources, which focuses on the interaction between the human system and the water resources system. Since the introduction of the concept of WRCC, a lot of fruitful research results have been published. However, it is difficult to find a well recognized definition of WRCC. (i) WRCC is the largest external activity that water resources can carry out [17], which only exists in theory, without relevant quantitative study. (ii) WRCC is a threshold, such as the largest amount of local water resources [18], the largest economic scale and the population that can be supported by local water resources [19,20]. At present, economic and population scales are often used as indicators of WRCC. Obviously, forces always appear in pairs in physics, so the carrying capacity can be divided into support ability and pressure ability. By comparing their values, the carrying states of the water resources can be diagnosed. Wang et al. judged that Tianjin 's water resources system was overloaded by comparing the WRCC with the actual population [21].

A complete and reasonable index system is essential for studying the WRCC [22,23]. At present, water environment, water resources, and social-economy that are closely related to the HWRS are usually included in the index system, and then three criteria layer layers are obtained [24–26]: (i) The water resources standard layer mainly reflects the ability of the local water resources subsystem to supply water to the socio-economic subsystem. (ii) The water environment standard layer is closely related to the water environment capacity. (iii) The socio-economic standard layer mainly contains elements in human production and living activities. In general, due to the concept of traditional WRCC, the indicators in the above-mentioned criterion layer are mostly related to water quantity and water quality, but the interaction in the HWRS is far more than them. Water intake and sewage discharge activities are generated by the socio-economic subsystem, causing damage and interference to the space of water ecological and hydrological processes of the water resources-environment subsystem, and thus generating pressure. In addition, the water resources-environment subsystem generates support through water supply and consumption of pollutants. It can be seen that in order to maintain the sustainable development of the HWRS, it is necessary to obtain multi-element support of water resources, such as quantity, quality, space, and flow.

The study of WRCC, which includes evaluation, prediction and regulation, is based on evaluation, using prediction as a starting point, and achieving the goal through regulation. The focus of the evaluation is to quantify the carrying capacity of a specific region [27]. Predictive research usually sets different development scenarios for the HWRS, and then predicts the WRCC of each scenario [28]. The research on regulation refers to planning, such as water environment protection planning and water resources development planning, and then the target of WRCC is set. The positive and negative feedback effects in the HWRS are used to regulate the activities of social and economic, and finally, the carrying capacity is in line with the expected target [29]. In general, whether it is evaluation, prediction or regulation, the results of WRCC research can provide reference for regional socio-economic development and water resources protection, but the scope is limited. The WRCC of 36 major Chinese cities has been evaluated [26,30]. These research results can reflect the past or current water resources carrying states. In order to study the WRCC of a specific region in the future, evaluation studies can be combined with predictive research. For example, Yang et al. predicted the WRCC of Xi'an in 2020 [31]. Usually, the prediction research needs to set up multiple scenarios and obtain the prediction results of WRCC and then get the regulation programs. These regulation programs are mostly derived from the parameter settings in the prediction scenario, which has the best simulation result. If the regulation is carried out according to these programs, there will often be an excess of support ability, resulting in high economic costs of regulation and low feasibility. In addition, the research on the development of HWRS shows that the carrying condition of water resources in a specific region is different in different historical stages and narrowing the evaluation of the carrying capacity in the long-term span is a key problem to be solved. Therefore, a three-stage hybrid model was established to study the WRCC in Jilin Province.

3. Study Area and Date Sources

3.1. Study Area

Jilin Province is located at longitude 121°38′–131°19′ E and latitude 40°50′–46°19′ N. It covers an area of 187,400 km^2, including eight prefecture-level cities and one autonomous prefecture, as shown in Figure 1. Jilin Province is an important industrial base and commodity grain base in China. With the strategy of revitalizing the northeast old industrial base and the modernization of water conservancy, socio-economic development has brought a lot of pressure on the water resources [32]. There are 19 major rivers in Jilin Province, which belong to five water systems [33]. These rivers transport a lot of water resources for Jilin Province. However, the volume of water resources per capita in Jilin Province is insufficient. As shown in the "Jilin Provincial Water Resources Bulletin (2017)", the regional volume of water resources per capita is only 1451.5 m^3, which is a water-stressed area [34]. In addition, approximately 1.12 billion tons of wastewater was discharged into water bodies in 2017. Therefore, surface water pollution has caused deterioration of the water environment. The shortage of water resources and pollution of the water environment have severely restricted the sustainable development of society and economy in Jilin [35].

The east of Jilin is rich in rainfall and rich in water resources, which is the birthplace of many rivers. The central part of Jilin is an important commodity grain production base in China, consuming a lot of water resources. The grasslands in the west of Jilin are vast, with large wetlands and abundant groundwater and transit water, but the excessive exploitation of groundwater extraction has caused wetland degradation and ground subsidence [36]. The spatial and temporal distribution of water resources in Jilin Province is uneven [37]. Thus, the water resource problems in Jilin Province are produced by both social and natural factors, which pose an urgent problem for the government and researchers.

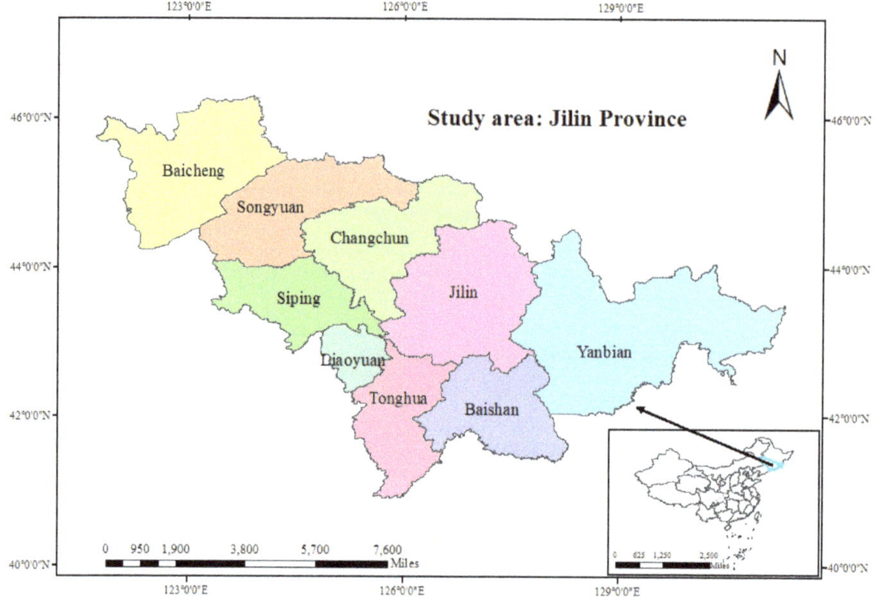

Figure 1. Location of Jilin Province.

3.2. Date Sources

The major datasets used in this study include water demand, water available, water conservation, population and economic. Table 1 presents all the data used in the study and their sources. In addition, the water area data in this paper are from the Resource Environment Cloud Platform of the Chinese Academy of Sciences and the Earthdata database.

Table 1. Data used in the study and their sources.

	Data Type	Source
1	Water demand	the Statistical Yearbook of Jilin Province (2006–2018)
2	Water available	Jilin Province long-term water supply and demand planning report
3	Water conservation	Statistical Bulletin of National Economic and Social Development of Jilin Province,
4	Water environment	Environmental Quality Bulletin of Jilin Province, Water Resources Bulletin of Jilin Province
5	Reservoir	Yearbook of Jilin Water Resources Province
6	Population and Economic	the Statistical Yearbook of Jilin Province (2006–2018)

4. Three-Stage Hybrid Model

Based on the existing research status, this paper establishes a three-stage hybrid model integrating evaluation, prediction and regulation. It is used to analyze the WRCC in Jilin Province, and to find out the factors affecting the HWRS, in order to improve the carrying capacity of the water resources. The overall research process of this paper is shown in Figure 2.

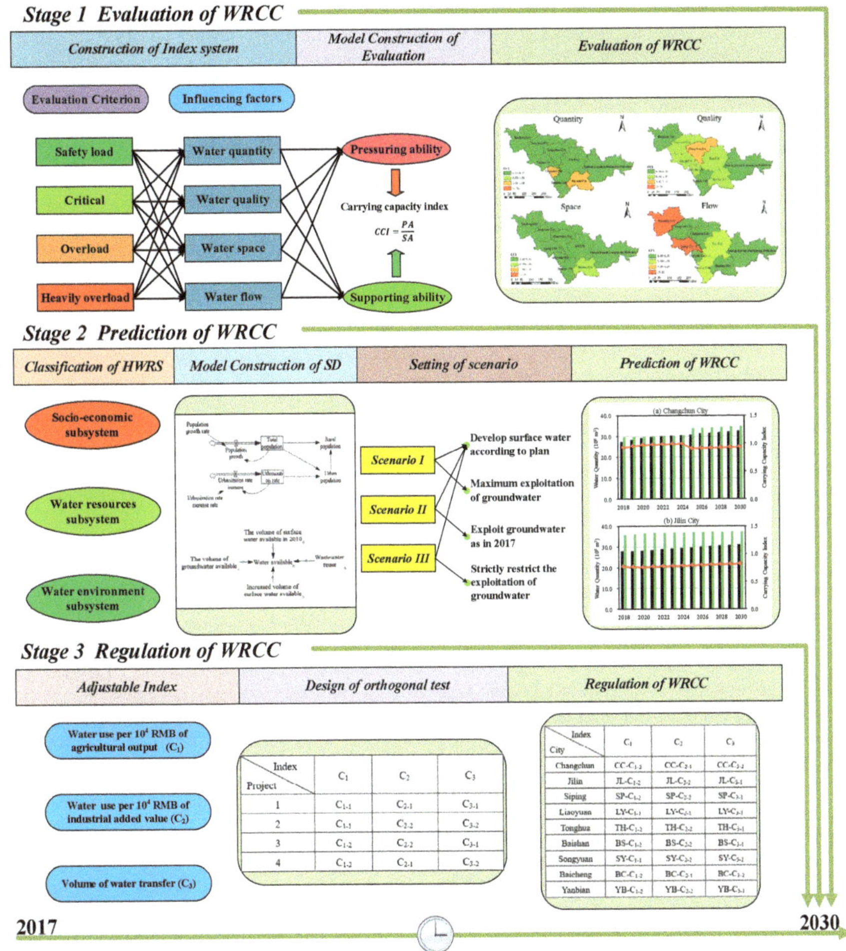

Figure 2. Framework of this work.

4.1. Evaluation of WRCC

4.1.1. Evaluation System for the WRCC

According to the system theory, the HWRS can be divided into the carrying body and the carrying object, which generate support and pressure, respectively. Between them, the carrying body is the water resources-environmental subsystem and the carrying object is the socio-economic subsystem. When the level of socio-economic activity does not exceed the threshold, the water resources-environmental subsystem can be sustainable. On the contrary, it will be damaged. It can be concluded that the accurate evaluation of the WRCC requires not only the calculation of support ability but also the calculation of pressure ability. Eventually, an evaluation system (Table 2) composed of eight indicators for the WRCC in Jilin Province was established [38].

Table 2. The index system for water WRCC.

Factors	Variables Symbol	Indexes Name	Attributes
Water quantity	PA_1	Water demand (billion m^3)	Pressuring ability
	SA_1	Water available (billion m^3)	Supporting ability
Water quality	PA_2	Discharge volume of pollutants (tons)	Pressuring ability
	SA_2	Water environment capacity (tons)	Supporting ability
Water space	PA_3	Area of actual water (km^2)	Pressuring ability
	SA_3	Area of natural water (km^2)	Supporting ability
Water flow	PA_4	Reservoir capacity (billion m^3)	Pressuring ability
	SA_4	Runoff (billion m^3)	Supporting ability

4.1.2. Evaluation Model of WRCC

As we all know, the various behaviors in the socio-economic subsystem are from human beings. Therefore, this paper uses the total population to quantitatively evaluate the WRCC. The calculation formula of carrying capacity is as follows [21]:

$$P_c = \frac{P_t}{CCI} \tag{1}$$

$$CCI_i = \frac{PA_i}{SA_i} \tag{2}$$

where P_c represents the carrying capacity at time t, including water quantity carrying capacity, water quality carrying capacity and comprehensive carrying capacity, where the comprehensive carrying capacity is equal to the minimum of the former two. P_t is the actual population of a region at time t. CCI is the carrying capacity index of four factors. PA, SA are pressure ability and support ability, respectively.

If the water quantity carrying capacity is bigger than the water quality carrying capacity, the limiting factor of the WRCC is water quality, which belongs to the water quality dominant area; otherwise, it belongs to the water quantity dominant area.

4.1.3. Evaluation Criterion of WRCC

According to the research results of the project and the opinions of relevant experts, a set of evaluation criteria was established to analyze the water resources carrying state. (i) If $CCI > 1.2$, the carrying state is heavily overload; (ii) if $1.2 \geq CCI > 1.0$, the carrying state is overload; (iii) if $1.0 \geq CCI > 0.9$, the carrying state is critical; (iv) if $0.9 \geq CCI$, the carrying state is safety load.

4.2. Prediction of WRCC

The system dynamics method was used to establish an SD model of WRCC, considering water resources, water environment and socio-economic. It simulates the development trend of HWRS in Jilin Province from the supply/demand of water resources, and carries out multi-scenario prediction of WRCC. The boundary of the model is the administrative boundary of Jilin Province, the time scale is 2005-2030, and the time step is 1 year. In addition, with reference to the administrative boundaries of 9 cities in Jilin Province, nine sub-models were established.

4.2.1. Flow Diagram of SD Model

The flow chart of the main variables in the sub-model is drawn (Figure 3).

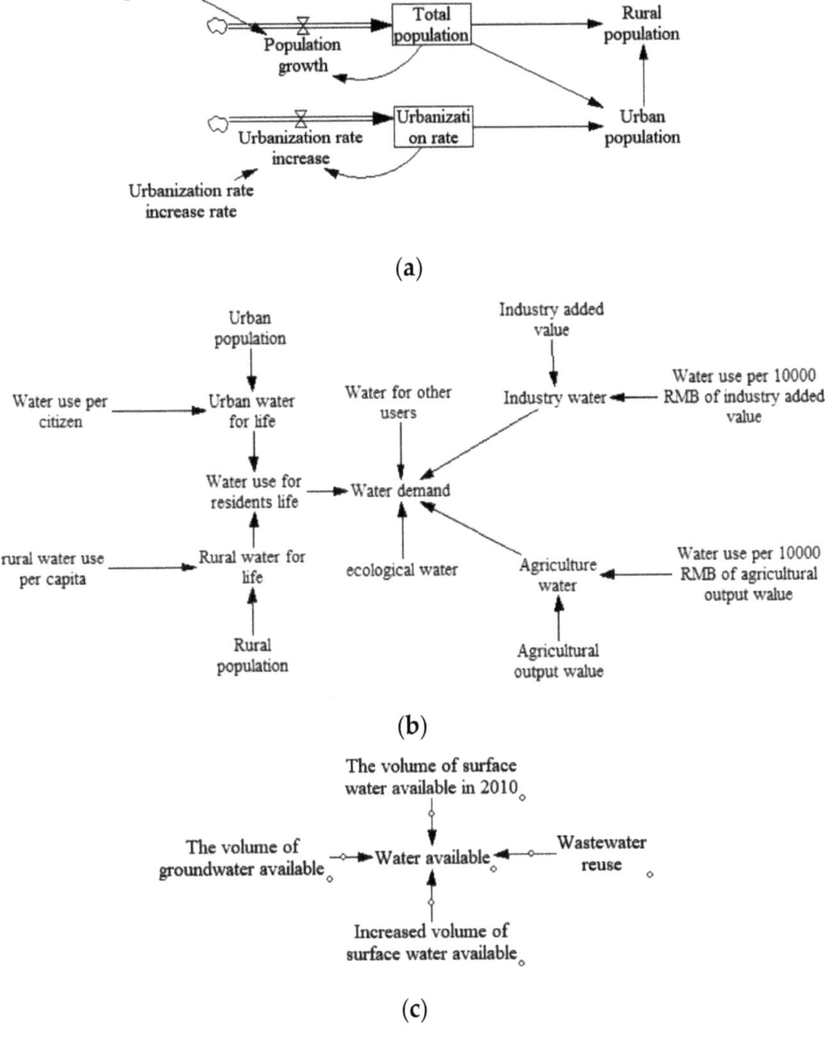

Figure 3. The flow diagram of main variables in the sub-model of WRCC: (**a**) total population, (**b**) water demand, (**c**) water available.

4.2.2. Equation of SD Model

In the SD model, mathematical equations are applied to quantify the water resources cycle. The indicators "available water quantity" and "water demand quantity" are calculated by Equations (3) and (4), respectively:

$$WA = WA_{t_1} + WA_{\Delta t} \tag{3}$$

$$WD = WD_{iw} + WD_{aw} + WD_{dw} + WD_{ew} + WD_{ow} \tag{4}$$

where WA represents the volume of water available. WA_{t_1} represents the volume of water available at time t_1. $WA_{\Delta t}$ represents the newly added volume of water from time t_1 to time t. WD represents regional water demand. WD_{iw}, WD_{aw}, WD_{ew}, WD_{ow} represent industrial water demand, agricultural water demand, ecological water demand, and other water demand, respectively. It should be mentioned that this study refers to the "Long-term Water Supply and Demand Planning of Jilin Province", and the time t_1 is set in 2010.

4.2.3. Scenario of HWRS

Three scenarios were set up for the development trend of the HWRS, and the WRCC under different scenarios was predicted separately.

Scenario I: The economic development goals, irrigation construction projects, and water transfer projects planned by Jilin Province were implemented as scheduled. These contents are derived from plans such as the "13th Five-Year Plan for National Economic and Social Development" and the "Water Conservancy Modernization Plan". In addition, groundwater is mined at maximum exploitable volume.

Scenario II: The volume of groundwater exploitation is the same as in 2017 during 2018–2030. It should be mentioned that if the groundwater exploitation in 2017 exceeds the maximum exploitable volume, the groundwater extraction will be maintained at maximum exploitable volume during the forecast period.

Scenario III: According to the "Water Conservancy Modernization Plan", the proportion of groundwater in the total water supply in Jilin Province fell to 18.8% in 2030, and the total groundwater supply decreased to less than 3.5 billion m^3. Therefore, in this paper, the groundwater exploitation in Changchun, Jilin, Siping, Liaoyuan, Tonghua, Baishan, Songyuan, Baicheng and Yanbian were adjusted to 60%, 60%, 70%, 80%, 60%, 50%, 70%, 70% and 50% of the maximum exploitable volume, respectively.

4.3. Regulation of WRCC

Two types of regulation were designed. The first is intra-regional regulation, that is, the regulation of the development trend of indicators in the sub-model; the second is cross-regional regulation, that is, the allocation of water resources between regions. When we design the regulation programs, we will focus on intra-regional regulation, supplemented by cross-regional regulation.

The HWRS is a complex system. The regulation of WRCC usually involves multiple variables and there are multiple adjustments to the variables, resulting in a large number of regulatory programs. The orthogonal test method is based on the probability theory, mathematical statistics and practical experience. Using the standardized orthogonal table to arrange the test plan, one can quickly find the optimization plan, which is often used to design multi-factor and multi-level optimization experiments [39,40]. In order to simplify the design of the scheme, the orthogonal test design method was applied to regulate the WRCC in Jilin Province.

5. Results

5.1. Analysis of The WRCC in 2017

Using the model described in the paper, the support ability, pressure ability and WRCC of Jilin was calculated. The WRCC of Jilin Province and its nine cities are described in Tables 3 and 4 and shown in Figure 4. It should be mentioned that the variables in the second line of Tables 3 and 4 are the acronyms of the names of Jilin Province. For example, "CC" is the acronym of Changchun City.

Table 3. The carrying capacity index (CCI) of Jilin.

CCI	Province	Prefecture-Level City and Autonomous Prefecture								
		CC	JL	SP	LY	TH	BS	SY	BC	YB
quantity	0.76	0.78	0.64	0.65	1.05	0.66	1.16	0.84	0.85	0.71
quality	0.91	1.01	0.93	0.93	0.94	0.67	0.91	0.96	0.89	0.87
domain	0.81	0.68	0.86	0.72	0.61	0.64	0.94	0.87	0.83	0.82
flow	0.45	0.43	0.95	1.47	0.20	0.95	0.08	0.76	1.54	0.06
Comprehensive	0.91	1.01	0.93	0.93	1.05	0.67	1.16	0.96	0.89	0.87

Table 4. The water resources carrying capacity (WRCC) of Jilin 10,000 people.

WRCC	Province	Prefecture-Level City and Autonomous Prefecture								
		CC	JL	SP	LY	TH	BS	SY	BC	YB
Current	2616	749	415	320	118	217	120	275	191	210
quantity	3496	965	645	492	112	329	103	330	224	295
quality	2855	739	448	345	125	322	131	287	215	242
Comprehensive	2855	739	448	345	112	322	103	287	215	242

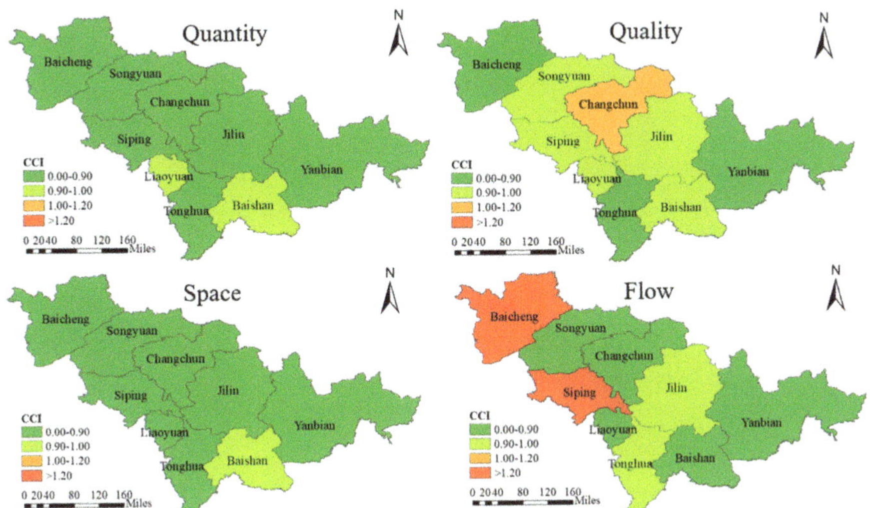

Figure 4. Carrying state of water resources.

(1) The WRCC is 28.13 million people in Jilin Province, bigger than the current population in 2017, and its comprehensive carrying capacity index (*CCI*) is 0.93, which is in a critical carrying state. Among the nine cities, Changchun, Liaoyuan and Baishan have smaller comprehensive carrying capacity than the current population, and their HWRS is in an overload state. Baishan's comprehensive *CCI* (1.16) is the largest of all cities, followed by Liaoyuan and Changchun. The comprehensive *CCI* of Baishan is equal to its water quantity *CCI*, which is significantly bigger than its water quality CCI, indicating that the limiting factor of WRCC of Baishan is water quantity, which belongs to the water quantity dominant area. Baishan belongs to Changbai Mountain and has abundant water resources. However, the reservoir capacity of reservoirs is insufficient. As shown in Table 3, its water flow *CCI* is 0.08. Therefore, new reservoirs can be built to increase water storage to alleviate the water shortage crisis. Liaoyuan is a water quantity dominated area and needs to increase the use of surface water. The water demand and the volume of pollutants discharged in Changchun are the largest among all cities, and its water quantity *CCI* and water quality *CCI* are 0.78 and 1.01, respectively, which are the water quality

dominant areas. As a provincial capital city, Changchun is the economic and cultural center of Jilin Province. Changchun is supposed to play a greater role in leading the industry, vigorously develop advanced manufacturing industries, and transfer some low-end industries. Furthermore, Changchun also needs to implement the principle of total pollutant control, increase investment in environmental protection. The comprehensive *CCI* of Jilin, Siping and Songyuan is higher than 0.9, and its HWRS is at a critical state, and it is necessary to adhere to the red line of water resources development and utilization. Jilin city, located in the middle of Jilin province, is the transition zone between the eastern mountain area and the western plain area. It has good water resources development conditions, making its water available rank first among all cities. However, similarly to Changchun, Jilin has developed industries and emits a large amount of pollutants every year. The water environment pollution crisis still exists. There is a huge shortage of water resources in Siping and Songyuan. In order to meet the water demand in the two cities, groundwater has to be used in large quantities, causing groundwater to fall and forming a groundwater level funnel. The Central City Songhua River Water Supply Project and the Hadashan Water Conservancy Project were constructed to alleviate the water crisis in Changchun, Siping, Liaoyuan and Songyuan. The comprehensive *CCI* of Tonghua, Baicheng and Yanbian are all below 0.9, and the HWRS is in a safe carrying state. The water quantity *CCI* of these three cities is higher than the water quality *CCI* and they belong to the water quality dominant areas. Therefore, while developing the economy in the future, they should always pay attention to environmental protection.

(2) The water quantity carrying state of Jilin Province is generally good. Except for the fact that Baishan and Liaoyuan are in an overload state, the remaining cities are in a safe carrying state. Among them, the reasons for the overload of Baishan and Liaoyuan are different. The main reason for the Baishan overload is the lack of water conservancy engineering facilities. The main reason for the Liaoyuan overload is the shortage of water resources. The water quality carrying state of Jilin Province is expected to gradually deteriorate from the periphery to the center. For example, the water quality *CCI* of Changchun and Jilin ranks first and third among all the cities. The water space carrying states of all cities are in good condition. Except for the fact that Baishan is in a critical state, the other eight cities are safely loaded. It can be concluded that each city has enough space to accommodate water resources. In terms of water mobility, the water flow carrying states of Baicheng and Siping are heavily overloaded, and the remaining seven cities are not. Baicheng is located in the western part of Jilin Province. The water resources are extremely scarce. A large amount of water needs to be stored to meet the demand for water. As a result, the area of wetlands and lakes in the territory is shrinking. As an ecological barrier zone in the west, Baicheng needs to improve its water use efficiency on the one hand and increase its water transfer from other regions on the other hand. The project to transfer water resources of Nenjiang into Baicheng was built to increase the volume of water available, and alleviate the crisis of ecological water shortage.

5.2. Future Simulation of the WRCC

5.2.1. Model Validation

Taking Jilin Province and Changchun City as examples, this model was used to estimate the values of major variables such as population, GDP, industrial water, and water demand in 2005–2017. The relative errors of these variables are shown in Table 5. The relative errors of the main variables in Jilin Province and Changchun City are small, and the maximum value is 2.44%. The SD model established in this study performed well and can be used for further analysis.

Table 5. Relative errors of the main variables between simulated data and historic data.

Year	Jilin Province				Changchun City			
	Population	GDP	Industry Water	Water Demand	Population	GDP	Industry Water	Water Demand
2005	0	0	0.48%	−1.45%	0	0	0.43%	0.58%
2006	0	0	0.05%	0.17%	0	0	1.25%	0.51%
2007	0	0	−0.10%	0.04%	0	0	0.80%	0
2008	0	0	0	0	0	0	0.40%	0.05%
2009	0	0	0	0	0	0	0.84%	0.10%
2010	0	0	0	0	0	0	−0.28%	0.12%
2011	0	0	0	0	0	0	0.28%	0.09%
2012	0	0	0	0	0	0	0.24%	0.08%
2013	0	0	0	0	0	0	0.48%	0.08%
2014	0	0	0	0	0	0	2.44%	0.04%
2015	0	0	2.41%	1.29%	0	0	0.28%	−0.13%
2016	0	0	0.47%	0.14%	0	0	0.35%	0.07%
2017	0	0	0.06%	0.03%	0	0	0.51%	0.16%

5.2.2. Provincial Simulation

Table 6 shows the simulation results of water demand/available in Jilin Province. As shown in Table 6, with the continuous development of the economy, the total water demand in Jilin Province increased year by year, but the growth rate slowed down. We predict that water demand in Jilin Province will reach 15.35 billion m^3 in 2020 and 17.05 billion m^3 in 2030. With the advancement of agricultural modernization and water conservancy modernization, Jilin Province will expand the area of agricultural planting, which will focus on expanding its rice planting area and its agricultural water consumption will also increase. We predict that it will exceed 11.5 billion m^3 in 2030. The total volume of industrial water is closely related to the total industrial added value and industrial water efficiency. It will increase slightly with the revitalization of industrial bases. We predict that it will reach 2.42 billion m^3 in 2020 and 2.85 billion m^3 in 2030. The living standards of the people are constantly improving. We assume that the per capita domestic water demand in urban areas and the per capita domestic water demand in rural areas will reach 160 L/capita*day and 95 L/capita*day in 2030, respectively. We predict that the total domestic water demand will reach 2.19 billion m^3. The water resources in the east are abundant, and the volume of ecological water is small. The ecological water demand will increase in the west, which will ensure the sustainable development of the ecosystem. We predict that it will reach 480 million m^3 in 2030.

Table 6. Simulation results of water demand/available in Jilin Province from 2018 to 2030.

Year	Water Demand/10^9 m^3					Water Available/10^9 m^3			CCI		
	Total	Agriculture	Industry	Domestic	Ecological	Scenario			Scenario		
						I	II	III	I	II	III
2018	145.4	101.6	24.4	15.2	4.2	168.8	160.5	151.1	0.86	0.91	0.96
2019	149.5	105.0	24.3	16.0	4.2	171.1	162.8	153.4	0.87	0.92	0.97
2020	153.2	108.1	24.2	16.8	4.1	182.3	174.0	164.6	0.84	0.88	0.93
2021	155.2	109.1	24.7	17.3	4.2	183.3	174.9	165.6	0.85	0.89	0.94
2022	157.2	110.0	25.1	17.8	4.2	189.4	181.1	171.7	0.83	0.87	0.92
2023	159.2	111.0	25.5	18.3	4.3	190.3	182.0	172.6	0.84	0.87	0.92
2024	161.2	112.0	25.9	18.9	4.3	192.1	183.8	174.4	0.84	0.88	0.92
2025	163.1	113.0	26.3	19.4	4.4	200.0	191.7	182.4	0.82	0.85	0.89
2026	164.7	113.5	26.8	19.9	4.4	200.5	192.1	182.8	0.82	0.86	0.90
2027	166.2	114.0	27.3	20.4	4.5	200.9	192.5	183.2	0.83	0.86	0.91
2028	167.7	114.4	27.8	20.9	4.5	201.3	193.0	183.6	0.83	0.87	0.91
2029	169.1	114.9	28.2	21.4	4.6	201.7	193.4	184.0	0.84	0.87	0.92
2030	170.5	115.4	28.5	21.9	4.6	202.1	193.8	184.5	0.84	0.88	0.92

The volume of water available is different in three scenarios in Jilin Province. Scenario I maximizes the development of groundwater, resulting in the largest volume of water available in Jilin Province.

We predict it will reach 18.23 billion m^3 in 2020 and 20.21 billion m^3 in 2030. Groundwater development in Scenario II is the same as in 2017, making groundwater supply less than the maximum developable amount, so the increase in water available will depend on surface water and reclaimed water. We predict that the water available will reach 17.4 billion m^3 in 2020 and 19.38 billion m^3 in 2030. Scenario III controls the groundwater development strictly and reduces the proportion of groundwater supply, which will increase the volume of groundwater in the emergency reserve water source and return the over-exploited groundwater in history. We predict that the water available will reach 16.46 billion m^3 in 2020 and 18.4 billion m^3 in 2030.

The water quantity CCI of Jilin Province has not exceeded one in three scenarios from 2018 to 2030. This shows that if Jilin Province can complete the modernization of water conservancy on schedule and transfer the water from the east to the west, the available water can generally meet the demand for water. In addition, the water quantity CCI in scenario III is the highest of all the scenarios during the same period. Groundwater development is strictly controlled, which is bound to reduce the volume of water available but it is still bigger than the water demand, indicating that scenario III is feasible.

It should be mentioned that the spatial distribution of water resources in Jilin Province is extremely uneven, so it is necessary to predict and analyze the water quantity carrying states of each city in scenario III.

5.2.3. City Simulation

Figure 5 shows the forecast results of water demand/available in all cities of Jilin Province from 2018 to 2030. As shown in Figure 5, the water demand of Changchun and Jilin is more than 2.5 billion m^3 during the forecast period and exceeds 3 billion m^3 in the 14th and 15th "five-year plans", respectively. Their GDP is located in the top two in Jilin Province, and the demand for water resources is enormous. However, there is a large gap in the volume of water available between the two cities. The volume of water resources in Jilin is significantly bigger than that in Changchun, which can fully meet the water demand. The water quantity CCI of Changchun exceeds 0.9 every year, and there is a risk of water shortage. It is necessary to improve the utilization rate of water resources. The water resources in Siping and Liaoyuan are extremely short, and it is urgent to transfer water across regions. We assume that The Central City Songhua River Water Supply Project will be completed in 2020. The water availability of the two cities will increase significantly and the water crisis will be basically solved. The water quantity CCI will decreased to 0.9 and 0.8.

Among the three cities in the east, Baishan is the main one engaged in the development of green economy, and its water resources development is low. Its water demand is close to that of Liaoyuan, ranking last in all cities. We predict that it will reach 0.33 billion m^3 in 2020 and 0.37 billion m^3 in 2030. Due to the insufficient water storage capacity of the water conservancy project, Baishan is unable to meet the water demand before 2024. During the 15th five-year plan period, after the completion of the proposed water conservancy project in Baishan, it will return to the safe carrying state, and the water crisis will be basically solved. Yanbian is located in the Chang-Ji-Tu Economic Zone and has a large demand for water. We predict it will reach 1.20 billion m^3 in 2020 and 1.25 billion m^3 in 2030. Several water storage projects and water transfer projects were planned in 2018-2030. Assuming that these projects are completed on schedule, we predict that the volume of water available for Yanbian will increase to 1.49 billion m^3 in 2020 and 1.57 billion m^3 in 2030, and its water quantity CCI will decrease to 0.81 and 0.79, respectively. The water demand of Tonghua is the highest among the three cities. We predict that it will reach 1.6 billion m^3 in 2020 and 1.68 billion m^3 in 2030. The degree of water resources development is significantly higher in Tonghua than in the other two cities, but the volume of water available of Tonghua is still lower than its water demand. Therefore, Tonghua needs to increase water conservation efforts.

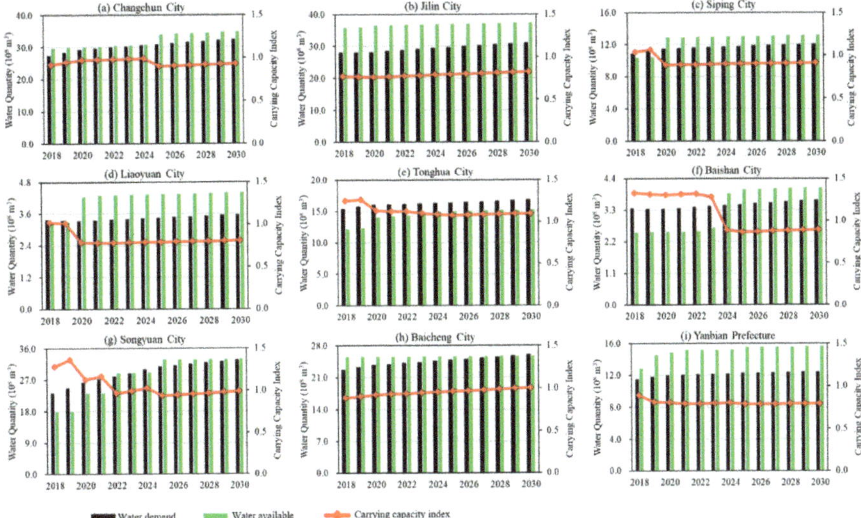

Figure 5. Simulation results of water demand/available in nine cities from 2018 to 2030.

Baicheng and Songyuan are located in the west and their water resources are extremely scarce. Due to fertile land resources, they are the most important agricultural bases in Jilin Province and even China. Their agricultural water use accounts for more than 80%. According to the plan, the area of newly built and expanded irrigation districts will exceed 3.5 million mu in Songyuan, resulting in an increase in the agricultural water demand. We predict that the water demand in Songyuan will reach 2.63 billion m^3 in 2020 and 3.27 billion m^3 in 2030. More than 2 million mu of irrigation districts are planned to be built in Baicheng. In order to meet the demand for water, the Hadashan Water Conservancy Project and the water diversion Project from Nenjiang river to Baicheng were constructed. They were used to transfer the water from the Songhua River and Nenjiang River to the two cities. The water shortage will be compensated, if the two projects meet the target in the two cities.

According to the above analysis, it can be found that the uneven distribution of water resources in Jilin Province will be alleviated by water saving and inter-regional water transfer. However, there is still a city with a water quantity CCI bigger than 1, so it is necessary to regulate the HWRS in Jilin Province.

5.3. Analysis of the Regulation Program and Results

5.3.1. Analysis of Regulatory Indicators

Based on the prediction results in Section 5.2, the three major water users in Jilin Province are agriculture, industry and domestic. Agricultural water use is closely related to agricultural output value and water use per 10,000 RMB of agricultural output value (C_1). Regulation of C_1 will directly affect the total volume of agricultural water. Industrial water is also the main target of regulation. Its value is determined by the industrial added value and the water use per 10,000 RMB of industrial added value (C_2). The modernization of agriculture and industry currently being implemented in Jilin Province will inevitably lead to an increase in agricultural output value and industrial added value. Therefore, it is crucial to control the water use by regulating C_1 and C_2. At present, the per capita domestic water use in Jilin Province is less than the average level in China. Scenario III has raised it to the average level, so this section will not regulate it. Furthermore, the forecast results show that some cities have not been able to meet their water demand after water transfer from other regions. Based on the above analysis, the variables will be used for the adjustment program of the next section, including the water use per 10,000 RMB of agricultural output value (C_1), the water use per 10,000 RMB of industrial added value (C_2) and the volume of water transfer (C_3).

5.3.2. Analysis of Regulatory Program

According to the national and local planning, two regulatory values were set for C1, C2 and C3, respectively. The most suitable regulatory project was selected through orthogonal tests. Based on the historical value of 2015, the adjusted value of this parameter is obtained in 2030. The value of 0.9 was chosen as the limit of the water quantity CCI in this section, that is, the HWRS reached a safe load state. The recommended projects for each city are described in Table 7.

Table 7. Regulation program of WRCC.

City	C_1 m^3/10^4 RMB	C_2 m^3/10^4 RMB	C_3 10^8 m^3	City	C_1 m^3/10^4 RMB	C_2 m^3/10^4 RMB	C_3 10^8 m^3
Changchun	+10%	−45%	11	Baishan	+10%	−45%	0
Jilin	+12%	−35%	0	Songyuan	+25%	−45%	15
Siping	+17%	−45%	2.6	Baicheng	+14%	−40%	12
Liaoyuan	0	−40%	0.7	Yanbian	+5%	−40%	0
Tonghua	−5%	−45%	4				

The water allocation data of Jilin Province (Table 8) was obtained, which regulates the HWRS according to the program shown in Table 7. The variables in Table 8 are composed of surface water (SW), groundwater (GW), transfer water (TW), reclaimed water (RW), agriculture water (AW), industry water (IW), domestic water (DW), and ecological water (EW). The volume of water available in Jilin Province is 18.85 billion m^3, the water demand is 16.68 billion m^3, and the water quantity CCI is 0.88, which is in a safe load state in 2030. This shows that the above regulation program is feasible. Fortunately, water conservancy modernization is currently being implemented in Jilin Province, which will increase the volume of water available in the central and western regions. In addition, the upgrading of industrial structure is underway, which is expected to reduce the water use per 10,000 RMB of industrial added value.

Table 8. Water allocation results of Jilin province in 2030.

Region	Water Available/10^8 m^3					Water Demand/10^8 m^3					CCI
	SW	GW	TW	RW	Total	AW	IW	DW	EW	Total	
CC	14.78	6.75	11	4.16	36.69	19.22	4.52	8	1.3	33.04	0.90
JL	29.85	3.66	0	3.73	37.24	14.37	13.59	4.03	1.1	33.09	0.89
SP	4.98	4.83	2.6	0.94	13.35	8.32	1.69	1.63	0.37	12.01	0.90
LY	2.03	0.99	0.7	0.41	4.13	1.75	1.07	0.55	0.3	3.67	0.89
TH	10.97	0.84	4	1.81	17.62	9.85	3.4	1.85	0.19	15.29	0.87
BS	3.35	0.44	0	0.39	4.18	0.58	1.73	1.12	0.23	3.66	0.88
SY	10.89	6.6	15	1.05	33.54	25.78	2.25	1.81	0.38	30.22	0.90
BC	5.96	7.93	12	0.54	26.43	21.53	0.79	1.01	0.24	23.57	0.89
YB	12.94	1.19	0	1.21	15.34	8.92	1.14	1.93	0.26	12.25	0.80
Total	95.74	33.23	45.30	14.24	188.51	110.32	30.18	21.93	4.37	166.80	0.88

6. Discussion

A three-stage hybrid model, including evaluation, prediction, and regulation, is established in this paper. It was used to analyze the WRCC of Jilin Province. According to the evaluation results in Section 5.1, the water resources problems faced by cities in Jilin Province in 2017 are different and there is inconsistency between the socio-economic subsystem and the water resources-environmental subsystem. For example, in order to meet the demand for water, surface water and groundwater have been extensively developed in Baicheng and Siping, causing their water flow factor to be overloaded. Due to the lack of water conservancy facilities, Baishan has less interference with rivers than the previous two cities, so that its water flow factor is in a safe load state, but the water quantity factor is in an overload state. The evaluation results of the above three cities indicate that there are advantages and disadvantages in reservoir construction. The advantage is that water can be stored to meet the demand for water. The disadvantage is that it destroys the fluidity of the water and causes damage the aquatic ecosystem [41,42].

According to the dynamic forecast of WRCC in Section 5.2, if the water transfer and water storage projects in the Water Conservancy Modernization Plan are completed on schedule, the volume of water resources transferred from the east to the central and western regions will increase, which can effectively alleviate the shortage of water resources in these regions. Groundwater resources are also protected. In addition, this study analyzes the impact of water conservancy engineering facilities on the water resources system of Jilin Province in the relevant plans, as shown in Table 9. After the modernization of water conservancy, the surface water in Changchun, Siping, Baicheng and Songyuan will increase significantly. In addition, the surface water of Baishan and Yanbian in the east also increased. This indicates that after the completion of the construction of the water conservancy project, it will not only transport surface water for the central and western regions, but also not cause surface water reduction in the eastern region. Water modernization construction will bring inestimable economic and ecological benefits to Jilin Province.

Table 9. The change in surface water available.

Year	CC	JL	SP	LY	TH	BS	SY	BC	YB
2017	14.78	29.85	4.98	2.03	9.99	1.83	10.89	12.8	10.91
2030	24.02	30.08	7.40	3.00	12.61	3.25	25.38	17.31	13.94

Combined with the regulation analysis of Section 5.3, the indicators need to be adjusted simultaneously, including the water use per 10,000 RMB of agricultural output, the water use per 10,000 RMB of industrial added value and the volume of water transfer, which will balance the water demand/available from 2018 to 2030. The water use efficiency of agriculture and industry will change through the regulation of the first two indicators. It can be concluded that the regulation of WRCC not only needs to strengthen water conservation, but also needs to rationally dispatch water resources by breaking the boundary of the region, which can solve the problem of water imbalance between regions. China's South-to-North Water Transfer Project is a successful case [43].

There is a plan in Jilin Province called "Main Pollutant Total Emission Reduction Plan", which sets the total control target for various pollutants. Referring to the plan, the value of the indicator "water environment capacity" is obtained. With the increasingly strict enforcement of environmental protection, all cities will strictly abide by the pollutant emission reduction plan, which is mandatory. We predict that there will be no more overloading of the water quality factor. In addition, it should be mentioned that the evaluation result of the carrying state of the water space has limitations. It can only reflect the ratio of the water area to the basin area, which is greatly affected by the water volume in the current year. In summary, this paper carries out the assessment of WRCC from four factors: water quantity, water quality, water space and water flow. It differs from previous studies that focused on the first two factors. Compared with the existing research, the SD model of this paper couples the city-scale models of each city into a provincial-scale model by connecting key indicators. It can simulate and analyze WRCC at different scales, which is complex and advanced. Furthermore, the results of water allocation are obtained in this paper, which will provide reference for water resources management in Jilin Province.

7. Conclusions

According to the characteristics of HWRS in Jilin Province, the load-balance method was used to establish an indexes system consisting of four factors and eight indicators. Then, the SD method was applied to construct a model that includes socio-economic, water resources, and water environment.

Through the validity test, it is verified that the SD model can the simulate the HWRS of Jilin Province. Three scenarios were set up to simulate the dynamic trend of WRCC. Using the simulation results of the all three scenarios, an orthogonal test method was adopted to select the best project. Finally, the results of water resources allocation in 2030 were obtained in Jilin Province. The current analysis allows us to draw the following conclusions.

(1) The evaluation results of WRCC indicate that the HWRS of Changchun, Liaoyuan and Baicheng are overloaded. The pressure of socio-economic activities exceeded the support of water resources-environmental in these cities. The HWRS of Jilin, Siping and Songyuan are in a critical state, very close to the warning line. The WRCC of other cities is surplus.

(2) The total water demand in Jilin Province continued to increase from 2018 to 2030. We predict that it will reach 15.35 billion m^3 in 2020 and 17.05 billion m^3 in 2030. The water available of the three scenarios can meet the future water demand. Among them, Scenario III imposes the most restrictive restrictions on groundwater development, resulting in the closest water quantity CCI to the warning line, but it is still feasible. Among the nine cities, Siping, Tonghua, Baishan, Songyuan and Baicheng will experience water shortages and the rest will not.

(3) In order to realize the sustainable development of the HWRS, Jilin Province needs to strictly control the growth rate of agricultural water consumption, improve the efficiency of industrial water use, and strengthen the cross-regional distribution of water resources.

The three-stage hybrid model proposed in this paper is reasonable for studying the WRCC and has application value for water resources management.

Author Contributions: Conceptualization, T.L. and X.Y.; methodology, T.L., B.S.; software, T.L., B.S.; validation, T.L.; formal analysis, X.Y.; investigation, T.L.; resources, X.Y.; data curation, T.L., L.G.; writing—original draft preparation, T.L.; writing—review and editing, X.Y.; visualization, X.Y.; supervision, X.Y.; project administration, X.Y.; funding acquisition, X.Y., L.G. All authors have read and agreed to the published version of the manuscript.

Funding: This work was supported by the National Key Research Program of China (No. 2016YFC0401305, No. 2017YFC0506603), the State Key Program of National Natural Science of China (No. 41530635), and the Project of National Natural Foundation of China (No. 51679007, 51379013).

Acknowledgments: Thanks to reviewers and editors for their valuable suggestions.

Conflicts of Interest: The authors declare no conflict of interest.

References

1. Chen, Y.L.; Wang, S.S.; Ren, Z.G.; Huang, J.F.; Wang, X.Z.; Liu, S.S.; Deng, H.J.; Lin, W.K. Increased evapotranspiration from land cover changes intensified water crisis in an arid river basin in north China. *J. Hydrol.* **2017**, *574*, 383–397. [CrossRef]
2. Beck, L.; Bernauer, L. How will combined changes in water demand and climate affect water availability in the Zambezi river basin. *Glob. Environ. Chang.* **2011**, *21*, 1061–1072. [CrossRef]
3. Chen, B.; Wang, M.; Duan, M.X.; Ma, X.T.; Hong, J.L.; Xie, F.; Zahng, R.R.; Li, X.Z. In search of key: Protecting human health and the ecosystem from water pollution in China. *J. Clean. Prod.* **2019**, *228*, 101–111. [CrossRef]
4. Gleick, P.H.; Palaniappan, M. Peak water limits to freshwater withdrawal and use. *Proc. Natl. Acad. Sci. USA* **2010**, *107*, 11155–11162. [CrossRef]
5. Wang, Y.Z.; Liu, L.; Guo, S.S.; Yue, Q.; Guo, P. A bi-level multi-objective linear fractional programming for water consumption structure optimization based on water shortage risk. *J. Clean. Prod.* **2019**, *237*, 117829. [CrossRef]
6. Wang, H.; Asefa, T.; Bracciano, D.; Adams, A.; Wanakule, N. Proactive water shortage mitigation integrating system optimization and input uncertainty. *J. Hydrol.* **2019**, *571*, 711–722. [CrossRef]
7. Yan, R.H.; Gao, Y.N.; Li, L.L.; Gao, J.F. Estimation of water environmental capacity and pollution load reduction for urban lakeside of Lake Taihu, eastern China. *Ecol. Eng.* **2019**, *139*, 105587. [CrossRef]
8. Wang, Q.R.; Liu, R.M.; Men, C.; Guo, L.J.; Miao, Y.X. Temporal-spatial analysis of water environmental capacity based on the couple of SWAT model and differential evolution algorithm. *J. Hydrol.* **2019**, *569*, 155–166. [CrossRef]

9. Wen, Z.G.; Meng, F.X.; Di, J.H.; Tan, Q.L. Technological approaches and policy analysis of integrated water pollution prevention and control for the coal-to-methanol industry based on Best Available Technology. *J. Clean. Prod.* **2016**, *113*, 231–240. [CrossRef]

10. Manzano-Solís, L.R.; Díaz-Delgado, C.; Gómez-Albores, M.A.; Mastachi-Loza, C.A.; Soares, D. Use of structural systems analysis for the integrated water resources management in the Nenetzingo river watershed, Mexico. *Land Use Policy* **2019**, *87*, 104029. [CrossRef]

11. Wang, K.; Davies, E.G.R.; Liu, J.G. Integrated water resources management and modeling: A case study of Bow river basin, Canada. *J. Clean. Prod.* **2019**, *240*, 118242. [CrossRef]

12. Arrow, K.; Bolin, B.; Costanza, R.; Dasgupta, P.; Folke, C.; Holling, C.S.; Jansson, B.O.; Levin, S.; Mäler, K.G.; Perrings, C.; et al. Economic growth, carrying capacity, and the environment. *Ecol. Econ.* **1995**, *15*, 91–95. [CrossRef]

13. Peng, B.H.; Li, Y.; Elahi, E.; Wei, G. Dynamic evolution of ecological carrying capacity based on the ecological footprint theory: A case study of Jiangsu province. *Ecol. Indic.* **2019**, *99*, 19–26. [CrossRef]

14. Li, C.; Li, H.J.; Feng, S.D.; Liu, X.Y.; Guo, S. A study on the spatiotemporal characteristics and change trend of the atmospheric environmental carrying capacity in the Jing-Jin-Ji region, China. *J. Clean. Prod.* **2019**, *211*, 27–35. [CrossRef]

15. Shi, Y.S.; Wang, H.F.; Yin, C.Y. Evaluation method of urban land population carrying capacity based on GIS—A case of Shanghai, China. *Comput. Environ. Urban. Syst.* **2013**, *39*, 27–38. [CrossRef]

16. Wang, D.L.; Shi, Y.H.; Wan, K.D. Integrated evaluation of the carrying capacities of mineral resource-based cities considering synergy between subsystems. *Ecol. Indic.* **2020**, *108*, 105701. [CrossRef]

17. Zeng, W.H.; Cheng, S.T. Preliminary suggestion on integrated planning of aquatic environment. *Shuili Xuebao* **1997**, *10*, 77–82. (In Chinese)

18. Falkenmark, M.; Lundqvst, J. Towards water security: Political determination and human adaptation crucial. *Nat. Resour. Forum* **1998**, *21*, 37–51. [CrossRef]

19. Yang, J.F.; Lei, K.; Khu, S.; Meng, W. Assessment of Water Resources Carrying Capacity for Sustainable Development Based on a System Dynamics Model: A Case Study of Tieling City, China. *Water Resour. Manag.* **2015**, *29*, 885–899. [CrossRef]

20. Wang, Z.G.; Luo, Y.Z.; Zhang, M.H.; Xia, J. Quantitative Evaluation of Sustainable Development and Eco-Environmental Carrying Capacity in Water-Deficient Regions: A Case Study in the Haihe River Basin, China. *J. Integr. Agric.* **2014**, *13*, 195–206. [CrossRef]

21. Wang, J.H.; Zhai, Z.L.; Sang, X.F.; Li, H.H. Study on index system and judgment criterion of water resources carrying capacity. *Shuili Xuebao* **2017**, *48*, 1023–1029. (In Chinese)

22. Munyaneza, C.; Kurwijila, L.R.; Mdoe, N.S.Y.; Baltenweck, I.; Twine, E.E. Identification of appropriate indicators for assessing sustainability of small-holder milk production systems in Tanzania. *Sustain. Prod. Consum.* **2019**, *19*, 141–160.

23. Lin, T.; Ge, R.B.; Huang, J.; Zhao, Q.J.; Lin, J.Y.; Huang, N.; Zhang, G.Q.; Li, X.H.; Ye, H.; Yin, K. A quantitative method to assess the ecological indicator system's effectiveness: A case study of the Ecological Province Construction Indicators of China. *Ecol. Indic.* **2016**, *62*, 95–100. [CrossRef]

24. Zhang, J.; Zhang, C.L.; Shi, W.L.; Fu, Y.C. Quantitative evaluation and optimized utilization of water resources-water environment carrying capacity based on nature-based solutions. *J. Hydrol.* **2019**, *568*, 96–107. [CrossRef]

25. Cheng, F.; Su, F.Z.; Chen, M.; Wang, Q.; Jiang, H.P.; Wang, X.G. An evolving assessment model for environmental carrying capacity: A case study of coral reef islands. *J. Environ. Manag.* **2019**, *233*, 543–552. [CrossRef]

26. Zhang, F.; Wang, Y.; Ma, X.J.; Wang, Y.; Yang, G.C.; Zhu, L. Evaluation of resources and environmental carrying capacity of 36 large cities in China based on a support-pressure coupling mechanism. *Sci. Total Environ.* **1991**, *688*, 838–854. [CrossRef]

27. Wang, Y.X.; Wang, Y.; Su, X.L.; Qi, L.; Liu, M. Evaluation of the comprehensive carrying capacity of interprovincial water resources in China and the spatial effect. *J. Hydrol.* **2019**, *575*, 794–809. [CrossRef]

28. Chi, M.B.; Zhang, D.S.; Fan, G.W.; Zhang, W.; Liu, H.L. Prediction of water resource carrying capacity by the analytic hierarchy process-fuzzy discrimination method in a mining area. *Ecol. Indic.* **2019**, *96*, 647–655. [CrossRef]

29. Li, T.H.; Yang, S.N.; Tan, M.X. Simulation and optimization of water supply and demand balance in Shenzhen: A system dynamics approach. *J. Clean. Prod.* **2019**, *207*, 882–893. [CrossRef]

30. Wei, X.M.; Wang, J.Y.; Wu, S.G.; Xin, X.; Wang, Z.L.; Liu, W. Comprehensive evaluation model for water environment carrying capacity based on VPOSRM framework: A case study in Wuhan, China. *Sustain. Cities Soc.* **2019**, *50*, 101640. [CrossRef]

31. Yang, Z.Y.; Song, J.X.; Cheng, D.D.; Xia, J.; Li, Q.; Ahamad, M.I. Comprehensive evaluation and scenario simulation for the water resources carrying capacity in Xi'an city, China. *J. Environ. Manag.* **2019**, *15*, 221–233. [CrossRef]

32. Ma, X.L.; Ma, Y.J. The spatiotemporal variation analysis of virtual water for agriculture and livestock husbandry: A study for Jilin Province in China. *Sci. Total Environ.* **2017**, *586*, 1150–1161. [CrossRef]

33. Yang, B.G.; Liu, C.H.; Zai, Y.Z.; Li, H.M.; Wang, J.; Shang, J.C.; Ji, Y.M. The Principal Component Analysis of the River in Jilin Province. *Sci. Geogr. Sinica* **1986**, *6*, 261–268. (In Chinese)

34. Wang, H. The Future Situation of Water Resources in China and its Management Requirements. *World Environ.* **2011**, *2*, 16–17. (In Chinese)

35. Chen, H.F.; Hu, Y.A.; Zhao, J.F. Meeting China's Water Shortage Crisis: Current Practices and Challenges. *Environ. Sci. Technol.* **2009**, *43*, 240–244. [CrossRef]

36. Yu, G.Q.; Du, B.J.; Liu, J.P. Ecosystem Vulnerability Assessment of Wetlands in Western Jilin Province from 1986 to 2012. *Wetland Sci.* **2016**, *14*, 439–445. (In Chinese)

37. Zhang, W.F.; Zhang, H.; Zhao, J.J. Discussion on Water Resources in Jilin Province. *J. China Hydrol.* **2009**, *29*, 68–70.

38. Bian, J.Y.; Huang, C.S.; Geng, L.H.; Fang, R.; Wang, Y.K. Diagnostic System Construction of Water Resources Carrying Capacity and the Key Factors Determination. *Water Sav. Irrig.* **2019**, *7*, 56–61. (In Chinese)

39. Wang, B.H.; Lin, R.; Liu, D.C.; Feng, B.W. Investigation of the effect of humidity at both electrode on the performance of PEMFC using orthogonal test method. *Int. J. Hydrog. Energy* **2019**, *44*, 13737–13743. [CrossRef]

40. Wang, B.H.; Jin, Y.; Luo, Y.G. Parametric optimization of EQ6110HEV hybrid electric bus based on orthogonal experiment design. *Int. J. Automot. Technol.* **2010**, *11*, 119–125. [CrossRef]

41. Wang, H.; Yang, Z.; Saito, Y. Inter-annual and seasonal variation of the Huanghe (Yellow River) water discharge over the past 50 years: Connections to impacts from ENSO events and dams. *Glob. Planet. Chang.* **2006**, *50*, 212–225. [CrossRef]

42. Jiang, T.; Zhang, Q.; Zhu, D.; Wu, Y. Yangtze floods and droughts (China) and teleconnections with ENSO activities (1470–2003). *Quat. Int.* **2006**, *144*, 29–37.

43. Office of the South-to-North Water Diversion Project Construction Committee, State Council, PRC. The South-to-North Water Diversion Project. *Engineering* **2016**, *2*, 265–267. [CrossRef]

Article

Role of Ion Chemistry and Hydro-Geochemical Processes in Aquifer Salinization—A Case Study from a Semi-Arid Region of Haryana, India

Gopal Krishan [1,*], Priyanka Sejwal [1], Anjali Bhagwat [2], Gokul Prasad [1], Brijesh Kumar Yadav [3], Chander Prakash Kumar [1], Mitthan Lal Kansal [4], Surjeet Singh [1], Natarajan Sudarsan [1], Allen Bradley [5], Lalit Mohan Sharma [6] and Marian Muste [5]

[1] Groundwater Hydrology Division, National Institute of Hydrology, Roorkee 247667, India; priyankasejwal6@gmail.com (P.S.); gokulghildiyal@gmail.com (G.P.); cpkumar@yahoo.com (C.P.K.); ssingh_sagar@yahoo.co.in (S.S.); sudarsan3010@gmail.com (N.S.)

[2] Environmental Hydrology Division, National Institute of Hydrology, Roorkee 247667, India; anjali.civil.iit@gmail.com

[3] Department of Hydrology, Indian Institute of Technology, Roorkee 247667, India; brijeshy@gmail.com

[4] Department of Water Resources and Development Management, Indian Institute of Technology, Roorkee 247667, India; mlkgkfwt@iitr.ac.in

[5] IIHR—Hydroscience & Engineering, the University of Iowa, Iowa City, IA 52242, USA; allen-bradley@uiowa.edu (A.B.); marian-muste@uiowa.edu (M.M.)

[6] Sehgal Foundation, Gurgaon 122003, India; lalit.sharma@smsfoundation.org

* Correspondence: drgopal.krishan@gmail.com

Citation: Krishan, G.; Sejwal, P.; Bhagwat, A.; Prasad, G.; Yadav, B.K.; Kumar, C.P.; Kansal, M.L.; Singh, S.; Sudarsan, N.; Bradley, A.; et al. Role of Ion Chemistry and Hydro-Geochemical Processes in Aquifer Salinization—A Case Study from a Semi-Arid Region of Haryana, India. *Water* **2021**, *13*, 617. https://doi.org/10.3390/w13050617

Academic Editor: Helder I. Chaminé

Received: 17 December 2020
Accepted: 3 February 2021
Published: 26 February 2021

Publisher's Note: MDPI stays neutral with regard to jurisdictional claims in published maps and institutional affiliations.

Abstract: In the present study, a total of sixty groundwater samples, twenty each for the pre-monsoon, monsoon and post monsoon seasons of 2018, were collected from selected locations in the Mewat district of Haryana, India. Electrical conductivity (EC) was measured at the site and total dissolved solids (TDS) were estimated. Samples were analysed for anions (chloride, sulphate, and bicarbonate) and cations (calcium, potassium, magnesium, and sodium). Multiple regression analysis was performed to analyse the data and report the dominant ions. Piper trilinear diagram and Gibbs plots were used to find out the water type and the factors controlling the chemistry of the groundwater, respectively. The saturation index of $CaCO_3$, $CaSO_4$ and $NaCl$ was determined, using the PHREEQC MODEL. Sodium and calcium among cations, and chloride among the anions, had the highest degree of affinity and strong significance for all three seasons. The calcium–chloride water type dominated for all three seasons and Gibbs plot depicted that most of the $Na^+/Na^+ + Ca^{2+}$ and $Cl^-/Cl^- + HCO_3^-$ ratios show the weathering of rocks to form minerals as the major reason behind the ionic chemistry of the groundwater. The highest level of dissolution is encountered in the case of $NaCl$, followed by $CaSO_4$, whereas $CaCO_3$ depicts precipitation. The geochemical aspects of weathering, evaporation and ion exchange are the major processes responsible for high salinity, and anthropogenic activities are leading to its expansion. The findings from this study will be useful in management and remediation of groundwater salinity of the region.

Keywords: salinity; ions; semi-arid region; Mewat; Haryana

1. Introduction

The groundwater crisis in the northwest region of India is in the spotlight reported in recent studies [1–4]. Apart from decreasing groundwater levels [5,6], aquifer salinization has become a serious concern [7,8].

Salinization in water resources is associated with high concentrations of some chemical elements such as sodium, calcium, magnesium, sulphate, chloride, boron, fluoride, selenium, and arsenic [7]. Salinity is dynamically correlated with the local geological factors and climatic conditions [9]. The various factors responsible for groundwater salinization are weathering, precipitation, ion exchanges, dissolution, leaching of fertilizers and manures and the subsurface biological activity. Since these processes depend upon the chemical and geological characteristics of Aquifer materials, the groundwater salinization varies both spatially and temporally [7,10,11]. Groundwater salinization is very rampant in state of Punjab and Haryana. Seven districts in the southwestern part of Haryana, namely Gurugram, Bhiwani, Rohtak, Kaithal, Mahindergarh, Mewat and Sonipat, cover approximately 30 percent of area of the state registers' saline groundwater [12].

Mewat, one of the 21 districts of Haryana, hovers on rain-fed agriculture due to the scanty irrigation sources. Despite unfavourable climatic conditions, agriculture is a significantly dominant contributor to livelihood. The growth and density of vegetation is relatively low due to the salinization of groundwater and limited amount of freshwater pockets; moreover, the TDS level of most of the locations reaches 35,000 mg/L [7]. Excessive amounts of specific dissolved ions outside their acceptance limits have detrimental effects on plant growth as well on the human health, which minimize agricultural production and reduce healthier practices. In order to address detrimental effects, it is vital to examine the associated hydro-geochemical process leading to the enhancement of salinity in groundwater as well as a need to maintain adequate planning, management and efficiency of groundwater resources. Keeping this in view, the present work aims to find the association between ion chemistry and hydro-geochemical processes in order to ascertain the role of ions in salinization of aquifer.

2. Study Area

2.1. Location and Climate

The research area is in Mewat district, Haryana, India, with a geographical extent of latitude 27° 39′, 28° 20′ N and longitude 76°51′, 77°20′ E (Figure 1). The study area covers about 1507 km^2 with a population of about one million [13]. The climate of the research area falls under semi-arid, with an annual rainfall of 594 mm with 75 percent received in monsoon season. In 2018, the annual rainfall in Nagina and Firozepur blocks was 650 and 724 mm, respectively, with more than 90 percent during June to September (Figure 2b). The daily maximum temperature of the study area is 40 °C (May and June) and daily minimum temperature is 5.1 °C (January). The recurrent shortage of rainfall during the monsoon season and limited availability of freshwater resources (Figure 2b) compel farmers to cultivate crops that require less water, such as wheat, millet and mustard.

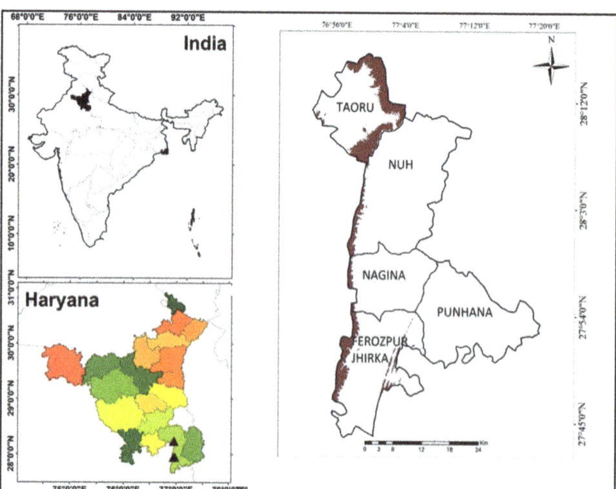

Figure 1. Location map of the study area depicting the district boundaries with Aravalli Hills.

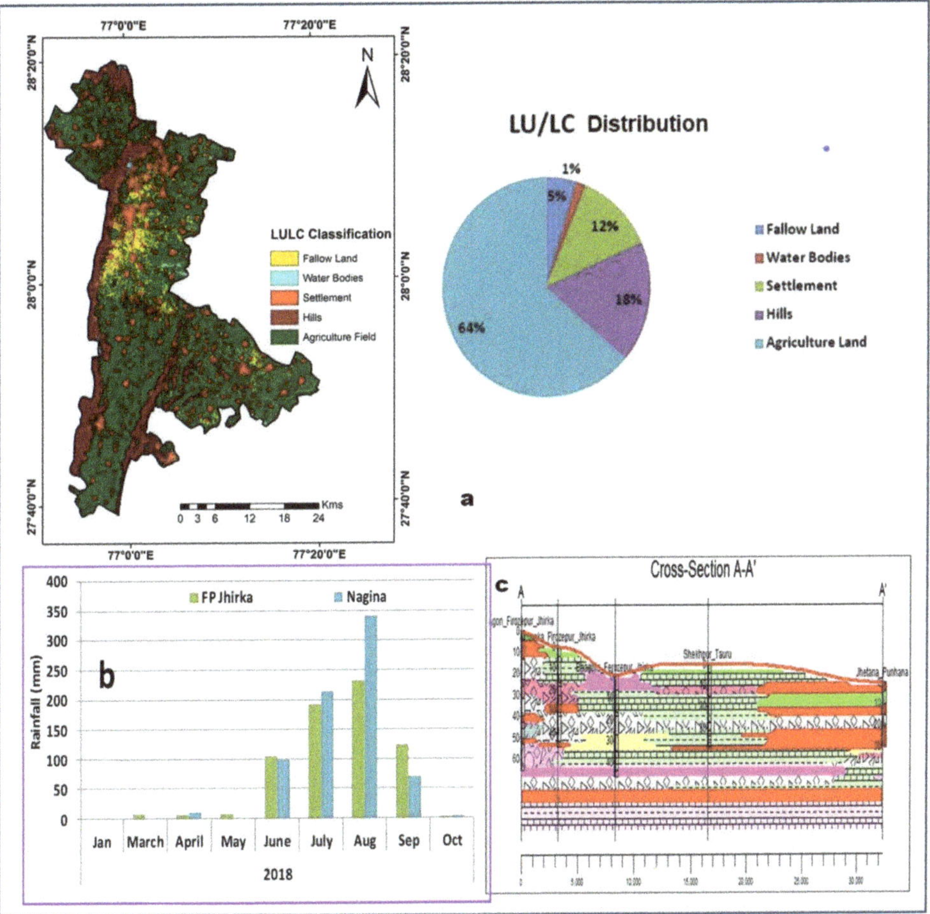

Figure 2. Landuse/land cover with class percentage distribution: (**a**) Landuse (**b**) rainfall; (**c**) lithology of study area.

2.2. Geomorphology and Soil

The district has a rolling topography with an urn-shaped structure. The central part is largely flat, but the western part slopes NW-SE and the northeastern parts slope NE-SW; altitude difference can be observed due to the presence of the Aravali Hills. The majority of the study area is dominated by the alluvial plains, with the northwestern area camouflaged by moderately deserted hills and valleys extending to the south-west and some parts of the south-east, whereas some parts of the mostly southeastern area are dominated by pediments and pediplain complexes with comparatively low water bodies.

Two major soil types, vertisols and salanchalks, are found in the study area. These soil types generally have medium textured loamy sand. The organic content of the soil ranges between 0.2 and 0.75 percent, while average electrical conductivity and the average pH of the soil are 0.80 µmhos/cm and 6.5–7, respectively. The top layer of the soil is significantly affected by salinization and salt crust can be found [14].

2.3. Hydrogeology

Topographically, the study area is mainly made up of alluvium of Quaternary and Paleoproterozoic age groups. The most dominant is the Quaternary age group alluvium, with a polycyclic sequence composed of sand, silt, and clay with kankar. The Paleoproterozoic age group consists of fledspathic gritty quartzite/amphibole/phyllite/schist of the Alwar group, Delhi (satellite data GSI). The groundwater has alluvial sediments with medium sand, clay and kankar. Good aquifer formations sand is intermittent, with multiple formations (Figure 2c). Overall, the clay ratio predominates at all depths throughout the districts. The depth of the bed rock in the central and eastern region lies within 300 mbgl, and the rest is around 90 mbgl in general [14]. The depth of water table ranges from 2 to 32 mbgl, with shallow water occurrences witnessed in the Nuh, Nagina and Punhana blocks.

2.4. Landuse and Land Cover

Agriculture is the main occupation of the people in the study area, as evident from landuse data, which consist of 64 percent agricultural land, 12 percent built-up area (human settlements including the residential sites), about 18 percent falls under hills, 1 percent under water bodies, and 5 percent under fallow area without perennial rivers and a normal semi-arid and arid drainage (Figure 2a). The major crops are Rabi, such as wheat, gram, and rice. The predominant economic activity includes agriculture, with no industrial settlement found in the area. Agriculture demands high irrigational water, but available water bodies do not meet these requirements. This creates a stress on the present water supplies in the study area.

3. Methodology

3.1. Sample Collection and Analysis

Groundwater samples from each of the 20 selected sites were collected in the pre-monsoon (April), monsoon (July), and post-monsoon seasons (October) in 2018 from hand pumps, open wells, and bore wells (Figure 3) with a depth range of 4–92 m. The water level of the open well in meters was recorded using a water-level indicator. GPS readings were taken to record latitude and longitude. Samples were taken in 125 mL capacity acid-washed Low-Density Polyethylene (LDPE) tarson bottles. Electrical Conductivity (EC) and pH were measured using a portable handheld Hach, HQ30d EC meter, and total dissolved solids (TDS) were estimated from measured EC values and expressed in mg/L. The samples collected in 125 mL bottles were analysed for cations (Ca^{2+}, Mg^{2+}, Na^+) and anions (HCO_3^-, SO_4^{2-}, Cl^-) at the water quality laboratory of the groundwater division of the institute, as per standard methodology [15].

The unpreserved 0.45 μm filtered water samples were used for the analysis of cations (Ca^{2+}, Mg^{2+}, Na^+) and anions (HCO_3^-, SO_4^{2-}, Cl^-). Ca^{2+} and Mg^{2+} were determined titrimetrically using standard EDTA. Cl^- was determined by standard $AgNO_3$ titration method. HCO_3^- was determined by titration with HCl. Na^+ was measured by flame photometry at a wavelength of 589 nm, and SO_4^{2-} by spectrophotometric turbidimetry by HACH spectrophotometer. All concentrations are expressed in milligrams per litre (mg/L).

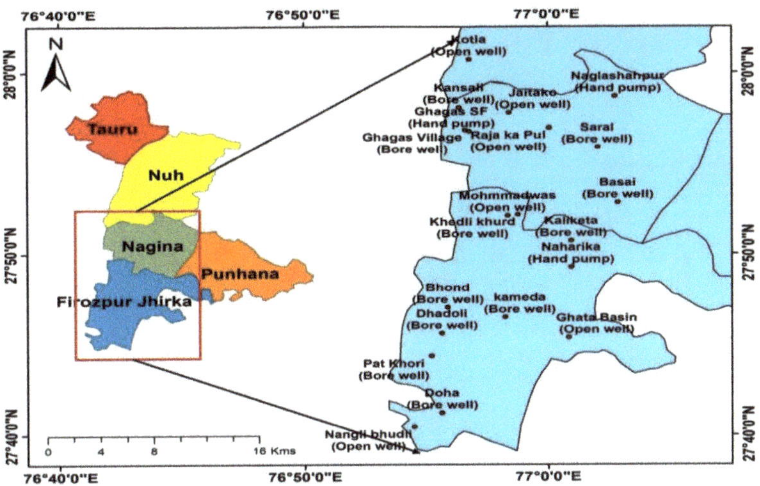

Figure 3. Sampling locations depicting sampled well type.

3.2. Data Analysis

Four trials were carried out in order to find the best-correlated value of TDS with respect to other ions. For all three seasons, the first trial included all six ions, namely Na^+, Cl^-, SO_4^{2-}, HCO_3^-, Mg^{2+} and Ca^{2+}. Regression Analysis was carried out using data analysis in Microsoft Excel to obtain the linear TDS equation and summative R^2 value. Multiple trails were performed and if the p-value did not satisfy the 95 percent level of significance, the respective ion was removed from the preceding trails.

AquaChem 2011.1 software was used to prepare piper trilinear diagram [16] in order to identify the hydro-geochemical facies and interpret the number of samples falling under the group sharing the same water qualities. Gibbs plots [17] was prepared using Sigma Plot 10 to interpret the factors controlling the ion chemistry of the groundwater. Arc GIS 10.4 was used for preparation of TDS maps using the inverse-distance weighting (IDW) interpolation technique.

4. Results and Discussion

Comparing all maps side-by-side shows that, in the southern part, the TDS in the pre-monsoon season is diluted in the monsoon season, which is evident from the increase in 0–1000 mg/L range values, which continues in the post-monsoon season with the withdrawal of rainfall. Similar observations were found using isotopes [18,19].

TDS level for all the seasons mostly lie in the 2000–12,000 mg/L range (Figure 4). Since TDS is a measure of all the dissolved salts that release cations and anions to water and has a direct relation with salinity, it is important to find out the dominant ion in the groundwater that elevates the level of salinity and its source.

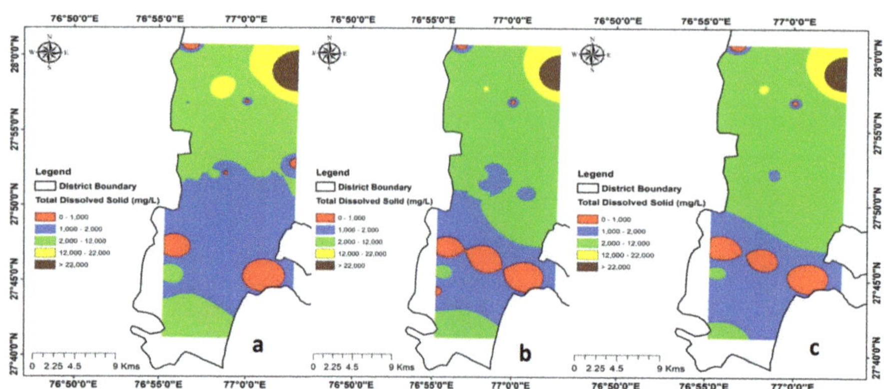

Figure 4. Total dissolved solids (TDS) maps (**a**) pre-monsoon (**b**) monsoon (**c**) post-monsoon.

4.1. Multiple Regression Analysis

In order to determine the dominant ions, multiple regression analysis was carried out. For all the seasons, three trails were carried out in accordance with the *p*-value with 95 percent significance level. The ions were removed if the value exceeded 0.05 and if all the ions fell under 95 percent significance level, then the one which had highest value relative to other ions was removed in the proceeding trail. Using this, two dominant cations and anions were determined. The dominant ions in the pre-monsoon were Cl^- and SO_4^{2-} from anions and Na^+ and Ca^{2+} from cations. The dominant ions in the monsoon season were Cl^- and HCO_3^- from anions, and Na^+ and Ca^{2+} from cations. The dominant ions in the post-monsoon season were Cl^- and SO_4^{2-} from anions and Na^+ and Mg^{2+} from cations. The compiled total of pre-monsoon and monsoon R^2 values was 0.99, and post-monsoon season value was 0.94 (Table 1).

Table 1. Different trails and dominant ion of each trail for all three seasons through Multiple Regression Analysis.

Parameters	Pre-Monsoon			Monsoon			Post-Monsoon		
	Trail	No. of Parameters	Dominant Ions	Trail	No. of Parameters	Dominant Ions	Trail	No. of Parameter	Dominant Ions
Na^+, Cl^-, Mg^{2+}, Ca^{2+}, SO_4^{2-}, HCO_3^-	1	6	Ca^{2+}, Na^+ & Cl^-, SO_4^{2-}	1	6	Ca^{2+}, Na^+ & Cl^-, HCO_3^-	1	6	Mg^{2+}, Na^+ Cl^-, SO_4^{2-}
	2	5 (*Exc. Mg^{2+})	Ca^{2+}, Na^+ & Cl^-, SO_4^{2-}	2	5 (*Exc. SO_4^{2-})	Ca^{2+} Na^+ & Cl^-, HCO_3^-	2	5 (*Exc. HCO_3^-)	Mg^{2+}, Na^+, Cl SO_4^{2-}
	3	4 (*Exc. HCO_3^- & Mg^{2+})	Ca^{2+}, Na^+ & Cl^-, SO_4^{2-}	3	4 (*Exc. SO_4^{2-} & Mg^{2+})	Ca^{2+} Na^+ & Cl^-, HCO_3^-	3	4 (*Exc. HCO_3^- & Cl^-)	Mg^{2+}, Na^+, Cl SO_4^{2-}
R^2	0.99			0.99			0.94		

*Exc.: excluding.

In order to understand the various processes responsible for the seasonal dominance of various ions, modeling of precipitation and dissolution process was carried out using PHREEQC model. SI was obtained for $CaCO_3$, $CaSO_4$ and $NaCl$ from the model, as shown in (Table 2 and Figure 5). The highest level of dissolution was encountered in the case of $NaCl$, followed by $CaSO_4$, whereas $CaCO_3$ depicted precipitation. In pre-monsoon, SI for $NaCl$ was -4.07 and for $CaSO_4$ was -1.38, which represents dissolution as a major reason for the dominance of Na^+ and Cl^- and Ca^{2+} and SO_4^{2-} ions.

Table 2. SI values of $CaCO_3$, $CaSO_4$ and NaCl using PHREEQC model.

Seasons	$CaCO_3$	$CaSO_4$	NaCl
Pre-monsoon	1.15	−1.38	−4.07
Monsoon	1.03	−0.99	−4.44
Post-monsoon	0.97	−0.88	−4.44

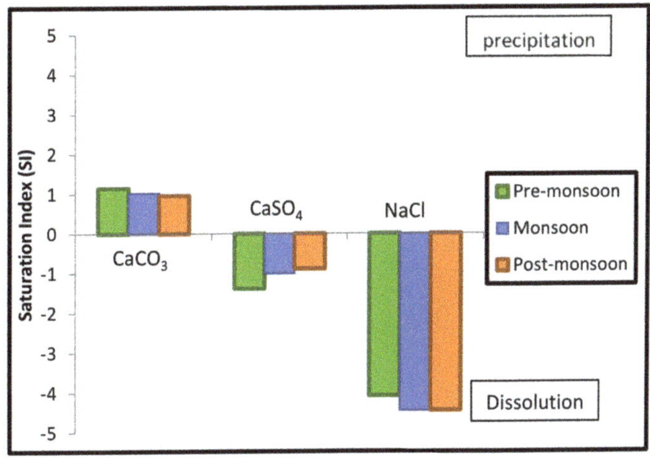

Figure 5. Season-wise trend of Saturation Index (SI) with respect to $CaCO_3$, $CaSO_4$ and NaCl in the groundwater of Mewat, Haryana, India.

In the monsoon season, SIs for NaCl and $CaSO_4$ are −4.44 and −0.99, which shows dissolution. Thus, Cl^-, Na^+, Ca^{2+} shows dominance; the SI for $CaSO_4$ reduced from −1.38 to -0.99 from the pre-monsoon to monsoon. HCO_3^- dissolution dominates over the SO_4^{2-} dissolution, making HCO_3^- dominant. This enhanced concentration is produced because, in case of monsoon season, the soil zone is the subsurface environment that contains elevated CO_2 pressure (produced as a result of the decay of organic matter and root respiration) which, in turn, combines with rainwater to form bicarbonate through the following reactions [20]:

$$CO_2 + H_2O \rightarrow H_2CO_3 \tag{1}$$

$$H_2CO_3 \rightarrow H^+ + HCO_3^- \tag{2}$$

Along with this, the monsoon rainfall results in the dissolution of carbonate and silicate minerals, as illustrated in Equation (3) [21]:

$$(Cations)(Silicates) + H_2CO_3 \rightarrow H_4SiO_4 + HCO_3 + cations + clay\ minerals \tag{3}$$

In case of post-monsoon season Cl^-, Na^+, SO_4^{2-} becomes dominant as a result of the dissolution of NaCl and $CaSO_4$ with SI = −4.44 and −0.88, respectively. Apart from this Mg^{2+} becomes the dominant cation in this season; because of the common ion effect, the value of Ca^{2+} is suppressed as $CaCO_3$ (SI = 0.97) undergoes precipitation, and $CaSO_4$ (SI = −0.88) undergoes dissolution at the same time [22].

To understand the correlation between the different ions contributing to the TDS level and the ions present in the groundwater of the study area, the piper trilinear is drawn as shown in Figure 6, with its two triangles depicting anions on the right and cations on the left side. Six different delineations can be identified: 1—calcium chloride type; 2—magnesium bicarbonate type; 3—sodium bicarbonate type; 4—sodium chloride type; 5 and 6—mixed type. The anion triangle on the right side illustrates that, for all three seasons, samples points fall in the H regions, which is the chloride dominant region, with some of the sample points shifting towards the mixed-type region E. However, for the cations in monsoon and post-monsoon seasons, approximately 50 percent of the sample points fall under the B region, that is, the Mg^{2+}-dominated region, and 50 percent of samples are distributed among the A, C and D regions with no clear delineation of any single cation's dominance. In the pre-monsoon season, 30 percent of the sample points fall under the B region, that is, the Mg^{2+}-dominated region, and the remaining 70 percent of samples are distributed among the A, C and D regions. In the diamond-shaped piper, for the pre-monsoon season, the dominant region was found to be the Ca^{2+} Cl^- type, with 55 percent of samples falling under this group, 25 percent of the samples falling under the mixed type and 20 percent under Na^+ Cl^- type water. A similar pattern was seen in the case of monsoon season, with 85 percent of sample points falling under the Ca^{2+} Cl^- type water type and 15 percent of samples falling under the Na^+ Cl^- type water. In the post-monsoon season, almost 90 percent of samples fall under the Ca^{2+} Cl^- type and only 10 percent under the Na^+ Cl^- type water. In all three seasons, the dominant water type is Ca^{2+} Cl^-. This dominance of Ca^{2+} Cl^- type water can be explained by looking at the hydrogeology of the area. The aquifer study of Mewat has alluvial sediments, with medium sand, clay and kankar. The dissolution of $CaCO_3$ present in the soil is dissolved with the infiltration of rainwater and elevates Ca^{2+} and HCO_3^- ions concentration in groundwater.

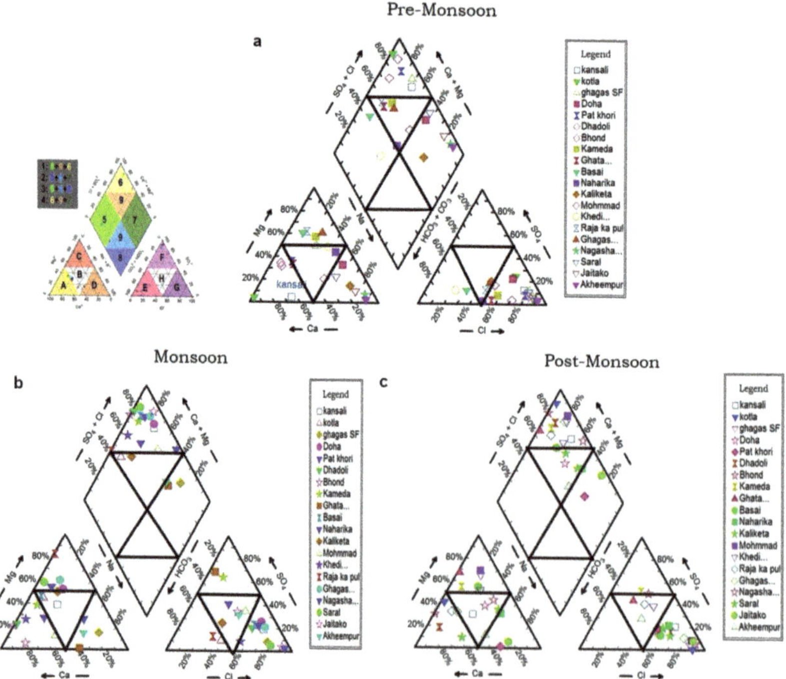

Figure 6. HREEQC modle. Piper trilinear diagram [16] of the groundwater water samples for (**a**) pre-monsoon (**b**) monsoon (**c**) post-monsoon. 1—calcium chloride type; 2—magnesium bicarbonate type; 3—sodium bicarbonate type; 4—sodium chloride type; 5 and 6—mixed type; A—mixed type; B—magnesium; C—sodium; D—calcium type, E—mixed, F—sulphate; G—bicarbonate; H—chloride.

The increase in ionic concentration of Ca^{2+} and HCO_3^- follows the groundwater level in the area, as shown in Figure 7. To justify the change in the ionic concentration with respect to the groundwater level, four different locations were taken randomly, and the ionic concentration was plotted against the groundwater level. As depicted in the graph, the groundwater level in the case of Mohmmadpur increased by 125 cm from pre-monsoon to post-monsoon and, as a result of increased groundwater level, the value of both Ca^{2+} (254 to 167 mg/L) and HCO_3^- (394 to 310) concentrations showed descending trends. Similarly, in the case of Bhond, the groundwater level increased by 25 cm from pre-monsoon to post-monsoon, and Ca^{2+} (176 to 167 mg/L) and HCO_3^- (165 to 84) concentrations showed a similar descending trend. However, in the case of Kotla, the reverse was noticed, i.e., the groundwater level showed a decreasing trend of 10 cm from pre- to post-monsoon, and the Ca^{2+} (138 to 1067) and HCO_3^- (34 to 365) concentrations showed an increasing trend. Therefore, as the groundwater level showed an increasing trend due to dissolution processes, the ionic concentration showed a decreasing trend and vice versa. Since the study area is dominated by agricultural practices, chloride in the groundwater is also contributed to by agricultural return flow, which also contributes to both the rise in groundwater level and the dissolution of ions contained in the soil [23,24].

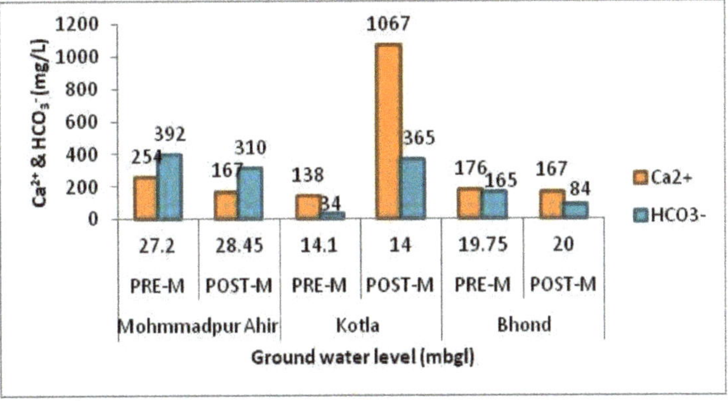

Figure 7. Seasonal change in Ca^{2+} and HCO_3^- ions with groundwater-level fluctuations.

4.2. Sources of Anions and Cations Causing Groundwater Salinization

The ion chemistry of the groundwater of the study area is described below to establish the correlation between the ions and the geochemical processes.

4.2.1. Weathering

To understand the dominance of cations and anions, plots are made between $Ca^{2+} + Mg^{2+}$ and $SO_4^{2-} + HCO_3^-$ (Figure 7). In the pre-monsoon and the monsoon, 20 percent of samples fall on the uniline and 80 percent fall below and near to the uniline. However, in the post-monsoon season, 25 percent of samples are lying on the uniline and 75 percent fall below and are not clustered near the uniline. As most of the sample points falls below the uniline (1:1), this depicts a higher concentration of Ca^{2+} and Mg^{2+} ions than that of $SO_4^{2-} + HCO_3^-$. In the pre-monsoon season, Ca^{2+} and Mg^{2+} ions lie in the range of 0–200 meq/L; in the monsoon, these range from 0 to 100 meq/L, and in the post-monsoon season, the range varies from 0 to 600 meq/L, making salinization more prominent in the post-monsoon season. The increased concentration of Ca^{2+} and Mg^{2+} ions can be attributed to rock (schist rock) and groundwater interaction, which is prominent in the hydrogeology of the study area [25]. Apart from Schist, Amphibole dominants the Paleoproterozoic age group of rocks found in the study area. By the process of Hydrolysis, the Amphiboles are broken down, causing an increase in the concentration of Ca^{2+}, Mg^{2+}, Na^+ ions [26]. Since the hydrogeology is dominated by fledspathic gritty quartzite, amphibole, phyllite and schist, it undergoes silicate weathering, making it one of the most prominent causes of enhanced concentrations of Mg^{2+} ions [27]. The enhanced concentration of Ca^{2+} and Mg^{2+} ions is further supported by the trilinear diagram (Figure 6), as it depicts calcium–chloride type water.

Na^+ becomes the dominant ion in the pre- and post-monsoon seasons, as mentioned in Table 1. The increased concentration is contributed by the silicate weathering and the cation exchange process [21]. The cation exchange process takes places as shown in Equation (4)

$$Ca^{2+} + Na\text{-}X \rightarrow Ca\text{-}X + Na^+ \tag{4}$$

where X denotes cation exchange sites.

When the infiltrating water interacts with clay lenses (fine-grained material that consists of hydrated aluminum silicate quartz, and organic fragments), the exchange between Ca^{2+} and Na^+ ions is triggered.

4.2.2. Ion Exchange

The plots of $Ca^{2+} + Mg^{2+}$ versus $HCO_3^- + SO_4^{2-}$ (Figure 8a) are used to examine the role of ion exchange. Since most of the points fall below the uniline, this indicates a reverse ion exchange process [28]. In order to understand the ion exchange process taking place between the groundwater and aquifer material Choloro-Alkaline indices, CAI1 and CAI2 were calculated using Equations (5) and (6) [29,30]. The comparisons between both indices for the sampling sites in all the three seasons were plotted in Figure 8b.

$$CAI1 = [Cl^- - (Na^+ + K^+]/Cl^- \tag{5}$$

$$CAI2 = [Cl^- - Na^+ + K^+]/SO_4^{2-} + HCO_3^- + NO^{3-} \tag{6}$$

Figure 8b illustrates that in the pre-monsoon, 75 percent of samples have positive, and 25 percent of samples have negative values. However, in the monsoon and the post-monsoon seasons, 90 percent of samples show positive values and only 10 percent have negative values. The positive value indicates that the Base Exchange takes place between Na^+ in the groundwater and Ca^{2+} or Mg^{2+} in the aquifer material. The negative value, on the other hand, indicates that ion exchange is taking place between Ca^{2+}–Mg^{2+} in the groundwater and Na^+ in the aquifer material. Since the samples are dominated by positive values, reverse ion exchange explains the high concentration of Na^+ in the groundwater.

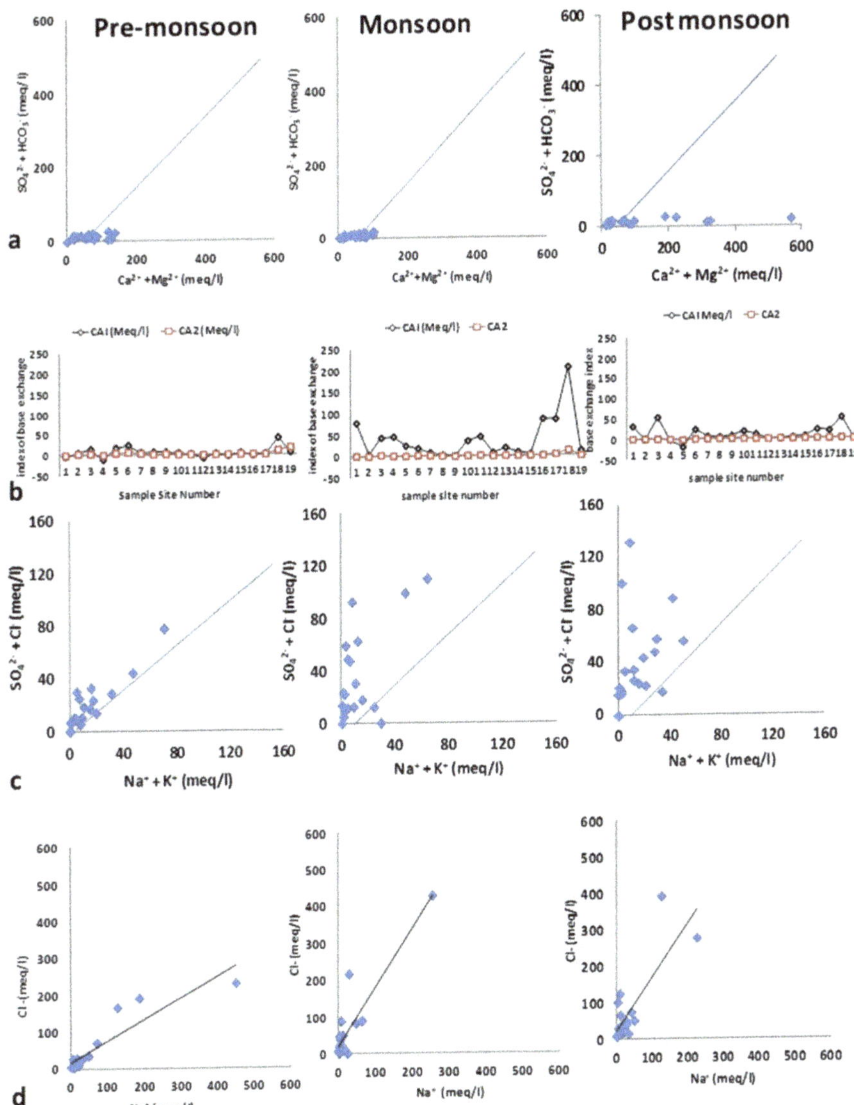

Figure 8. (**a**) $Ca^{2+} + Mg^{2+}$ v/s $SO_4^{2-} + HCO_3^-$ (**b**) Base exchange indices (**c**) $Na^+ + K^+$ v/s $SO_4^{2-} + Cl^-$ (**d**) Na^+ v/s Cl^- for all 3 seasons.

4.2.3. Anthropogenic Sources

The results obtained from multiple linear regressions (Table 1) illustrate Cl^- as one of the dominant ions in the study area. These help to understand the sources responsible for the increase in the concentration of chloride, Na^+, K^+ v/s $Cl^- + SO_4^{2-}$, as shown in Figure 8c. In case of the pre-monsoon, 100 percent of sample points fall above the uniline. For the monsoon season, 90 percent fall above the uniline and 5 percent below the uniline, and 95 percent of sampling points are above the uniline and 5 percent below the uniline for the post-monsoon. Irrespective of season, the majority of sample plots above the uniline indicate the dominance of $Cl^- + SO_4^{2-}$. The hydrogeology of the region is dominated by silicate minerals, but these silicate-bearing strata lack sodalite and chlorapatite minerals, which are known to be the sources of chloride and sulphate. Since no geological sources for chloride and sulphate are found in the region, anthropogenic sources dominated by agricultural activities can be thought as the major contributor of Cl^- and SO_4^{2-} ions [31].

The sulphate concentration is low in the pre-monsoon season, but the concentration increases in the monsoon season to the maximum in the post-monsoon season, as shown in Figure 8a–c. In the absence of any known geological reasons, the application of inorganic fertilizers such as potash (KCl) and gypsum ($CaSO_4 \cdot 2H_2O$) can be considered as the main sources of this increase [31]. The fertilizers are intensively applied for kharif crops during the monsoon and the post-monsoon period, which are also periods of maximum rainfall in Mewat; therefore, the leaching of minerals to groundwater becomes the main cause of sulphate contamination, which is also witnessed in the sulphate contour maps for the study area (Figure 9).

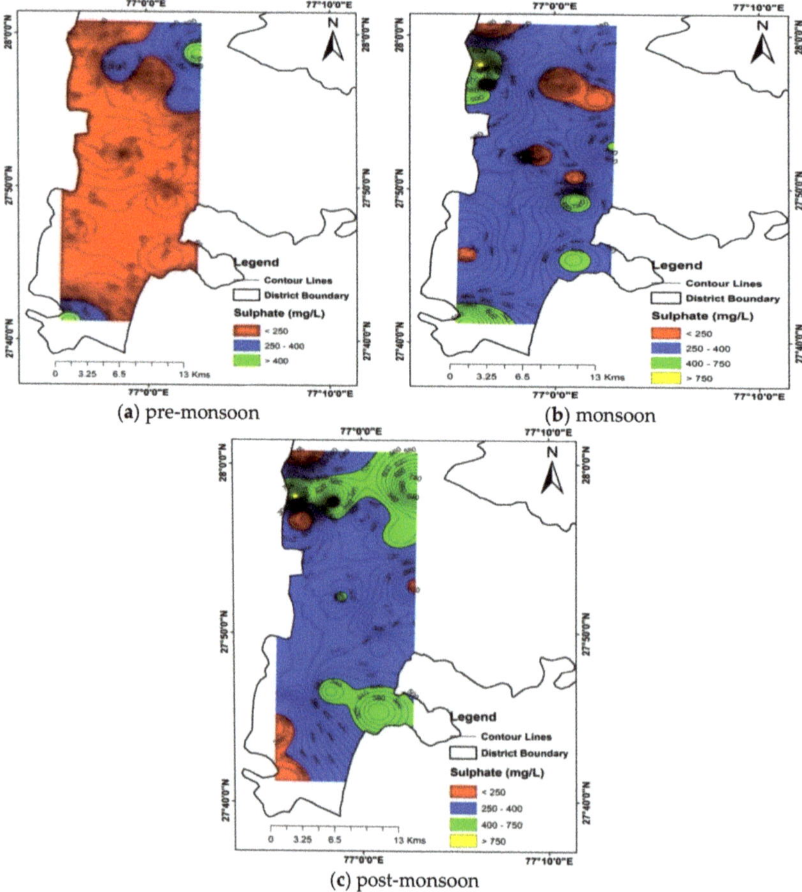

Figure 9. Contour map of sulphate for (**a**) pre-monsoon (**b**) monsoon (**c**) post-monsoon.

4.2.4. Evaporation

Na$^+$ v/s Cl$^-$ plots (Figure 8d) show that the points falling near to the uniline (1:1) are considerable, due to the process of evaporation. The Quaternary age group of the study area is dominated by sand, silt, and clay with kankar (CaCO$_3$), which reinforce the vigorous acquaintance of evaporation in the study area and are elevated by the presence of a semi-arid climate [21,32]. The sample points fall above the uniline in all three seasons, depicting the higher concentrations of chloride ion. The above-mentioned reasons, along with the natural geography, Aravalli ranges and rise in temperature, play a major role in groundwater salinity.

Gibbs plots illustrate (Figure 10) the factors influencing the ion chemistry, i.e., rock, precipitation and evaporation dominance. The plot shows the chemical weathering of rocks to form minerals as the major reason governing the groundwater chemistry in the study area. Higher anthropogenic activities increase the concentration of TDS which, in turn, shifts some of the sample points from rock- to evaporation-dominant regions [33] Similar results are found for semi-arid areas [34–36] where, through tracer techniques, it has been observed that the mineral dissolution is major cause of salinity.

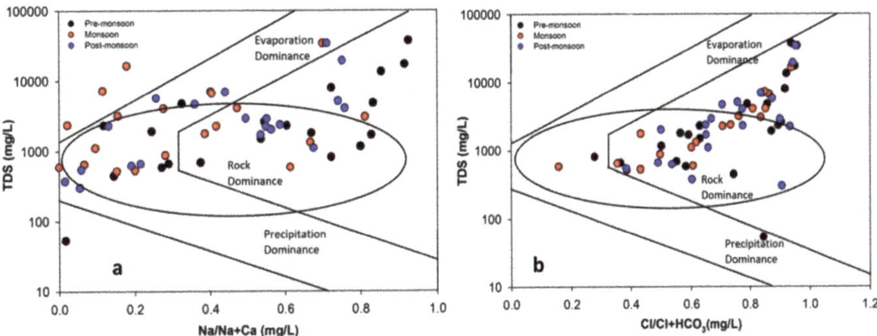

Figure 10. Gibbs plots (**a**) TDS v/s Na$^+$+/Na$^+$+ Ca^{2+} and (**b**) TDS v/s Cl$^-$/Cl$^-$ + HCO$_3^-$.

5. Conclusions

In the semi-arid Mewat region of Haryana, where more than half of groundwater samples that were collected in 2018 for pre-monsoon, monsoon, and post-monsoon seasons were categorized as saline water, this study has led to an understanding of the region in terms of its high groundwater salinity. The salinity is caused due to the dominance of Na$^+$, Cl$^-$ and Ca^{2+} ions, and these ions have a strong positive correlation with TDS. The processes responsible for the dominance of these ions are weathering, evaporation and ion exchange. The primary source of salinity in the study area is found to be geogenic, further enhanced by anthropogenic activities. These results will be useful in the management of groundwater salinity and planning for remediation measures in the region. The Aquifer Storage and Recovery (ASR) technique for storage of fresh water in saline water has been tested under controlled conditions and will now be transferred to field conditions at a large scale.

Author Contributions: Conceptualization, methodology, funding acquisition, review and editing: G.K.; Writing, original draft preparation: P.S.; writing—review and editing: A.B. (Anjali Bhagwat); Formal analysis: G.P.; Data curation: B.K.Y.; project administration: C.P.K.; visualization: M.L.K.; validation: S.S.; software: N.S.; review and editing: A.B. (Allen Bradley); investigation: L.M.S.; supervision: M.M. All authors have read and agreed to the published version of the manuscript.

Funding: This work was carried out under Purpose Driven Study under National Hydrology Project funded by World Bank.

Institutional Review Board Statement: Not applicable.

Informed Consent Statement: Not applicable.

Data Availability Statement: Available on request from corresponding author.

Acknowledgments: GK thanks Director, National Institute of Hydrology, Head, GWHD, Nodal officer, NHP, PDS Coordinator, NHP training coordinator, NPMU-NHP, TAMC, Funding received from National Hydrology Project is duly acknowledged.

Conflicts of Interest: The authors declare that they have no conflict of interest.

References

1. Bonsor, H.C.; MacDonald, A.M.; Ahmed, K.M.; Burgess, W.G.; Basharat, M.; Calow, R.C.; Dixit, A.; Foster, S.S.D.; Gopal, K.; Lapworth, D.; et al. Hydrogeological typologies of the Indo-Gangetic basin alluvial aquifer, South Asia. *Hydrogeol. J.* **2017**, *25*, 1377–1406. [CrossRef]
2. Krishan, G.; Bisht, M.; Ghosh, N.C.; Prasad, G. Groundwater salinity in north west of India: A critical appraisal. In *Environment Management*; WSTL Book Series; Springer: Berlin/Heidelberg, Germany, 2019; Chapter 19.
3. Lapworth, D.J.; MacDonald, A.M.; Krishan, G.; Rao, M.S.; Gooddy, D.C.; Darling, W.G. Groundwater recharge and age-depth profiles of intensively exploited groundwater resources in northwest India. *Geophys. Res. Lett.* **2015**, *42*. [CrossRef]
4. MacDonald, A.; Bonsor, H.; Ahmed, K.; Burgess, W.; Basharat, M.; Calow, R.; Dixit, A.; Foster, S.; Krishan, G.; Lapworth, D.; et al. Groundwater depletion and quality in the Indo-Gangetic Basin mapped from in situ observations. *Nat. Geosci.* **2016**, *9*, 762–766. [CrossRef]
5. Malik, A.; Bhagwat, A. Modelling groundwater level fluctuations in urban areas using artificial neural network. *Groundw. Sustain. Dev.* **2021**, *12*, 100484. [CrossRef]
6. Rodell, M.; Velicogna, I.; Famiglietti, J.S. Satellitebased estimates of groundwater depletion in India. *Nature* 2009. [CrossRef]
7. Krishan, G. Groundwater Salinity. *Curr. World Environ.* **2019**, *14*, 186–188. [CrossRef]
8. Lapworth, D.; Krishnan, G.; Macdonald, A.; Rao, M. Groundwater quality in the alluvial aquifer system of northwest India: New evidence of the extent of anthropogenic and geogenic contamination. *Sci. Total Environ.* **2017**, *599–600*, 1433–1444. [CrossRef] [PubMed]
9. Misra, A.K.; Mishra, A. Study of quaternary aquifer in Ganga Plains, India: Focus on groundwater salinity, fluoride and fluorosis. *J. Hazard. Mater.* **2006**, *144*, 438–448. [CrossRef]
10. Pazand, K.; Ardeshir, H. Investigation of hydrochemical characteristics of groundwater in the Bukan basin, North. *Iran. Appl. Water Sci.* **2012**, *2*, 309–315. [CrossRef]
11. Lakshmanan, A.R.; Kannan, M.; Senthil, K. Major ion chemistry and identification of hydrogeochemical processes of groundwater in a part of Kancheepuram district, Tamil Nadu, India. *Environ. Geosci.* **2003**, *10*, 157–166. [CrossRef]
12. Anjali, P.; Saravanan, V.S.; Jayanti, C. *Interlacing Water and Human Health: Case Studies from South Asia*; SAGE Publishing Pvt. Ltd.: Thousand Oaks, CA, USA, 2012; pp. 201–207.
13. CENSUS. 2011. Available online: http://www.census2011.co.in/census/district/226-mewat.html (accessed on 1 August 2020).
14. Central Ground Water Board. Ground Water Information Booklet, Mewat District, Haryana. 2012. Available online: www.cgwb.gov.in/district_profile/haryana/mewat.pdf (accessed on 1 August 2020).
15. APHA. *Standard Methods for The Examination of Water and Wastewater*, 21st ed.; American Public Health Association: Washington, DC, USA; New York, NY, USA, 2005.
16. Piper, A.M. A graphical procedure in the geochemical interpretation of water analysis. *Eos Trans. Am. Geophys. Union* **1944**, *25*, 914–928. [CrossRef]
17. Gibbs, R.J. Mechanisms Controlling World Water Chemistry. *Science* **1970**, *170*, 1088–1090. [CrossRef]
18. Krishan, G.; Ghosh, N.C.; Kumar, C.P.; Sharma, L.M.; Yadav, B.; Kansal, M.L.; Singh, S.; Verma, S.K.; Prasad, G. Understanding stable isotope systematics of salinity affected groundwater in Mewat, Haryana, India. *J. Earth Syst. Sci.* **2020**, *129*, 109. [CrossRef]
19. Krishan, G.; Kumar, C.P.; Prasad, G.; Kansal, M.L.; Yadav, B.; Verma, S.K. Stable Isotopes and Inland Salinity Evidences for Mixing and Exchange. In Proceedings of the Roorkee Water Conclave (RWC-2020), Uttarakhand, India, 26–28 February 2020.
20. Singh, A.K.; Mondal, G.C.; Kumar, S.; Singh, T.B.; Tewary, B.K.; Sinha, A. Major ion chemistry, weathering processes and water quality assessment in upper catchment of Damodar River basin, India. *Environ. Geol.* **2007**, *54*, 745–758. [CrossRef]
21. Subba Rao, N.; Surya Rao, P. Major ion chemistry of groundwater in a River basin: A study of India. *Environ. Earth Sci.* 2009. [CrossRef]
22. Freeze, R.A.; Cherry, J.A. *Groundwater*; Presentice Hall: Upper Saddle River, NJ, USA, 1979.
23. Hem, J.D. *Study and Interpretation of the Chemical Characteristics of Natural Water*; Scientific Publisher: Jodhpur, India, 1991; p. 2254.
24. Todd, D.K. *Groundwater Hydrology*, 2nd ed.; John Wiley: New York, NY, USA; Chichester, UK; Brisbane, Australia; Toronto, ON, Canada, 1980; Volume xiii, p. 535.
25. Craw, D. Water-rock interaction and acid neutralization in a large schist debris dam, Otago, New Zealand. *Chem. Geol.* **2000**, *171*, 17–32. [CrossRef]

26. Malov, A.I. Water–rock interaction in vendian sandy–clayey rocks of the Mezen Syneclise. *Lithol. Miner. Res.* **2004**, *39*, 345–356. [CrossRef]
27. Holland, H.D. *The Chemistry of the Atmosphere and Ocean*; Wiley—Intersciences: New York, NY, USA, 1978; p. 351.
28. Fisher, R.S.; Mullican, W.F. Hydrochemical evolution of sodium sulphate and sodium-chloride groundwater beneath the Northern Chihuahuan desert, Trans-Pecos, Texas, USA. *Hydrogeol J.* **1997**, *5*, 4–16. [CrossRef]
29. Schoeller, H. Qualitative evaluation of groundwater resources. In *Methods and Techniques of Groundwater Investigations and Development*; UNESCO Water Resources Series 33; UNESCO: Paris, France, 1965; pp. 44–52.
30. Schoeller, H. Geochemistry of groundwater. In *Groundwater Studies—An International Guide for Research and Practice*; UNESCO: Paris, France, 1977; Chapter 15; pp. 1–18.
31. Rajmohan, N.; Elango, L. Hyrogeochemistry and its relation to groundwater level fluctuation in the Palar and Cheyyar River basins, southern India. *Hydrol. Process* **2006**, *20*, 2415–2427. [CrossRef]
32. Gurdak, J.J.; Hanson, R.T.; McMohan, P.B.; Bruce, B.W.; McCray, J.E.; Thyne, G.D.; Reedy, R.C. Climate variability controls on unsaturated water and chemical movement, High Plans Aquifer, USA. *Vadose Zone J.* **2007**, *6*, 533–547. [CrossRef]
33. Subba Rao, N. Groundwater quality in crystalline terrain of Guntur district, Andhra Pradesh. *Visakhapatnam J. Sci.* **1998**, *2*, 51–54.
34. Krishan, G.; Prasad, G.; Anjali; Kumar, C.; Patidar, N.; Yadav, B.; Kansal, M.; Singh, S.; Sharma, L.; Bradley, A.; et al. Identifying the seasonal variability in source of groundwater salinization using deuterium excess- A case study from Mewat, Haryana, India. *J. Hydrol. Reg. Stud.* **2020**, *31*, 100724. [CrossRef]
35. Krishan, G.; Vashisth, R.; Sudersan, N.; Rao, M.S. Groundwater salinity and isotope characterization: A case study from south-west Punjab, India. *Environ. Earth Sci.* **2021**, (in press).
36. Krishan, G.; Ghosh, N.C.; Kumar, B.; Kumar, C.P.; Rao, M.S.; Sudarsan, N.; Singh, S.; Sharma, A.; Kumar, S.; Jain, S.K. *Aquifer Salinization in Punjab*; Report-CS-175/GWHD/2019-20 Submitted to PSFC; Government of Punjab: Chandigarh, India, 2021; p. 178.

Article

Proposing the Optimum Withdrawing Scenarios to Provide the Western Coastal Area of Port Said, Egypt, with Sufficient Groundwater with Less Salinity

Mohamed Abdelfattah [1,*], Heba Abdel-Aziz Abu-Bakr [2], Ahmed Gaber [1], Mohamed H. Geriesh [3], Ashraf Y. Elnaggar [4], Nihal El Nahhas [5] and Taher Mohammed Hassan [2]

[1] Geology Department, Faculty of Science, Port Said University, Port Said 42522, Egypt; ahmedgaber_881@hotmail.com

[2] National Water Research Centre, Research Institute for Groundwater, Cairo 13621, Egypt; heba_rigw@yahoo.com (H.A.-A.-B.); drtahermh@yahoo.com (T.M.H.)

[3] Geology Department, Faculty of Science, Suez Canal University, Ismailia 41522, Egypt; helmi_mohamed@hotmail.com

[4] Department of Food Science and Nutrition (Previously Chemistry), College of Sciences, Taif University, P.O. Box 11099, Taif 21944, Saudi Arabia; aynaggar@tu.edu.sa

[5] Department of Botany and Microbiology, Faculty of Science, Alexandria University, Alexandria 21526, Egypt; nihal.elnahhas@alexu.edu.eg

* Correspondence: mohamed_abdelfatah@sci.psu.edu.eg; Tel.: +20-1012409010

Citation: Abdelfattah, M.; Abu-Bakr, H.A.-A.; Gaber, A.; Geriesh, M.H.; Elnaggar, A.Y.; Nahhas, N.E.; Hassan, T.M. Proposing the Optimum Withdrawing Scenarios to Provide the Western Coastal Area of Port Said, Egypt, with Sufficient Groundwater with Less Salinity. *Water* **2021**, *13*, 3359. https://doi.org/10.3390/w13233359

Academic Editors: Maurizio Barbieri, Helder I. Chaminé, Maria José Afonso and Aldo Fiori

Received: 4 October 2021
Accepted: 24 November 2021
Published: 26 November 2021

Publisher's Note: MDPI stays neutral with regard to jurisdictional claims in published maps and institutional affiliations.

Abstract: Recently, groundwater resources in Egypt have become one of the important sources to meet human needs and activities, especially in coastal areas such as the western area of Port Said, where seawater desalination cannot be used due to the problem of oil spill and the reliance upon groundwater resources. Thus, the purpose of the study is the sustainable management of the groundwater resources in the coastal aquifer entailing groundwater abstraction. In this regard, the Visual MODFLOW and SEAWAT codes were used to simulate groundwater flow and seawater intrusion in the study area for 50 years (from 2018 to 2068) to predict the drawdown, as well as the salinity distribution due to the pumping of the wells on the groundwater coastal aquifer based on field investigation data and numerical modelling. Different well scenarios were used, such as the change in well abstraction rate, the different numbers of abstraction wells, the spacing between the abstraction wells and the change in screen depth in abstraction. The recommended scenarios were selected after comparing the predicted drawdown and salinity results for each scenario to minimize the seawater intrusion and preserve these resources from degradation.

Keywords: groundwater sustainable management; groundwater abstraction; seawater intrusion; numerical modelling; coastal aquifer

1. Introduction

Water scarcity has recently become an important problem, especially in arid and semi-arid areas such as Egypt [1]. The continuing need for water due to the population increase in Egypt has led to an increase in the use of groundwater resources, since these needs are not being met by surface water sources [2]. Population growth increases the water requirements leading to increased pumping of the aquifers [3]. Therefore, groundwater is considered an important natural resource and the main source of water supply in many coastal regions [4]. In such regions, groundwater resources are over-exploited to meet the development and urbanization of coastal zones and the excessive pumping of groundwater in coastal aquifers has reduced the freshwater flux to the sea and allowed seawater to migrate inland [5].

The western area of Port Said is one of the coastal regions located in the north of Egypt, northeast of the Nile Delta, where the increase in population and human needs has resulted in the development of projects that depend on water sources and need a permanent source of water. Water desalination is one of the ways to obtain a continuous water source and, in such coastal areas, the desalination plants rely on seawater [6]. Despite the Mediterranean Sea is located north of the study area, it cannot be relied upon and desalinated due to the problem of an oil spill which was previously monitored using satellite images [7–10]. Therefore, desalination has relied on saline groundwater, using appropriate solutions that reduce the resulting salinity, thus reducing the effect of seawater intrusion [11].

The Nile Delta aquifer is among the largest underground reservoirs in the world [12] with a total capacity of 500 Bm^3 [3]. About 20% of groundwater comes from conventional water resources where the total abstraction from the Nile Delta Aquifer in Egypt was estimated to be 7 billion cubic meters (BCM) in 2016 [4,13,14]. Annual groundwater abstraction from the Nile aquifer system has been estimated to be 4.6 BCM [15]. The Nile Delta aquifer was affected by sea water intrusion, especially in the coastal areas [16], due to the combined effects of climate change-induced sea-level rise (SLR) [17] and excessive groundwater extractions for reclamation development projects [18]. Overexploitation of groundwater also occurred in the Nile Delta, as the rapidly growing population depends increasingly upon groundwater extraction for domestic water needs [19].

Seawater intrusion (SWI) is a worldwide problem in coastal aquifers where groundwater resources are significantly threatened by saline water [20]. Several studies have been performed to study aquifer salinization; see, for example, [12,21–25]. Due to major causes of SWI [3,26,27], groundwater withdrawals create a new hydraulic gradient, whereby the water level could be lowered to the extent that the piezometric head of a freshwater body becomes less than that of the adjacent saline water body [28]. This change in the hydraulic gradient of aquifers accelerates the progressive landward invasion of seawater toward the abstraction wells and consequently results in the degradation of the chemical quality of abstracted water and surrounding groundwater followed by other problems, such as decrease in fresh water availability, as well as human health and ecosystem damage [27,29].

Several strategies were put forward to minimize seawater intrusion into coastal aquifers, such as the construction of subsurface barriers and the installation of injection wells [30–32]. The optimization of abstraction wells can target the following: maximizing the abstraction rate [33], minimizing aquifer salinity [34], or minimizing seawater intrusion [35]. Artificial recharge helps to raise groundwater levels in aquifers [14] and the recharge can be used in the coastal aquifers to manage seawater intrusion [36–39]. The abstraction of saline water and its disposal in the sea helps to decrease the amount of saline water in coastal aquifers [14]. This method was applied to control saltwater intrusion in coastal aquifers by a number of researchers; see, for example [40,41]. A number of studies have been directed to identify SWI in the NDA using different numerical models [12,22–24,42–45]. A balance between abstraction rates and aquifer salinity should be considered [30].

Recently, many research studies investigated the evaluation of the aquifer by groundwater modeling [28,46–55], used MODFLOW [56,57], SEAWAT and a combined version of MODFLOW [58] and MT3DMS [59].

Therefore, the overall objective of this paper is to propose optimum withdrawing scenarios to provide the western coastal area of Port Said, Egypt, with sufficient groundwater with less salinity to be used in the desalination plant. The groundwater modelling technique was performed with Visual MODFLOW, with the MT3D and SEAWAT codes, to simulate groundwater flow and solute transport, as well as seawater intrusion on groundwater quality due to over pumping.

2. The Study Area

The study area is located 150 m south of the Mediterranean coast in the northeast of the Nile Delta and west of Port Said, Egypt (Figure 1a). As shown in Figure 1b, c, it extends between the latitudes of 31°21′ to 31°21′12″ east and longitudes of 32°4′57″ to 32°5′18″ north. The study area belongs to the northeastern part downstream of the Nile watershed [60], covers an area of about 23.5 km^2 and is characterized by mild topography, as it does not exceed 5 meters above sea level, either to the north and south where the Mediterranean Sea and Manzala Lake are 0 and less than 0 meters, respectively. The main geomorphic units along the study area are the coastal sand dunes, which are concentrated on the north side of the study area (on the coast), and the sabkha, which is concentrated in the southern part in the direction of Manzala [61].

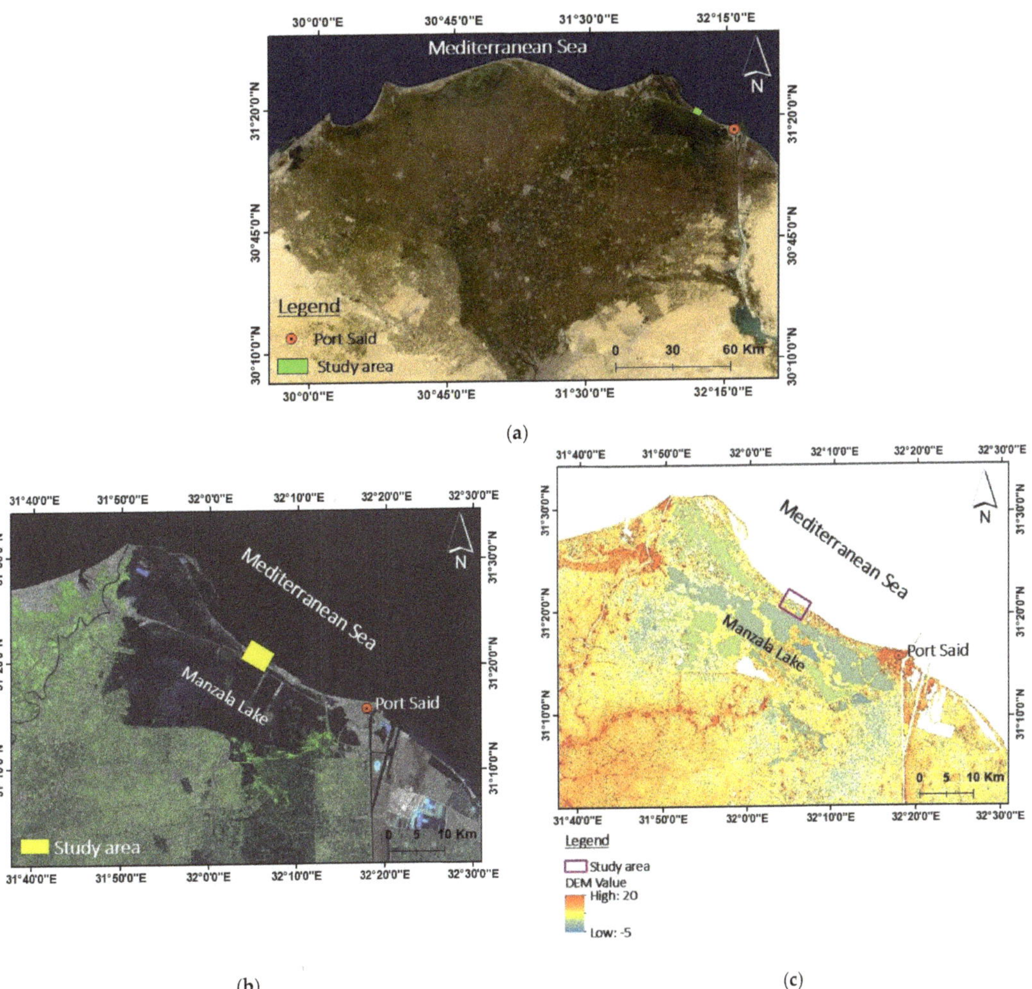

Figure 1. (**a**) A satellite image showing the Nile Delta region. (**b**) The location of the study area relative to Port Said city. (**c**) Digital elevation model for the study area and surrounding.

2.1. Meteorological Data of the Study Area

The study area is located in an arid climate region; the average daily temperature varies between 17 °C and 20 °C at the Mediterranean Sea coast [62]. The annual mean values for relative humidity in the morning and the evening are between 60% and 80%, respectively [4]. The highest average monthly rainfall in the study area is 50 mm in November (Figure 2a) and the highest average annual rainfall amount in the last 12 years was about 64 mm in 2015 (Figure 2b). The average values of the surface evaporation and evapotranspiration are about 180 mm/year and 126 mm/year, respectively, in the study area [63]. Due to the high evaporation rates and limited precipitation in the study area, local rainfall provides negligible replenishment of groundwater [2,60]. The analysis of the time series of rainfall during the period 2009–2021 (Figure 2b) led us to conclude that rainfall will not notably change and will not contribute to the aquifer within the study area. Therefore, it is ignored.

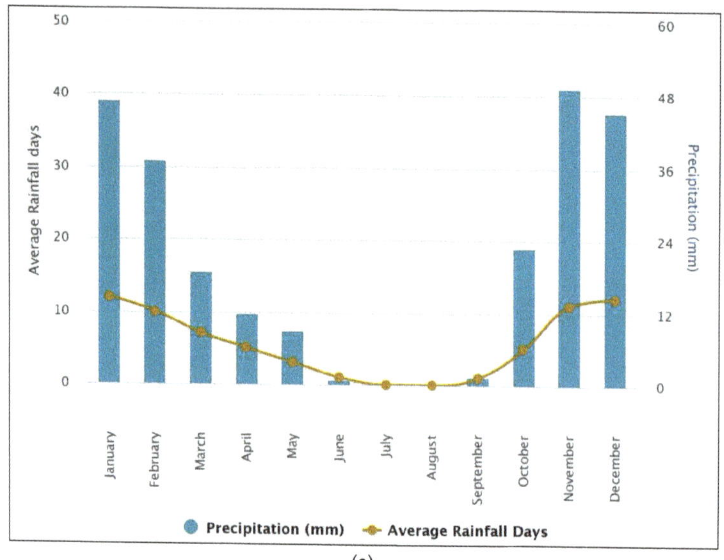

(a)

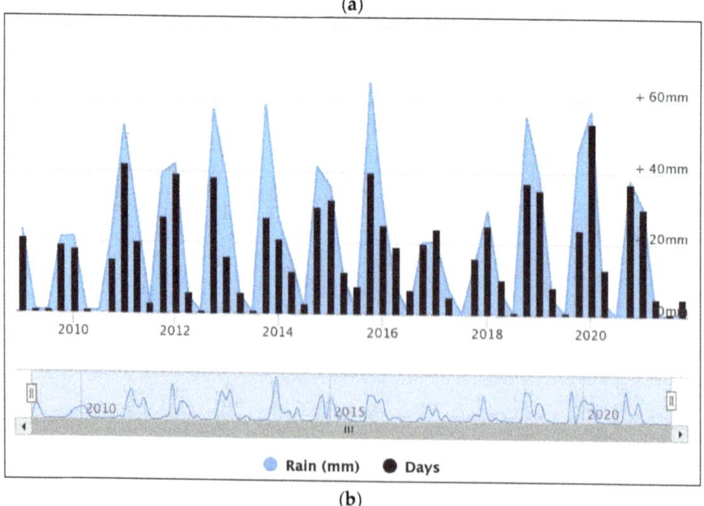

(b)

Figure 2. (**a**) Monthly average rainfall. (**b**) Average annual rainfall amount (mm) and rainy days during the period 2009–2021 for Port Said [64].

2.2. Geological Setting

Generally, two main geological components in the ND region are Quaternary deposits and Tertiary deposits [65]. The Quaternary includes the Holocene and Pleistocene sediments. Holocene deposits are widely spread with a maximum thickness of about 77 m [66].

Moreover, the thickness of Quaternary deposits increases in a northward direction to reach 250 m in the south and 1000 m in the north [67].

The study area clearly consists of quaternary deposits [65]. These deposits are gravel and sand with some clay lenses belonging to Bilqas, Mit Ghamr and Wastany formations from the Holocene and Pleistocene periods.

2.3. Hydrogeological Setting

Hydro-geologically, these quaternary strata are very important with a significantly amount of water stored. According to well data, the groundwater aquifer system is classified into three aquifers (unconfined, leaky and deep), which have about 90 m, 200 m and 400 m thickness, respectively. Figure 3a shows the subsurface geological cross-sections in (Figure 3b) and around the study area (Figure 4). Five test wells are located in the study area to carry out full hydrogeological studies of the aquifer, rely upon the groundwater of the aquifer and desalinate it to improve its quality. The static water levels in the five wells within the study area were measured and indicated that the shallow wells (GW 01 and GW 05) range from −1.4 to −0.9 m and from −1.4 to −1.1 m, respectively, while the groundwater levels for the deep wells (GW 02, GW 03 and GW 04) range from 0.6 to 1 m, from 0.8 to 1.1 m and from 0.5 to 0.8, respectively. The aquifers' salinities were monitored and the hydraulic parameters of the aquifers were estimated through pumping tests of wells.

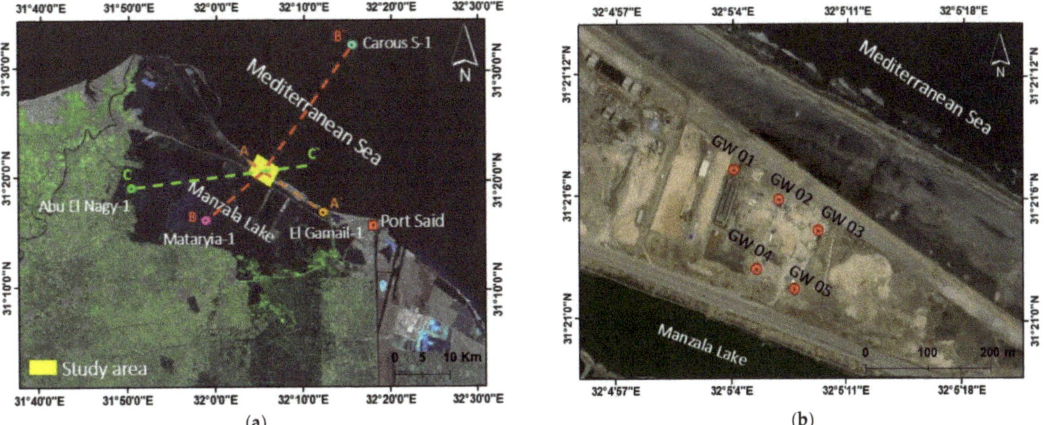

Figure 3. (a) The directions of geological cross sections using wells around the study area. (b) The locations of five test wells within the study area.

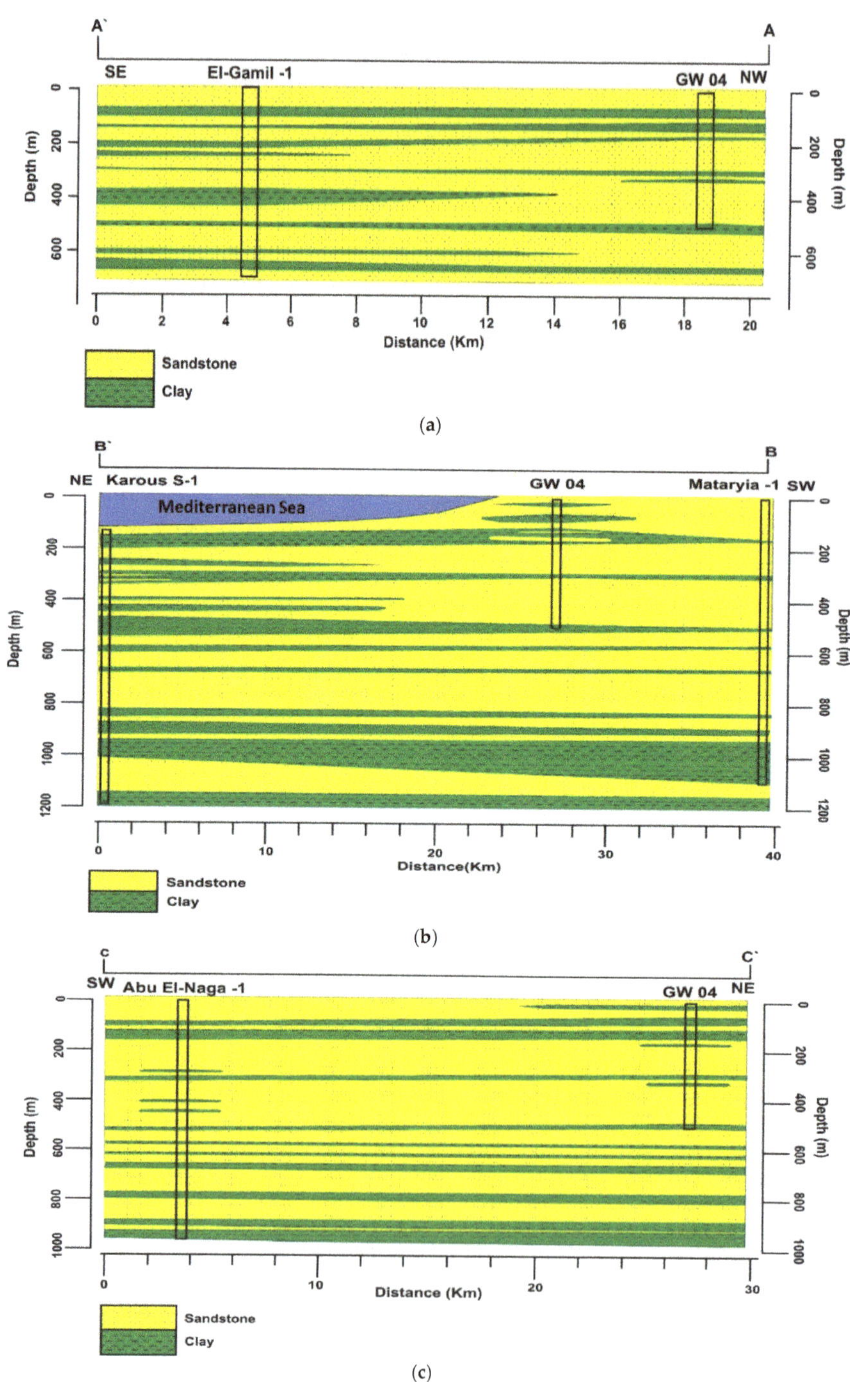

Figure 4. Subsurface geological cross sections: (**a**) A–A′, (**b**) B–B′ and (**c**) C–C′.

3. Methodology

Groundwater modeling is the most widely used method for understanding the flow regime of groundwater aquifers, as well as for assessing the amount of water that will be used to know the potential of the aquifer and the effect of seawater intrusion due to over-pumping.

To achieve the research objectives, field data were collected to identify the hydraulic properties and salinity of the aquifer. Groundwater flow and solute transport (seawater intrusion) were simulated using the MODFLOW and SEAWAT codes and the testing scenarios were developed to simulate flow and transport in the western Port Said coastal area for 50 years (from 2018 to 2068) to predict the drawdown as well as the salinity distribution due to the pumping of the wells on the groundwater coastal aquifer according to the methodology in Figure 5. This study defined the characteristics of the aquifer system and the baseline salinity of the coastal aquifer using field data and a numerical model. Eighteen scenarios were developed to minimize the effect of seawater intrusion on the Nile Delta aquifer within the study area. The simulation results were compared to the initial salinity values (baseline condition) in the aquifer at the control points, which indicated a change in the different scenarios.

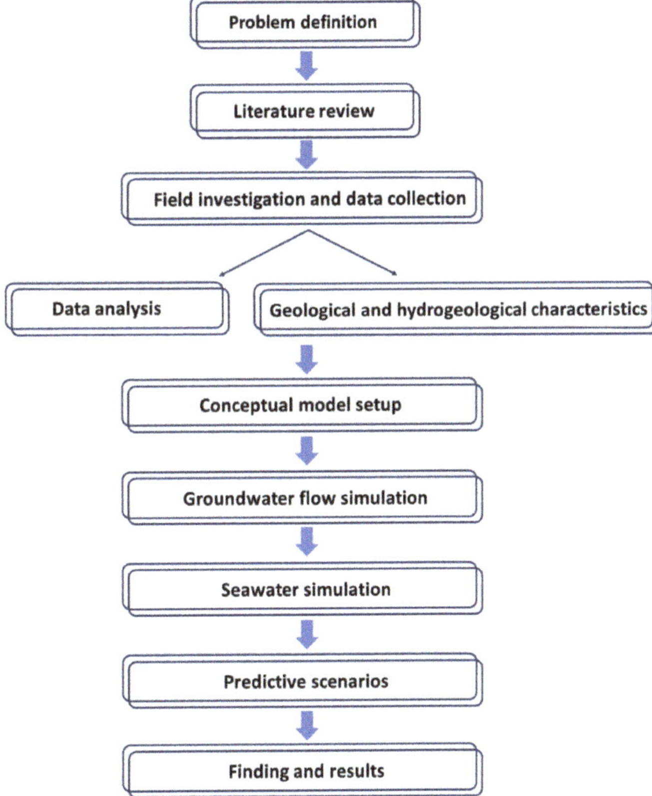

Figure 5. Flow chart explaining the development of the groundwater modelling study.

The scenarios were utilized to predict future responses of the aquifer using the following principles:

- The change in well abstraction rate;
- The different numbers of abstraction wells;

- The spacing between the abstraction wells;
- The change in screen depth in the abstraction wells.

As mentioned in Figure 2b, there are five of wells in the study area on which the above-mentioned scenarios of the 2018 (steady state condition)–2068 projection were based; the different scenarios for each arrangement of the wells are discussed as follows:

a. The first arrangement of the wells (Figure 6a) consists of five wells located in the northern part of the study area being used with two screen depths (unconfined aquifer and leaky confined aquifer); for each screen depth, different well abstraction rates are considered:

  - Sc a.1: screen depth in aquifer A, with total pumping of 50,000 m^3/day;
  - Sc a.2: screen depth in aquifer A, with total pumping of 100,000 m^3/day;
  - Sc a.3: screen depth in aquifer A, with total pumping of 150,000 m^3/day;
  - Sc a.4: screen depth in aquifer B, with total pumping of 50,000 m^3/day;
  - Sc a.5: screen depth in aquifer B, with total pumping of 100,000 m^3/day;
  - Sc a.6: screen depth in aquifer B, with total pumping of 150,000 m^3/day.

b. The second arrangement of the wells (Figure 6b) consists of five wells on the southern part of the study area with different well abstraction rates from aquifer B (leaky confined aquifer):

  - Sc b.1: total pumping from 5 proposed wells of 50,000 m^3/day;
  - Sc b.2: total pumping from 5 proposed wells of 100,000 m^3/day;
  - Sc b.3: total pumping from 5 proposed wells of 150,000 m^3/day.

c. The third arrangement of the wells (Figure 6c) consists of ten wells with different well abstraction rates from aquifer B (leaky confined aquifer):

  - Sc c.1: total pumping of 50,000 m^3/day;
  - Sc c.2: total pumping of 100,000 m^3/day;
  - Sc c.3: total pumping of 150,000 m^3/day.

d. The fourth arrangement of five wells (Figure 6d) considers a spacing between wells of 100 m and different abstraction rates from aquifer B (leaky confined confined):

  - Sc d.1: total pumping of 50,000 m^3/day;
  - Sc d.2: total pumping of 100,000 m^3/day;
  - Sc d.3: total pumping of 150,000 m^3/day.

e. The fifth arrangement of five wells (Figure 6e) considers a spacing between wells of 200 m and different abstraction rates from aquifer B (leaky confined confined):

  - Sc e.1: total pumping of 50,000 m^3/day;
  - Sc e.2: total pumping of 100,000 m^3/day;
  - Sc e.3: total pumping of 150,000 m^3/day.

These criteria for the scenarios (the change in well abstraction rate, the different numbers of abstraction wells, the spacing between the abstraction wells and the change in screen depth in abstraction wells) were taken into consideration as the requirements for the desalination plant. For example, the quantities of discharge were determined for what this plant would meet in terms of the quantities of water required to be used. In addition, the locations of the proposed wells were determined in accordance with the available places in the study area, taking into account the distances between wells and engineering constructions. The comparison of the scenarios was made using five observation points to observe aquifer salinity. The best and most appropriate scenario is the scenario with the lowest predicted salinity concentrations.

The hydrogeological conceptual model was developed to describe the site hydrogeological conditions. The flow conceptual model was built based on the findings of field investigations and the geological and hydrogeological characteristics of the subsurface layers, while the transport conceptual model describes the transport of seawater and parameters of concern to the groundwater aquifers and, especially, into the abstraction wells.

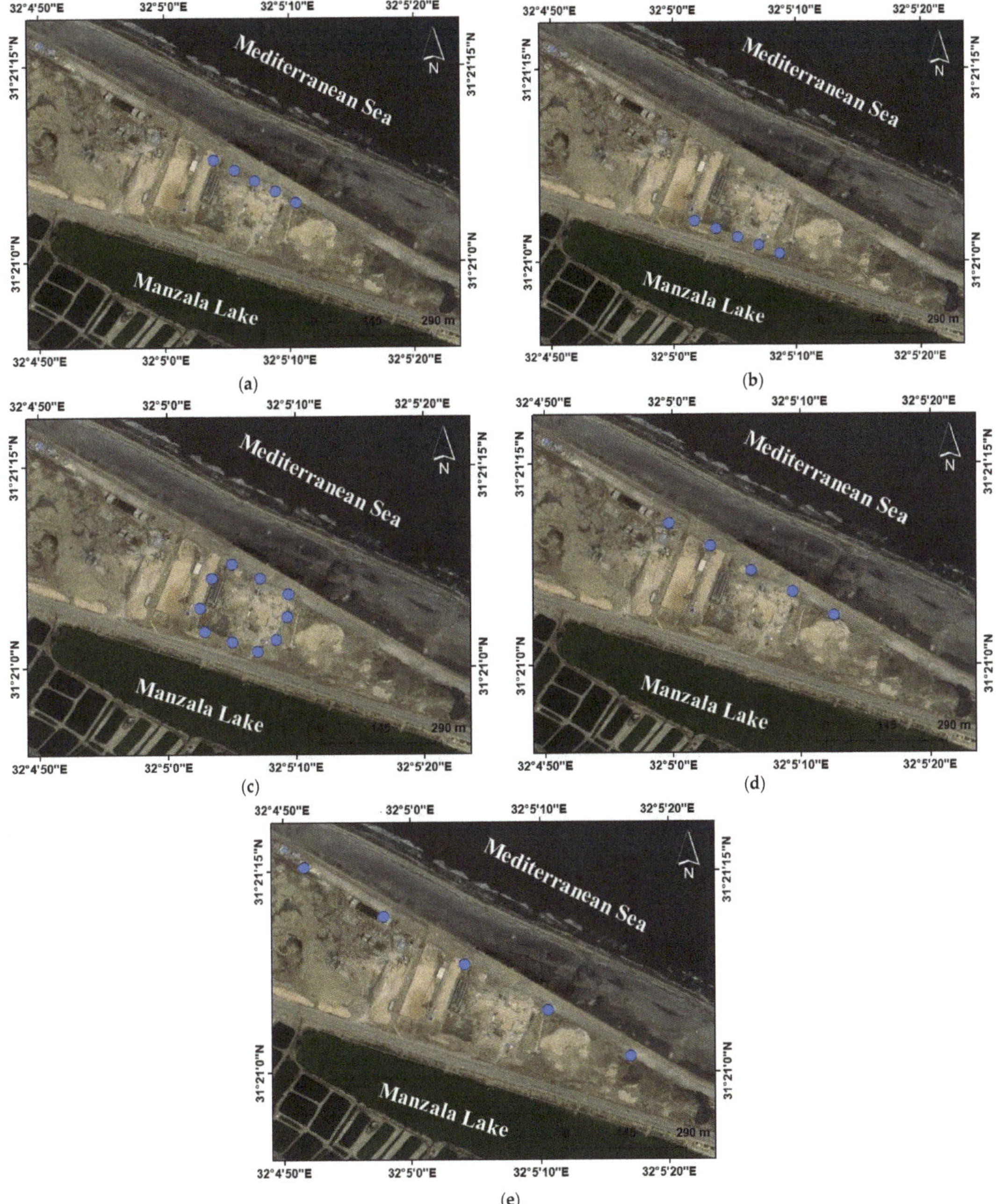

Figure 6. Arrangement of proposed pumping wells for different scenarios: (**a**) first arrangement, (**b**) second arrangement, (**c**) third arrangement, (**d**) fourth arrangement and (**e**) fifth arrangement.

3.1. Model Geometry

The numerical model of the study area was carried out using a mesh of 48 rows and 48 columns with an area of about 23.5 km², as shown in Figure 7, in which the aquifer system defined in the conceptual model is composed of three aquifers: the first unit is unconfined shallow aquifer A, which belongs to the quaternary age, with an average thickness of 90 m; the second unit is leaky confined aquifer B, which belongs to the Mit Ghamr formation of the Pleistocene age, with an average thickness of 200 m; the third unit is confined deep aquifer C, which belongs to the Wastany formation of the upper Pleistocene age, with an average thickness of 200 m. There are two layers of clay with a thickness of about 20 meters; the first separates the shallow aquifer and the leaky aquifer B, while the other separates the aquifers B and C. All aquifers consist of sand and gravel with intercalated clay lenses.

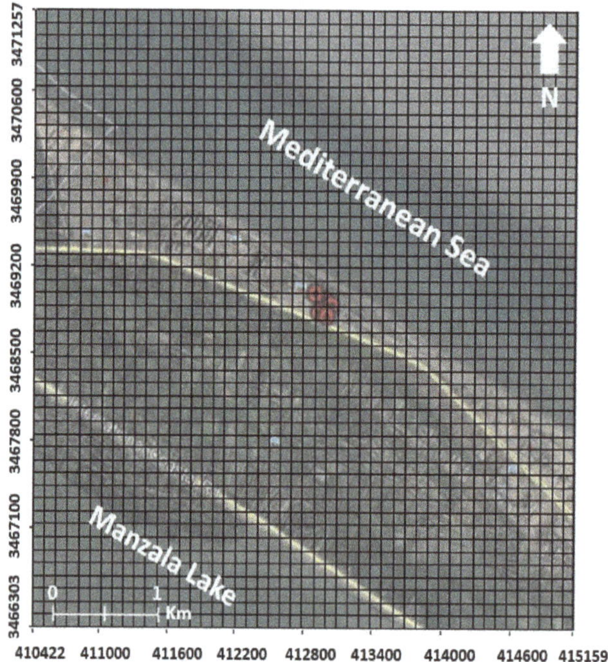

Figure 7. The grid cells for conceptual groundwater model.

3.2. Boundary Condition

The hydrogeological boundary conditions were selected according to water level data from drilled wells and the hydrogeological map of the Nile Delta aquifer [1], as follows:

1. Northern boundary: Mediterranean Sea representing the northern constant head and constant salinity concentration (40,600 mg/L), according to water analysis.
2. Southern boundary: Manzala Lake representing the southern constant head with a -3 value and constant salinity concentration (50,760 mg/L), according to water analysis.

3.3. Hydrogeological Parameters

To identify the hydraulic properties of the aquifer, pumping tests were conducted on the five exploratory wells. After completion of well construction and development, step and continuous pumping tests were performed on each well.

Pumping tests were conducted in two wells in aquifer A (GW 01 and GW 05), in three wells in aquifer B (GW 02, GW 03 and GW 04) and reviewed from previous works [18,42,68].

According to the pumping test analysis, the transmissivity value ranges were determined as follows:

- GW 01: According to the pumping tests analysis by Theis, with Jacob correction for unconfined aquifer, the transmissivity values ranged from 586 to 1150 m^2/d.
- GW 05: According to the pumping tests analysis by Theis, with Jacob correction for unconfined aquifer, the transmissivity values ranged from 1010 to 1270 m^2/d.
- GW 02: According to the pumping tests analysis by Theis, for confined aquifer, the transmissivity values ranged from 1140 to 1170 m^2/d.
- GW 03: According to the pumping tests analysis by Theis, for confined aquifer, the transmissivity values ranged from 763 to 890 m^2/d.
- GW 04: According to the pumping tests analysis by Theis, for confined aquifer, the transmissivity values ranged from 1110 to 1950 m^2/d.

Four-step pumping tests were conducted at each well with different pumping rates for each well with a six-hour duration in each step. Then, continuous pumping tests were conducted at each well with the highest pumping rate and different time durations, such as 24, 48 and 72 h. Table 1 summarizes the pumping rate and drawdown for the steps and continuous pumping tests.

Table 1. Pumping test field data for three aquifers.

Well	Step Pumping Tests		Continuous Pumping Test			Aquifer	Hydraulic Conductivity (m/day)
	Discharge Rate (m^3/hr)	Drawdown (m)	Discharge Rate (m^3/hr)	Drawdown (m)	Time (Hour)		
GW 01	150	7.39	250	13.38	72	Sandstone (Unconfined)	13–28
	183	9.32					
	215	11.12					
	250	13.28					
GW 05	182	13.8	300	24.34	24		
	215	16.44					
	265	21.37					
	280	23.48					
GW 02	150	13.59	300	31.33	72	Sandstone (Leaky Confined)	4–11
	200	18.39					
	257	24.04					
	300	28.35					
GW 03	65	0.95	120	4.9	48		
	85	1.83					
	105	3.2					
	120	4.25					
GW 04	180	9.82	300	16.87	24,72	Sandstone (Deeper confined)	3–5.5
	220	12.14					
	260	14.6					
	300	17.04					

The Boulton 1963, Hantush 1960 and Thesis 1935 pumping test analysis methods [69–71] were used to estimate the hydraulic properties (transmissivity and storativity of pumped aquifer; vertical hydraulic conductivity and storage coefficient of aquitard) of unconfined, leaky confined (semi-confined) and confined deep aquifers, respectively (Figure 8).

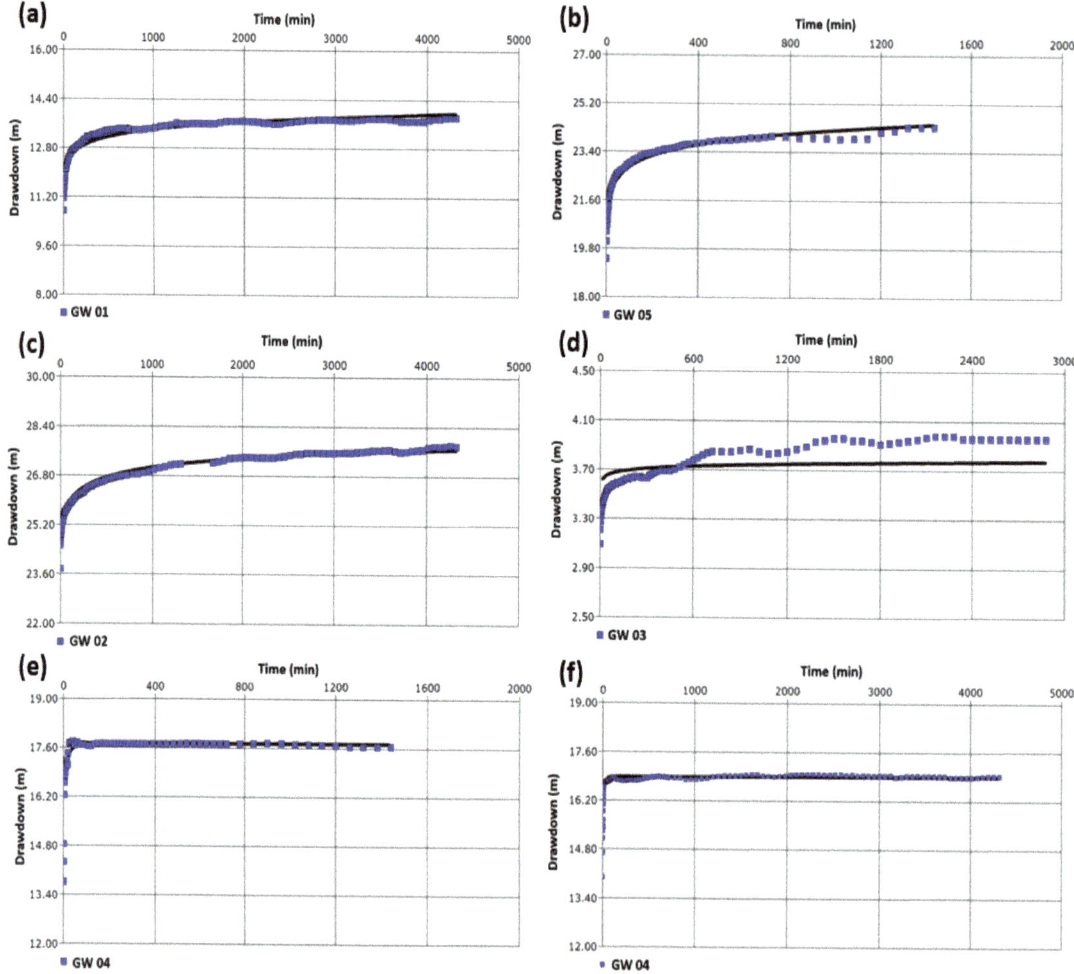

Figure 8. Examples of analyzing the pumping tests in (**a**,**b**) unconfined aquifer, (**c**,**d**) leaky confined aquifer and (**e**,**f**) deep confined aquifer.

The results show that the values of hydraulic conductivity of unconfined sandstone ranging from 13 to 28 m/day were obtained from the pumping tests of two wells (GW 01 and GW 05), while the values of hydraulic conductivity of confined sandstone ranging from 4 to 11 m/day were obtained from the pumping tests of two wells (GW 02 and GW 03) and the values of horizontal hydraulic conductivity of confined deeper sandstone ranging from 3 to 5.5 m/day were obtained from the pumping tests of the GW 04 well. The vertical hydraulic conductivity value is one-tenth of the value of horizontal hydraulic conductivity [72].

3.4. Model Calibration

Boundary conditions, hydraulic properties and initial conditions are classic prerequisite parameters for the aquifer domain and the accuracy of the final output is highly dependent on these parameters [2], especially when simulating groundwater flow in such heterogeneous media [73]. To achieve an accurate simulation, the spatial distribution of the aquifer hydraulic properties conditioned with physical measurement (e.g., hydraulic conductivity, porosity, storativity and dispersivity) should be determined [73].

The model was calibrated by trial and error by changing the hydraulic parameters and the calculated water levels and comparing them with the measured levels. Hydraulic conductivity was used as a calibration parameter. In model calibration, trial-and-error approaches were used to manually match the field and simulated data. A sensitivity analysis was conducted on hydraulic conductivity and porosity. The hydraulic conductivity ranged from 3 to 28 m/d and the porosity ranged from 0.2 to 0.45. The parameter of sensitivity was evaluated using the root mean square error (RMSE) method, which demonstrated that the model was sensitive to hydraulic conductivity but insensitive to porosity for the test range.

Regarding field data, the model was calibrated using 2018 field data. The year 2018 was chosen as the calibration year to calibrate the model under equilibrium (steady state) conditions. Figure 9 shows the groundwater flow in the study area where the flow direction gradually decreases from the Mediterranean Sea in the north to Manzala Lake in the south. The model was calibrated until the lowest possible error values were reached for five observation points and the results show that the residual varied between 0.098 m (GW 03) and 0.67 m (GW 01) with a root mean square (RMS) of 0.362 m and a normalization root mean square of 9.675%, as shown in Figure 10a.

The computed water budget consists of two parts, inflow and outflow. In the model studied, the inflow budget included one component, inflow across boundaries, while the outflow budget consisted of two components, outflow across boundaries and groundwater withdraw of wells, which was 0 m^3/day in a steady state but changes according to discharge rates in the transient state of the model and is used in domestic and water supply. Table 2 shows that the calculated total inflow was 89,910.4297 m^3/day and the total outflow was 89,919.2109 m^3/day with a net flow of about 9 m^3/day, which indicates that the discrepancy percentage was 0.01%.

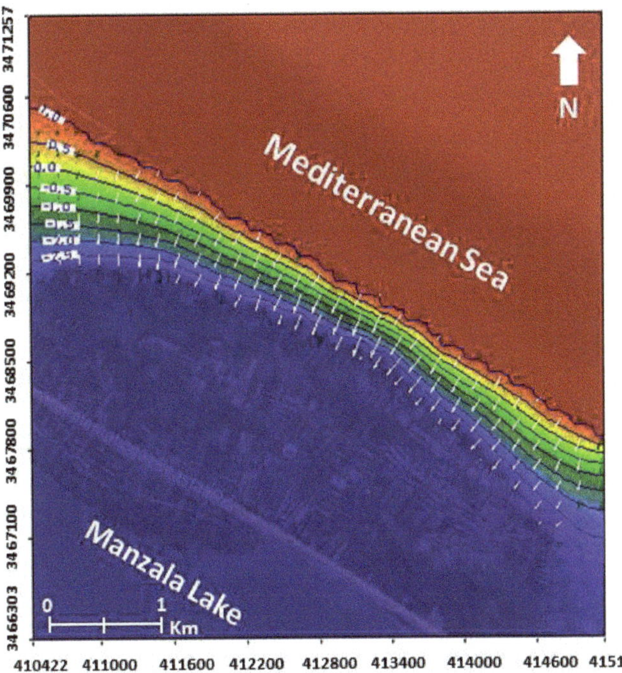

Figure 9. Calculated groundwater levels in 2018.

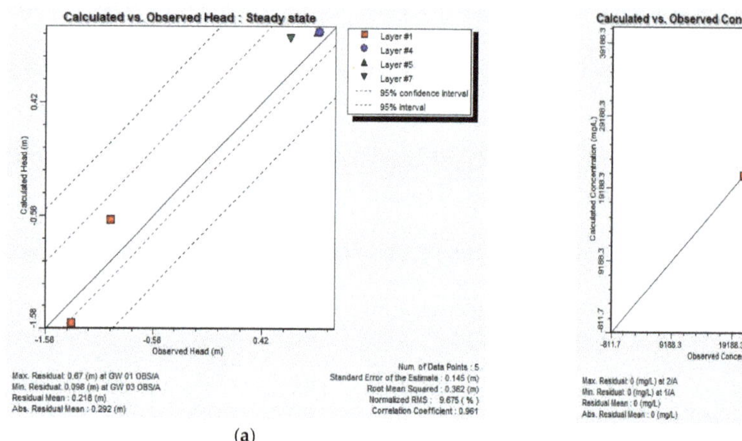
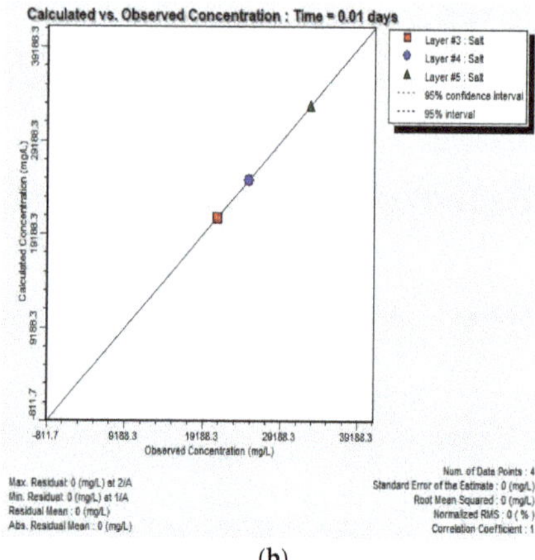

Figure 10. (**a**) Calibration of the head at steady state (2018) and (**b**) calibration of the TDS (mg/L) at steady state (2018).

Table 2. Groundwater balance of the study area.

Component	Recharge (m^3/day)	Discharge (m^3/day)
Subsurface drainage	0	-
Seepage from the Nile and main canals	0	-
Inflow across boundaries	89,910.4297	-
Discharge by tile drains		0
Inflow to Damietta		0
Discharge into drains		0
Groundwater withdrawals		0
Evaporation		0
Outflow across boundaries		89,919.2109
Balance	89,910.4297	89,919.2109

The SEWAT code was calibrated using the existing salinity field data of 2018. The three main aquifers were simulated to study the effect of seawater intrusion. The initial conditions of TDS in the three aquifers were assigned according to the water samples analysis and the results of the EC logs. Seawater salinity was 40,600 mg/L TDS. For aquifer A, TDS was 32,410 mg/L. Aquifer B was divided into three sub-layers with salinities of 21,000 mg/L, 25,000 mg/L and 33,000 mg/L, respectively. Salinity of aquifer C ranged between 50,000 mg/L and 60,000 mg/L. This aquifer was divided into two sub-layers and the sensitivity analysis was performed to predict the impact of this bottom aquifer on the salinity of abstracted water.

Longitudinal dispersivity was used as a calibration parameter. Figure 10b shows the calibration curve between the field and simulated salinity values in 2018 for three observation points.

4. Results

4.1. Groundwater Flow Model

Groundwater flow and solute transport (seawater intrusion) were simulated under transient conditions for a prediction period using the MODFLOW and SEAWAT codes for 50 years (18,250 days), from 2018 to 2068, to predict the drawdown in groundwater levels and changes in groundwater salinity as a result of extraction scenarios. The stress period of the calibration run was set to one year (365 days). The prediction period was divided into 50 stress periods.

Table 3 and Figure 11 explain the results of predicted drawdown in 18 scenarios as follows:

- The first three scenarios (Sc a.1, Sc a.2 and Sc a.3) refer to extraction from the first aquifer, with wells located at a distance of about 165 m from the sea, showing the drawdown of the water level would range, over time, from 2.5 m in Sc a.1 (50,000 m^3/day) to 6.7 m in Sc a.3 (150,000 m^3/day); according to the remaining three scenarios (Sc a.4, Sc a.5 and Sc a.6), in which extraction is performed from the second aquifer, the drawdown of the water level would range, over time, from 9.7 m in Sc a.4 (50,000 m^3/day) to 29.3 m in Sc a.6 (150,000 m^3/day).
- For scenarios Sc b.1, Sc b.2 and Sc b.3, where extraction is performed from the second aquifer, with wells located at a distance of about 300 m from the sea, the drawdown of the water level would range, over time, from 7.5 m in Sc b.1 (50,000 m^3/day) to 22.6 m in Sc b.3 (150,000 m^3/day).
- For scenarios Sc c.1, Sc c.2 and Sc c.3, where extraction is divided into 10 wells instead of 5, from the second aquifer, the drawdown of the water level would range, over time, from 9.2 m in Sc c.1 (50,000 m^3/day) to 27.7 m in Sc c.3 (150,000 m^3/day).
- For scenarios Sc d.1, Sc d.2 and Sc d.3, where extraction is performed from the second aquifer, with wells located at a distance of about 165 m from the sea with distance of 100 m between wells, the drawdown of the water level would range, over time, from 7.2 m in Sc d.1 (50,000 m^3/day) to 21.6 m in Sc d.3 (150,000 m^3/day).
- For scenarios Sc e.1, Sc e.2 and Sc e.3, where extraction is performed from the second aquifer, with wells located at a distance of about 165 m from the sea with a distance of 200 m between wells, the drawdown of the water level would range, over time, from 3.5 m in Sc e.1 (50,000 m^3/day) to 10.7 m in Sc e.3 (150,000 m^3/day).

Table 3. Summary of testing scenarios.

Scenario	Wells Arrangement	Screen Depth	Discharge (m^3/d)	Drawdown after 50 Years (m)
Sc a.1			50,000	2.5
Sc a.2		Aquifer A	100,000	4.6
Sc a.3			150,000	6.7
Sc a.4	a		50,000	9.7
Sc a.5		Aquifer B	100,000	19.5
Sc a.6			150,000	29.3
Sc b.1			50,000	7.5
Sc b.2	b	Aquifer B	100,000	15.1
Sc b.3			150,000	22.6
Sc c.1			50,000	9.2
Sc c.2	c	Aquifer B	100,000	18.4
Sc c.3			150,000	27.7
Sc d.1			50,000	7.2
Sc d.2	d	Aquifer B	100,000	14.4
Sc d.3			150,000	21.6
Sc e.1			50,000	3.5
Sc e.2	e	Aquifer B	100,000	7
Sc e.3			150,000	10.5

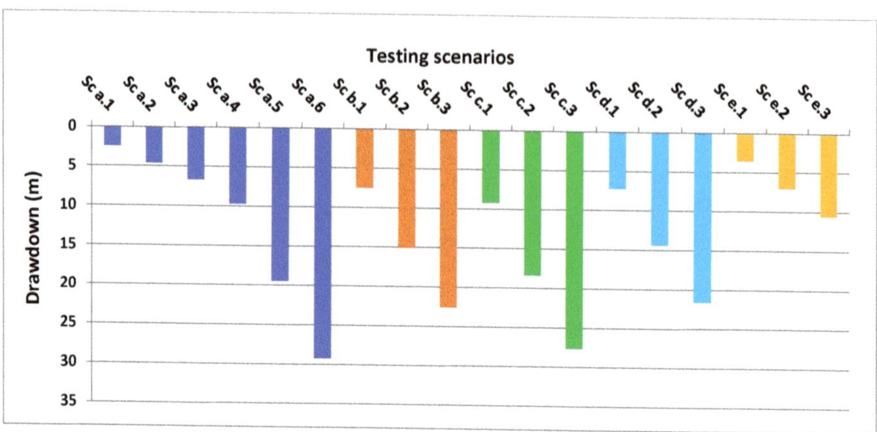

Figure 11. Predicted drawdown after 50 years for 18 scenarios.

4.2. Simulation of Saltwater Intrusion

Here, the scenarios relative to future changes in seawater due to over-pumping of groundwater resources in the study area are discussed. Figures 12 and 13 show the predicted curves of salinity concentrations for 18 scenarios at one control point (GW 02).

Based on the results of the salinity of the monitoring well from the results of the representation of water salinity and the test of sea interference with groundwater, the following become clear:

- According to scenarios Sc a.1, Sc a.2 and Sc a.3 (Figure 12a), which refer to extraction from the first aquifer, over time, the salinity of the water would reach about 40,000 mg/L when the largest amount is extracted, which is 150,000 m^3/day; however, these scenarios were excluded because they required studying the effect of the extraction of this amount of groundwater on the facilities, foundations and infrastructure of the station, as well as on the nearby facilities. According to scenarios Sc a.4, Sc a.5 and Sc a.6 (Figure 12a′), in which extraction is performed from the second aquifer, with wells located at a distance of about 165 m from the sea, the salinity would range from 38,000 to 40,000 mg/L.
- For scenarios Sc b.1, Sc b.2 and Sc b.3 (Figure 12b), in which extraction is performed from the second aquifer, with wells located at a distance of about 300 m from the sea, the salinity would range from 34,000 to 38,000 mg/l.
- For scenarios Sc c.1, Sc c.2 and Sc c.3 (Figure 12c), in which extraction is performed from 10 wells instead of 5, from the second aquifer, the salinity would range from 35,000 to 38,000 mg/L.
- For scenarios Sc d.1, Sc d.2 and Sc d.3 (Figure 12d), in which extraction is performed from the second aquifer, with wells located at a distance of about 165 m from the sea with a distance of 100 m between wells, the salinity would range from 36,000 to 38,500 mg/L.
- For scenarios Sc e.1, Sc e.2 and Sc e.3 (Figure 12e), in which extraction is performed from the second aquifer, with wells located at a distance of about 165 m from the sea with a distance of 200 m between wells, the salinity would range from 38,500 to 40,000 mg/L.

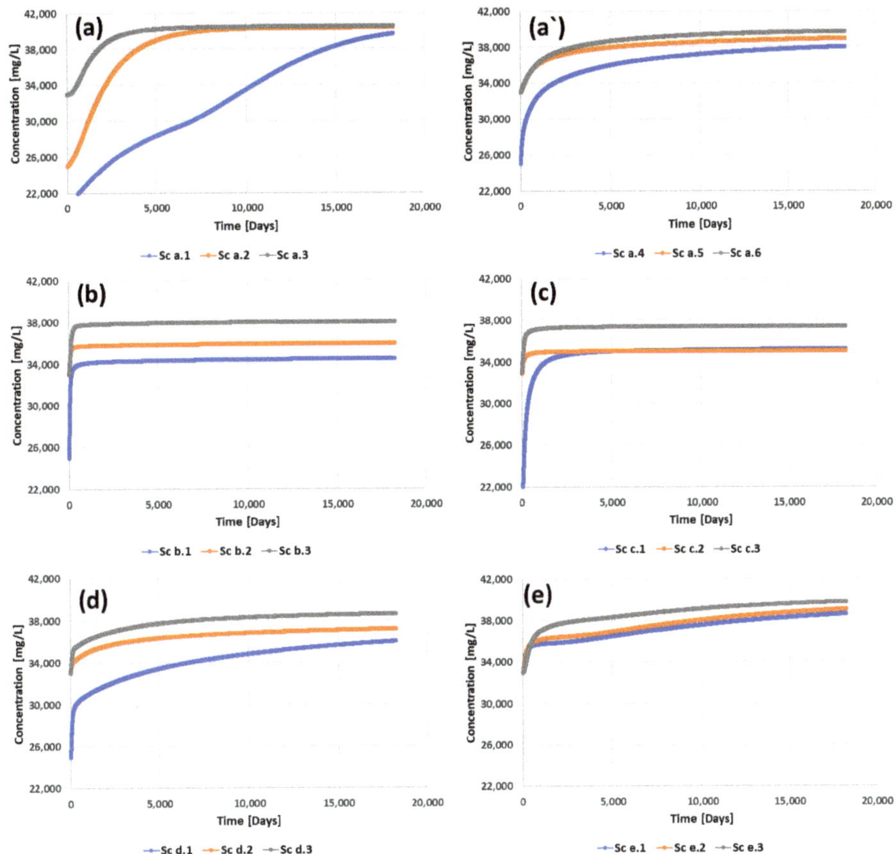

Figure 12. Predicted salinity curves over 50 years for scenario a (**a**,**a′**), scenario b (**b**), scenario c (**c**), scenario d (**d**) and scenario e (**e**) at one control point (GW 02).

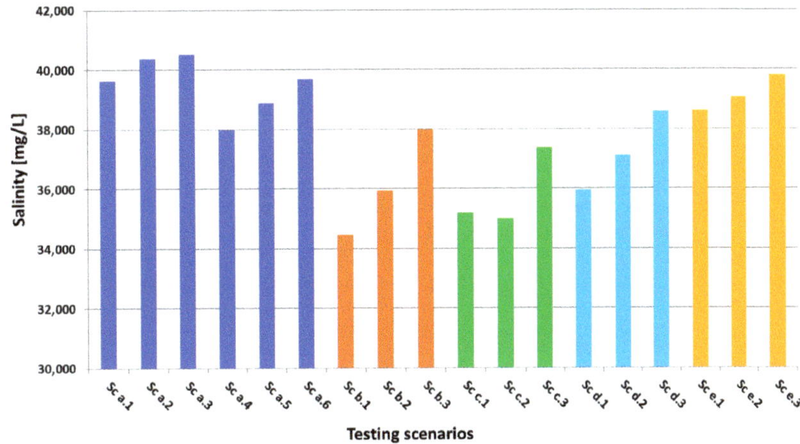

Figure 13. Predicted maximum salinity after 50 years for 18 scenarios.

5. Conclusions

This study presents the sustainable management of groundwater resources in the coastal aquifer at the western area of Port Said, Egypt, due to human expansion activity in this area. The main objective is proposing the optimum withdrawing scenarios to provide the study area, with sufficient groundwater with less salinity. In this regard, a groundwater model was developed. The Visual MODFLOW and SEAWAT codes were used in the study area to simulate, over 50 years (from 2018 to 2068), groundwater flow and groundwater salinity and predict the drawdown and, consequently, the impact of seawater intrusion according to the testing scenarios. The eighteen testing scenarios included the change in well abstraction rate, the different numbers of abstraction wells, the spacing between the abstraction wells and the change in screen depth in the abstraction wells. After comparing, the results recommend the groundwater abstraction to be performed from aquifer B and, preferably, a time of 25 years in the scenarios Sc b.1, Sc b.2, Sc c.1, Sc c.2, Sc d.1 and Sc d.2 to minimize the highest values of drawdown and salinity concentration due to seawater intrusion. The average value of salinity would be about 35,000 mg/L if the groundwater abstraction was used in the quantities illustrated in the recommended scenarios in a shorter period of time, such as 25 years, in order to obtain a sufficient and permanent source of water required to be utilized as a main source of water. On the other hand, taking into consideration the groundwater salinity, this would achieve the goal of sustainable development of renewable resources such as groundwater.

Author Contributions: Conceptualization, M.A. and H.A.-A.A.-B.; methodology, M.A., N.E.N. and H.A.-A.A.-B.; validation, M.A., N.E.N., A.Y.E. and H.A.-A.A.-B.; writing—original draft preparation, M.A. and H.A.-A.A.-B.; supervision, writing—review and editing, A.G., A.Y.E., N.E.N., M.H.G. and T.M.H. All authors have read and agreed to the published version of the manuscript.

Funding: This research study received no external funding.

Institutional Review Board Statement: Not applicable.

Informed Consent Statement: Not applicable.

Data Availability Statement: Not applicable.

Acknowledgments: We thank Taif University Researchers Supporting Project Number (TURSP-2020/32), Taif University, Taif, Saudi Arabia.

Conflicts of Interest: The authors declare no conflict of interest.

References

1. RIGW. *Hydrogeological Map of Nile Delta, Scale 1: 500,000*, 1st ed.; MWRI Publishing Unit: Cairo, Egypt, 1992.
2. Eltarabily, M.G.A.; Negm, A.M. Groundwater management for sustainable development east of the nile delta aquifer. In *Groundwater in the Nile Delta*; Springer: Berlin/Heidelberg, Germany, 2017; pp. 687–708.
3. Bear, J.; Cheng, A.H.-D.; Sorek, S.; Ouazar, D.; Herrera, I. *Seawater Intrusion in Coastal Aquifers: Concepts, Methods and Practices*; Springer Science & Business Media: Berlin/Heidelberg, Germany, 1999; Volume 14.
4. Negm, A.M.; Sakr, S.; Abd-Elaty, I.; Abd-Elhamid, H.F. An overview of groundwater resources in Nile Delta aquifer. *Groundw. Nile Delta* **2018**, *73*, 3–44.
5. Abd-Elhamid, H.; Javadi, A.; Abdelaty, I.; Sherif, M. Simulation of seawater intrusion in the Nile Delta aquifer under the conditions of climate change. *Hydrol. Res.* **2016**, *47*, 1198–1210. [CrossRef]
6. Boumaiza, L.; Chesnaux, R.; Drias, T.; Walter, J.; Huneau, F.; Garel, E.; Knoeller, K.; Stumpp, C. Identifying groundwater degradation sources in a Mediterranean coastal area experiencing significant multi-origin stresses. *Sci. Total. Environ.* **2020**, *746*, 141203. [CrossRef] [PubMed]
7. Elbeih, S.F.; Negm, A.M.; Kostianoy, A. *Environmental Remote Sensing in Egypt*; Springer Nature: Basingstoke, UK, 2020.
8. Abou El-Magd, I.; El-Zeiny, A. Quantitative hyperspectral analysis for characterization of the coastal water from Damietta to Port Said, Egypt. *Egypt. J. Remote Sens. Space Sci.* **2014**, *17*, 61–76. [CrossRef]
9. El-Magd, A.; Zakzouk, M.; Abdulaziz, A.M.; Ali, E.M. The potentiality of operational mapping of oil pollution in the mediterranean sea near the entrance of the suez canal using sentinel-1 SAR data. *Remote Sens.* **2020**, *12*, 1352. [CrossRef]
10. El-Magd, I.A.; Zakzouk, M.; Ali, E.M.; Abdulaziz, A.M. An open source approach for near-real time mapping of oil spills along the mediterranean coast of Egypt. *Remote Sens.* **2021**, *13*, 2733. [CrossRef]

11. Abd-Elaty, I.; Shahawy, A.E.; Santoro, S.; Curcio, E.; Straface, S. Effects of groundwater abstraction and desalination brine deep injection on a coastal aquifer. *Sci. Total. Environ.* **2021**, *795*, 148928. [CrossRef]

12. Sherif, M.M.; Singh, V.P. Effect of climate change on sea water intrusion in coastal aquifers. *Hydrol. Process.* **1999**, *13*, 1277–1287. [CrossRef]

13. Molle, F.; Gaafar, I.; El-Agha, D.E.; Rap, E. *Irrigation Efficiency and the Nile Delta Water Balance*; Water and salt management in the Nile Delta: Report No.9; IWMI: Anand, India, 2016; p. 7.

14. Abd-Elaty, I.; Javadi, A.A.; Abd-Elhamid, H. Management of saltwater intrusion in coastal aquifers using different wells systems: A case study of the Nile Delta aquifer in Egypt. *Hydrogeol. J.* **2021**, *29*, 1767–1783. [CrossRef]

15. Abdel-Shafy, H.I.; Kamel, A.H. Groundwater in Egypt issue: Resources, location, amount, contamination, protection, renewal, future overview. *Egypt. J. Chem.* **2016**, *59*, 321–362.

16. El-Raey, M.; Nasr, S.; Frihy, O.; Desouki, S.; Dewidar, K. Potential impacts of accelerated sea-level rise on Alexandria Governorate, Egypt. *J. Coast. Res.* **1995**, *14*, 190–204.

17. IPCC; Field, C.B.B.; Barros, V.R.; Dokken, D.J.; Mach, K.J.; Mastrandrea, M.D.; Bilir, T.E.; Chatterjee, M.; Ebi, K.L.E.; Anokhin, Y.; et al. *Climate Change 2014–Impacts, Adaptation and Vulnerability: Regional Aspects*; Cambridge University Press: Cambridge, UK, 2014.

18. Morsy, W.S. Environmental Management to Groundwater Resources for Nile Delta Region. Ph.D. Thesis, Faculty of Engineering, Cairo University, Cairo, Egypt, 2009.

19. Mabrouk, M.; Jonoski, A.; Solomatine, D.; Uhlenbrook, S. A review of seawater intrusion in the Nile Delta groundwater system–the basis for assessing impacts due to climate changes and water resources development. *Nile Water Sci. Eng. J.* **2017**, *10*, 46–61.

20. Chang, C.-M.; Yeh, H.-D. Spectral approach to seawater intrusion in heterogeneous coastal aquifers. *Hydrol. Earth Syst. Sci.* **2010**, *14*, 719–727. [CrossRef]

21. Wilson, J.L.; Townley, L.R.; Da Costa, A.S. *Mathematical Development and Verification of a Finite Element Aquifer Flow Model AQUIFEM-1*; Technological Planning Program; Tap Report; Massachusetts Institute of Technology: Cambridge, MA, USA, 1979; p. 114.

22. Amer, A.; Farid, M. About sea water intrusion phenomenon in the Nile Delta aquifer. In Proceedings of the International Workshop on Management of the Nile Delta Groundwater Aquifer, CU/Mit, Cairo, Egypt, 1981.

23. Darwish, M. Effect of Probable Hydrological Changes on the Nile Delta Aquifer System. Ph.D. Thesis, Cairo University, Cairo, Egypt, 1994.

24. Amer, A.; Sherif, M. *An Integrated Study for Seawater Intrusion in the Nile Delta Aquifer*; Working Paper for SRP; NWRC-MPWWR: Cairo, Egypt, 1996.

25. Sherif, M.; Singh, V. *Groundwater Development and Sustainability in the Nile Delta Aquifer*; Final Report; Binational Fulbright Commission: Giza Governorate, Egypt, 1997.

26. Mahlknecht, J.; Merchán, D.; Rosner, M.; Meixner, A.; Ledesma-Ruiz, R. Assessing seawater intrusion in an arid coastal aquifer under high anthropogenic influence using major constituents, Sr and B isotopes in groundwater. *Sci. Total Environ.* **2017**, *587*, 282–295. [CrossRef] [PubMed]

27. Howard, K.W. Beneficial aspects of sea—Water intrusion. *Groundwater* **1987**, *25*, 398–406. [CrossRef]

28. Hussain, M.S.; Javadi, A.A.; Sherif, M.M. Three dimensional simulation of seawater intrusion in a regional coastal aquifer in UAE. *Procedia Eng.* **2015**, *119*, 1153–1160. [CrossRef]

29. Patel, A.S.; Shah, D.L. *Water Management: Conservation, Harvesting and Artificial Recharge*; New Age International (P) Limited, Publishers: Hyderabad, India, 2008.

30. Abarca, E.; Vázquez-Suñé, E.; Carrera, J.; Capino, B.; Gámez, D.; Batlle, F. Optimal design of measures to correct seawater intrusion. *Water Resour. Res.* **2006**, *42*. [CrossRef]

31. Tsanis, I.K.; Song, L.F. Remediation of sea water intrusion: A case study. *Groundw. Monit. Remediat.* **2001**, *21*, 152–161. [CrossRef]

32. Essink, G.H.O. Improving fresh groundwater supply—problems and solutions. *Ocean Coast. Manag.* **2001**, *44*, 429–449. [CrossRef]

33. Shamir, U.; Bear, J.; Gamliel, A. Optimal annual operation of a coastal aquifer. *Water Resour. Res.* **1984**, *20*, 435–444. [CrossRef]

34. Das, A.; Datta, B. Development of multiobjective management models for coastal aquifers. *J. Water Resour. Plan. Manag.* **1999**, *125*, 76–87. [CrossRef]

35. Emch, P.G.; Yeh, W.W. Management model for conjunctive use of coastal surface water and ground water. *J. Water Resour. Plan. Manag.* **1998**, *124*, 129–139. [CrossRef]

36. Narayan, K.; Schleeberger, C.; Charlesworth, P.; Bistrow, K. Effects of roundwater pumping on saltwater intrusion in the lower Burdekin Delta, North Queensland. In Proceedings of the MODSIM 2003 International Congress on Modelling and Simulation, Townsville, Australia, 14–17 July 2003; pp. 212–217.

37. Trichakis, Y. Modeling the saltwater intrusion phenomenon in coastal aquifers-a case study in the industrial zone of herakleio in crete. *Glob. NEST J.* **2005**, *7*, 197–203.

38. Kashef, A.-A.I. Control of salt-water intrusion by recharge wells. *J. Irrig. Drain. Div.* **1976**, *102*, 445–457. [CrossRef]

39. Mahesha, A. Transient effect of battery of injection wells on seawater intrusion. *J. Hydraul. Eng.* **1996**, *122*, 266–271. [CrossRef]

40. Maimone, M.; Fitzgerald, R. Effective modeling of coastal aquifer systems. In Proceedings of the First International Conference on Saltwater Intrusion and Coastal Aquifers-Monitoring, Modeling, and Management, Essaouira, Morocco, 23–25 April 2001; pp. 23–25.
41. Kacimov, A.; Sherif, M.; Perret, J.; Al-Mushikhi, A. Control of sea-water intrusion by salt-water pumping: Coast of Oman. *Hydrogeol. J.* **2009**, *17*, 541–558. [CrossRef]
42. Farid, M.S. Nile Delta Groundwater Study. Master's Thesis, Faculty of Engineering, Cairo University, Cairo, Egypt, 1980.
43. Farid, M.S. Management of Groundwater System in the Nile Delta. Ph.D. Thesis, Faculty of Engineering, Cairo University, Cairo, Egypt, 1985.
44. Mohsen, M.S.; Singh, V.P.; Amer, A.M. A note on saltwater intrusion in coastal aquifers. *Water Resour. Manag.* **1990**, *4*, 123–134. [CrossRef]
45. Sherif, M.; Sefelnasr, A.; Javadi, A. Incorporating the concept of equivalent freshwater head in successive horizontal simulations of seawater intrusion in the Nile Delta aquifer, Egypt. *J. Hydrol.* **2012**, *464*, 186–198. [CrossRef]
46. Abu-Bakr, H.A.e.-A.; Elkhedr, M.; Hassan, T.M. Optimization of abstraction wells near coastal zone. *J. Earth Sci. Res.* **2016**, *4*, 30–42. [CrossRef]
47. Abd-Elaty, I.; Zeleňáková, M.; Krajníková, K.; Abd-Elhamid, H.F. Analytical solution of saltwater intrusion in costal aquifers considering climate changes and different boundary conditions. *Water* **2021**, *13*, 995. [CrossRef]
48. Mastrocicco, M.; Busico, G.; Colombani, N.; Vigliotti, M.; Ruberti, D. Modelling actual and future seawater intrusion in the Variconi Coastal Wetland (Italy) due to climate and landscape changes. *Water* **2019**, *11*, 1502. [CrossRef]
49. Abdelhalim, A.; Sefelnasr, A.; Ismail, E. Numerical modeling technique for groundwater management in Samalut city, Minia Governorate, Egypt. *Arab. J. Geosci.* **2019**, *12*, 124. [CrossRef]
50. Hagagg, K.H. Numerical modeling of seawater intrusion in karstic aquifer, Northwestern Coast of Egypt. *Modeling Earth Syst. Environ.* **2019**, *5*, 431–441. [CrossRef]
51. Mabrouk, M.; Jonoski, A.; Oude Essink, G.H.; Uhlenbrook, S. Assessing the fresh–saline groundwater distribution in the Nile delta aquifer using a 3D variable-density groundwater flow model. *Water* **2019**, *11*, 1946. [CrossRef]
52. Mabrouk, M.; Jonoski, A.; HP Oude Essink, G.; Uhlenbrook, S. Impacts of sea level rise and groundwater extraction scenarios on fresh groundwater resources in the Nile Delta governorates, Egypt. *Water* **2018**, *10*, 1690. [CrossRef]
53. Abd-Elaty, I.; Abd-elhamid, H.; Fahmy, M.; Abd-Elaal, G. Study of Impact of some changes on groundwater system in Nile delta aquifer. *Egypt. J. Eng. Sci. Technol.* **2014**, *17*, 10–11. [CrossRef]
54. Abd-Elaty, I.; Abd-Elhamid, H.F.; Nezhad, M.M. Numerical analysis of physical barriers systems efficiency in controlling saltwater intrusion in coastal aquifers. *Environ. Sci. Pollut. Res.* **2019**, *26*, 35882–35899. [CrossRef]
55. Abd-Elaty, I.; Abd-Elhamid, H.F.; Qahman, K. Coastal aquifer protection from saltwater intrusion using abstraction of brackish water and recharge of treated wastewater: Case study of the gaza aquifer. *J. Hydrol. Eng.* **2020**, *25*, 05020012. [CrossRef]
56. Langevin, C.D.; Shoemaker, W.B.; Guo, W. *Modflow-2000, the US Geological Survey Modular Ground-Water Model–Documentation of the SEAWAT-2000 Version with the Variable-Density Flow Process (VDF) and the Integrated MT3DMS Transport Process (IMT)*; US Geological Survey: Tallahassee, FL, USA, 2003; ISSN 2331-1258.
57. Bear, J. Seawater intrusion into coastal aquifers. In *Encyclopedia of Hydrological Sciences*; Anderson, M.G., Ed.; Wiley: Devon, UK, 2005.
58. McDonald, M.G.; Harbaugh, A.W. *A Modular Three-Dimensional Finite-Difference Ground-Water Flow Model*; US Geological Survey: Reston, VA, USA, 1988.
59. Zheng, C.; Wang, P.P. *MT3DMS: A Modular Three-Dimensional Multispecies Transport Model for Simulation of Advection, Dispersion, and Chemical Reactions of Contaminants in Groundwater Systems; Documentation and User's Guide*; Aquaveo Reseller Network: Provo, UT, USA, 1999.
60. Abd-Elaty, I.; Said, A.M.; Abdelaal, G.M.; Zeleňáková, M.; Jandora, J.; Abd-Elhamid, H.F. Assessing the impact of lining polluted streams on groundwater quality: A case study of the eastern Nile delta aquifer, Egypt. *Water* **2021**, *13*, 1705. [CrossRef]
61. Abdelfattah, M.; Gaber, A.; Geriesh, M.H.; Hassan, T.M. Investigating the less ambiguous hydrogeophysical method in exploring the shallow coastal stratified-saline aquifer: A case study at West Port Said Coast, Egypt. *Environ. Earth Sci.* **2021**, *80*, 1–14. [CrossRef]
62. EEAA. *Egypt Second National Communication under the United Nations Framework Convention on Climate Change (UNFCCC)*; Egyptian Environmental Affairs Agency, Ministry of State for Environmental Affairs: Cairo, Egypt, 2010.
63. El Haddad, I. Hydrogeological Studies and Their Environmental Impact on Future Management and Sustainable Development of the New Communities and Their Surroundings, East of the Nile Delta, Egypt. Ph.D. Thesis, Mansoura University, Mansoura, Egypt, 2002. Unpublished.
64. Available online: https://www.worldweatheronline.com/port-said-weather-averages/bur-said/eg.aspx (accessed on 8 November 2021).
65. Sestini, G. Nile Delta: A review of depositional environments and geological history. *Geol. Soc. Lond. Spec. Publ.* **1989**, *41*, 99–127. [CrossRef]
66. Zaghloul, Z.; Taha, A.; Hegab, O.; El Fawal, F. The neogene-quaternary sedimentary basins of the Nile delta. *Egypt. J. Geol.* **1977**, *21*, 1–19.

67. El-Fayoumy, I.F. Geology of Groundwater Supplies in the Eastern Region of the Nile Delta and its Extension in North Sinai. Ph.D. Thesis, Faculty of Science, Cairo University, Cairo, Egypt, 1968.
68. Dahab, K. Hydrogeological Evolution of the Nile Delta after the High Dam Construction. Ph.D. Thesis, Menoufia University, Al Minufiyah, Egypt, 1993.
69. Boulton, N. Analysis of data from non-equilibrium pumping tests allowing for delayed yield from storage. *Proc. Inst. Civ. Eng.* **1963**, *26*, 469–482. [CrossRef]
70. Theis, C.V. The relation between the lowering of the piezometric surface and the rate and duration of discharge of a well using ground—water storage. *Eos Trans. Am. Geophys. Union* **1935**, *16*, 519–524. [CrossRef]
71. Hantush, M.S. Modification of the theory of leaky aquifers. *J. Geophys. Res.* **1960**, *65*, 3713–3725. [CrossRef]
72. Bear, J. *Hydraulics of Groundwater*; Dover Publications: Mineola, NY, USA, 2007.
73. Pasquier, P.; Marcotte, D. Steady-and transient-state inversion in hydrogeology by successive flux estimation. *Adv. Water Resour.* **2006**, *29*, 1934–1952. [CrossRef]

MDPI

St. Alban-Anlage 66

4052 Basel

Switzerland

Tel. +41 61 683 77 34

Fax +41 61 302 89 18

www.mdpi.com

Water Editorial Office

E-mail: water@mdpi.com

www.mdpi.com/journal/water